Lecture Notes Series on Computing - Vol. 17

Algorithmic Aspects of Domination in Graphs

LECTURE NOTES SERIES ON COMPUTING

Editor-in-Chief: D T Lee (*Academia Sinica, Taiwan*)

Published

Lecture Notes Series on Computing - Vol. 17

Algorithmic Aspects of Domination in Graphs

Gerard Jennhwa Chang

National Taiwan University, Taiwan

World Scientific

NEW JERSEY • LONDON • SINGAPORE • BEIJING • SHANGHAI • HONG KONG • TAIPEI • CHENNAI • TOKYO

Published by

World Scientific Publishing Co. Pte. Ltd.

5 Toh Tuck Link, Singapore 596224

USA office: 27 Warren Street, Suite 401-402, Hackensack, NJ 07601

UK office: 57 Shelton Street, Covent Garden, London WC2H 9HE

Library of Congress Cataloging-in-Publication Data

Names: Chang, Gerard J. author
Title: Algorithmic aspects of domination in graphs / Gerard Jennhwa Chang,
    National Taiwan University, Taiwan.
Description: New Jersey : World Scientific, [2026] | Series: Lecture notes series on computing ;
    vol. 17 | Includes bibliographical references and index. | Contents: Introduction to domination in
    graphs -- Design and analysis of algorithms -- Trees -- Chordal graphs -- Interval graphs --
    Strongly chordal graphs -- Cocomparability graphs and asteroidal triple-free graphs --
    Permutation graphs -- Distance-hereditary graphs.
Identifiers: LCCN 2025026671 | ISBN 9789819817283 hardcover |
    ISBN 9789819817290 ebook for institutions | ISBN 9789819817306 ebook for individuals
Subjects: LCSH: Graph algorithms | Domination (Graph theory)
Classification: LCC QA166.245 .C43 2026
LC record available at https://lccn.loc.gov/2025026671

British Library Cataloguing-in-Publication Data
A catalogue record for this book is available from the British Library.

For any available supplementary material, please visit
https://www.worldscientific.com/worldscibooks/10.1142/14419#t=suppl

Desk Editors: Kannan Krishnan/Amanda Yun

Typeset by Stallion Press
Email: enquiries@stallionpress.com

Dedicated to my wife Chain-Chin Yen

and our daughters Katharine and Karen.

Preface

Domination in graph theory is a natural model for many location problems in operations research. As an example, consider the following fire station problem. Suppose a county has decided to build some fire stations, which must serve all towns in the county. The fire stations are to be located in some towns so that every town either has a fire station or is a neighbor of a town that has a fire station. To save money, the county wants to build the minimum number of fire stations that satisfy the above requirements.

The fire station problem can be abstracted into the concept of domination in graphs. In a graph, a dominating set is a subset of vertices such that every vertex of the graph not in this set is adjacent to some vertex of the set. For the fire station problem, consider the graph having all towns of the county as its vertices and a town is adjacent to its neighbor towns. The problem is just to find a minimum sized dominating set as the set of towns where the fire stations are to be built. For more applications of domination, see the books *Fundamentals of Domination in Graphs* by Haynes, Hedetniemi and Slater [1998] and *Domination in Graphs: Core Concepts* by Haynes, Hedetniemi and Henning [2023].

From the graph theoretical point of view, domination is a covering type problem. In other words, minimum number of closed neighborhoods of vertices are used to cover all vertices of the graph. Covering problems are extensively studied in graph theory. For instance, covering vertices by cliques and covering vertices by stable sets play important roles in the perfect graph theory. In fact, closed neighborhood covering has close relation to clique covering, as a closed neighbor in a graph is a clique in the square of the graph, for some detailed discussions, see the chapter on strongly chordal graphs. Of course, a clique in the square of a graph is not always back to a closed neighborhood of the original graph. This makes these two covering problems in fact not the same.

The concept of domination originated from defensive and offensive strategies of the Roman Empire in the fourth century, the war-oriented board game *Chaturanga* developed in India during the sixth century and domination for different chess pieces in the nineteenth century. Although many theorems for the domination problem have been established for a long time, the first algorithmic result on this topic was given by Cockayne, Goodman and Hedetniemi in 1975. They gave a linear-time algorithm for the domination problem in trees by the labeling approach. On the other hand, David Johnson at a graph theory conference held in Qualicum Beach, Vancouver Island, British Columbia sometime during 1975–1976 showed that the domination problem is NP-complete. Since then, many algorithmic results have been studied for variations of the domination problem in different classes of graphs.

The purpose of this book is to introduce the algorithms of variations of domination during the past half century. In the process of achieving this, we examine the structures of various graph classes, including trees, chordal graphs, strongly chordal graphs, interval graphs, cocomparability graphs, permutation graphs and distance-hereditary graphs. Proof techniques for NP-completeness results on domination are also mentioned, see Section 4.4. In fact, from Chapter 3 on each chapter discusses a class of graphs mentioned above. Each chapter begins with properties of the graphs in one or two sections and then follows algorithms for variations of the domination problem.

While there are a lot of algorithms for variations of domination, the attempt of this book is not to list all of them. Instead, three commonly used approaches are discussed in the chapter for trees: labeling approach, dynamic programming approach and primal-dual approach.

The first two approaches are in fact implementations of mathematical induction. Induction and algorithm are, in a sense, two sides of the same concept. To prove a theorem by induction, it is very often easier to prove a stronger statement. This is because one has a stronger induction hypothesis to use, see examples in Section 1.4. When designing algorithms for domination, it is easier to consider boundary conditions as seen in the labeling and the dynamic programming approaches, see Sections 3.2 and 3.3 for some illustrations. The primal-dual approach is a popular and powerful method in mathematics. To our concern, linear programming and the primal-dual relation in the perfect graph theory are closely related to domination, see Sections 1.5 and 3.4.

We try to make the arguments as concise as possible, sometimes new proof methods are established, such as the introduction of strongly chordal graphs in Sections 6.1 and 6.2. When presenting existing algorithms, alternative ones are sometimes offered. For instance, besides the original labeling algorithms for domination, total domination and pair-domination in trees, primal-dual solutions are also provided, see Sections 3.4 and 3.6.

I am grateful to Professor Der-Tsai Lee for his encouragement on writing this book. Special thanks to Professors Stephen T. Hedetniemi and Xuding Zhu for their detailed comments and many useful references which greatly enrich the book. I also thank my son-in-law Wei-Ting Yen for help checking the grammar and spelling of the book using ChatGPT.

Gerard Jennhwa Chang

About the Author

Gerard Jennhwa Chang is Professor Emeritus of Mathematics at National Taiwan University in Taipei. He received his Ph.D. in Operations Research from Cornell University in Ithaca, New York, USA and has held academic appointments at National Central University, National Chiao Tung University, and National Taiwan University in Taiwan. His research focuses on discrete mathematics and combinatorial optimization, with particular emphasis on graph theory and its algorithms. Over the course of his career, he has published more than 200 scholarly articles and authored three academic books, two of which are in Chinese. He also played a key role in revising Taiwan's 2019 Mathematics Curriculum Guidelines (commonly known as the "108 Curriculum") and has written numerous introductory articles on the subject.

Contents

Contents

List of Figures

Chapter 1

Introduction to Domination
in Graphs

1.1 The star of the show

It is often said that graph theory was founded by Euler (1736) as a generalization of the solution to the famous problem of Königsberg bridges. Here's the story.

The city of Königsberg was located on the Pregel river in Prussia. The city occupied the island of Kneiphopf plus areas on both banks. These regions were linked by seven bridges shown on the left in Figure 1.1. The citizens wondered whether they could leave home, cross every bridge exactly once, and return home. The problem reduces to traversing the figure on the right, with heavy dots representing the four land masses and curves representing the seven bridges.

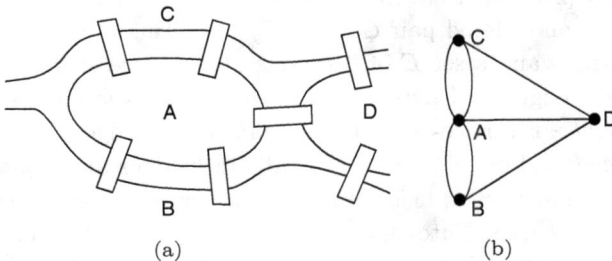

(a)   (b)

Fig. 1.1. The problem of Königsberg bridges.

1

The model on the right makes it easy to argue the desired traversal does not exist. Each time one enters and leaves a land mass, two bridges ending at it are used. One can also pair the first bridge and the last bridge on the land mass where he/she begins and ends. Thus, the existence of the desired traversal requires that each land mass be involved to an even number of bridges. This necessary condition does not hold for the problem of Königsberg bridges.

Euler's paper in fact established a theorem for the general case. A necessary and sufficient condition for the existence of such a traversal for a general configuration was given. That is, the figure in the model must be connected, and each land mass must be involved in an even number of bridges. In case of a positive answer, an algorithm for finding such a traversal tour was also illustrated.

From 1736 to 1936, concepts of graphs—though sometimes under different names—were used in various scientific fields as models of real world problems, see the historical book by Biggs *et al.* (1986). Exactly two centuries after Euler's paper, Kőnig (1936) wrote the first book on graphs, officially announcing the birth of graph theory. For the past ninety years, various books on graph theory are published at an exponential rate. This book is one of them.

Domination in graph theory is a natural model for many location problems in operations research. As an example, consider the following fire station problem. Suppose a county has decided to build some fire stations, which must serve all towns in the county. The fire stations are to be located in some towns so that every town either has a fire station or is a neighbor of a town that has a fire station. To save money, the county wants to build the minimum number of fire stations that satisfy the above requirements. Graph theory provides a natural model for this problem.

A *graph* is an ordered pair $G = (V, E)$ consisting of a finite nonempty set V of *vertices* and a set E of 2-subsets of V, whose elements are called *edges*. For any edge $e = \{u, v\}$, it is said that vertices u and v are *adjacent* and are *neighbors*; vertices u and v are *incident* to edge e; and vertices u and v are *end vertices* of edge e. Two distinct edges are *adjacent* if they contain a common vertex. It is convenient to henceforth denote an edge by uv rather than $\{u, v\}$. Note that uv and vu represent the same edge in a graph. In this book, it is common to say that G is a graph with vertex set $V(G)$ and edge set $E(G)$.

Two graphs G and H are *isomorphic* if there is a bijection f from $V(G)$ to $V(H)$ such that $uv \in E(G)$ if and only if $f(u)f(v) \in E(H)$. Two isomorphic graphs are essentially the same as one can be obtained from the other by renaming vertices.

For visual convenience it is useful to express a graph diagrammatically. To do this, each vertex of the graph is represented by a point (or a small circle) in the plane and each edge by a curve joining the points (or small circles) corresponding to the two vertices incident to the edge. It is convenient to refer to such diagram of a graph as the graph itself. Figure 1.2 shows a graph G with

$$\text{vertex set } V(G) = \{t, u, v, w, x, y, z\} \text{ and}$$
$$\text{edge set } E(G) = \{tw, tx, uv, vw, wx, xy, wz, xz\}.$$

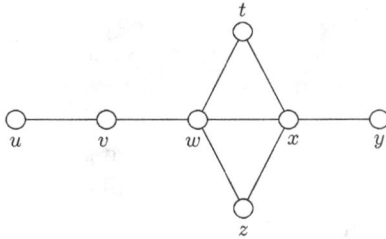

Fig. 1.2. A graph G of 7 vertices and 8 edges.

The graphs described above are often called *simple graphs*. For practical applications, there are many variations of simple graphs including *infinite graphs*, *multigraphs*, *pseudographs*, *directed graphs* (*digraphs*) and *mixed graphs*:

an infinite graph allows the vertex set to be infinite,
a multigraph allows *parallel edges* (edges with the same end vertices),
a pseudograph further allows *loops* (edges of the form vv),
an edge in a directed graph has a direction, and
a mixed graph contains both directed edges and undirected edges.

The graph model for the problem of Königsberg bridges is a multigraph, see the figure on the right in Figure 1.1.

Directed graphs (*digraphs*) can be defined similar to graphs except that a (directed) edge (u, v) is now an ordered pair rather than a 2-subset. Note that now uv stands for (u, v) and vu for (v, u); they are different ordered pairs. To emphasize the direction, a directed edge uv is also written as $\vec{uv}$ or $u \to v$.

An *orientation* of a graph G is a digraph $\vec{G}$ with vertex set $V(\vec{G}) = V(G)$ such that for each edge $\{u, v\}$ of $E(G)$ exactly one of $\vec{uv}$ and $\vec{vu}$ is in $E(\vec{G})$ and all edges in $E(\vec{G})$ are chosen in this way.

This book mainly deals with simple graphs, and may consider their orientations sometimes.

The fire station problem can be abstracted into the concept of domination in terms of graphs as follows. In a graph, a vertex is said to *dominate* itself and all its neighbors. A *dominating set* of a graph G is a subset D of $V(G)$ such that every vertex in the graph is dominated by some vertex in D, or equivalently every vertex not in D is adjacent to some vertex in D. The *domination number* $\gamma(G)$ of a graph G is the minimum size of a dominating set of G. As an example, $D = \{v, x\}$ is a dominating set of the graph G in Figure 1.2 and so $\gamma(G) \leq 2$. In fact, $\gamma(G) > 1$ and so $\gamma(G) = 2$, since no vertex can dominate all vertices in G.

For the fire station problem, consider the graph G having all towns of the county as its vertices and a town is adjacent to its neighbor towns. The fire station problem is just the domination problem, as $\gamma(G)$ is the minimum number of fire stations needed.

Historically speaking, the concept of domination originated from defensive and offensive strategies of the Roman Empire in the fourth century, the war-oriented board game *Chaturanga* developed in India during the sixth century and domination for different chess pieces in the nineteenth century. Beyond the fire station problem, domination theory has numerous real-world applications. The book by Haynes *et al.* (2003) illustrates many examples, including radio broadcasting, computer communication networks, sets of representatives, school bus routing and bus stop selection, electrical power domination, influence in social networks, topological maps, transportation of hazardous materials. For other examples, see the book by Haynes *et al.* (1998). For some other books and some Ph.D./Master's theses on domination, see the references marked by (1.1 book) and (1.1 thesis), respectively.

The domination problem for different chess pieces was widely studied in the nineteenth century. These include the eight queens problem posted by Bezzel (1848), the queens domination problem, the queens independent domination problem and the queens total domination problem. For further studies on problems in chessboards, see the references marked (1.1 chess).

The following illustrates an example. The problem is to determine the minimum number of kings dominating the entire chessboard. It is not hard to see that the answer to an $m \times n$ chessboard is $\lceil \frac{m}{3} \rceil \lceil \frac{n}{3} \rceil$. In the Chinese chess game, a king only dominates the four neighbor cells having common sides with the cell the king lies in. The solution to the Chinese king domination problem for an $m \times n$ chessboard is more complicated. Figure 1.3 shows optimal solutions to the king and the Chinese king problems on a 3×6 chessboard.

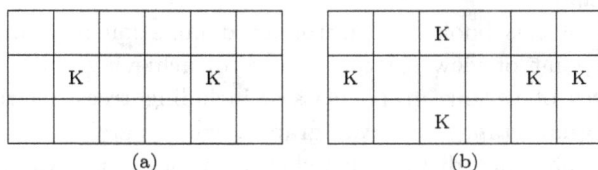

(a)   (b)

Fig. 1.3. King domination on a 3×6 chessboard. (a) chessboard and (b) Chinese chessboard.

For the king domination problem on an $m \times n$ chessboard, consider the king's graph G, where vertices correspond to the mn squares of the board, and two vertices are adjacent if and only if their corresponding squares share a common point. For the Chinese king domination problem, the vertex set is the same but two vertices are adjacent if and only if their corresponding squares have a common side. Figure 1.4 shows the corresponding graphs for the king and the Chinese king domination problems on a 3×6 chessboard. The king/Chinese king domination problem is just the domination problem in the corresponding graph G, and $\gamma(G)$ is the minimum number of kings needed. Black vertices in the graph form a minimum dominating set of the graph.

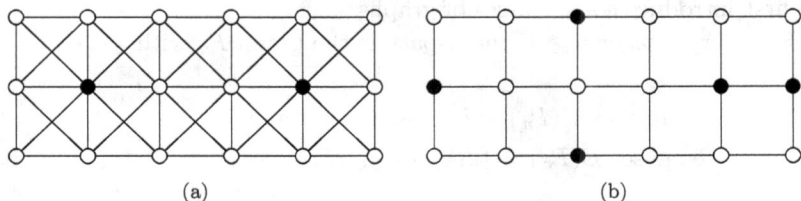

(a)   (b)

Fig. 1.4. King's graphs for chess and Chinese chess. (a) chess and (b) Chinese chess.

Although many theorems for the domination problem have been established for a long time, the first algorithmic result on this topic was given by Cockayne *et al.* (1975) who presented a linear-time algorithm for the domination problem in trees by the labeling approach. On the other hand, David Johnson at a graph theory conference held in Qualicum Beach, Vancouver Island, British Columbia sometime during 1975–1976 showed that the domination problem is NP-complete using a reduction from 3-SAT, see the book by Haynes *et al.* (2023, p. 52). For more information, see the book by Garey and Johnson (1979, p. 190). Since then, many algorithmic results have been studied for variations of the domination problem in different classes of graphs.

The aim of this book is to introduce domination in graphs from an algorithmic point of view. In the process of achieving this, we examine the structures of various graph classes, including trees, chordal graphs, strongly chordal graphs, interval graphs, cocomparability graphs, permutation graphs and distance-hereditary graphs. Surveys on this line were given by Hedetniemi and Laskar (1990, Part IV), Haynes *et al.* (1998, Chapters 8 and 9; 1998a, Chapter 12; 2021, Part III; 2023, Chapter 3), Chang (1998, 2013), Kang (2013), and Henning and Yeo (2013, Chapter 3).

1.2   Graph terminology

This section introduces graph terminology that is frequently used in this book. For more information on graphs, see general books on graph theory, such as the book by West (2001). For more books on graph theory, see the references at the end of the book marked by (1.2 book).

Graph classes
We first introduce some classes of graphs.

For every integer $n \geq 1$, the n-path is the graph P_n with

$$\text{vertex set } V(P_n) = \{v_1, v_2, \ldots, v_n\} \text{ and}$$
$$\text{edge set } E(P_n) = \{v_1v_2, v_2v_3, \ldots, v_{n-1}v_n\}.$$

Figure 1.5 shows graphs P_1, P_2, P_3 and P_4.

Fig. 1.5. Graphs P_1, P_2, P_3 and P_4.

For every integer $n \geq 3$, the *n-cycle* is the graph C_n with

vertex set $V(C_n) = \{v_1, v_2, \ldots, v_n\}$ and

edge set $E(C_n) = \{v_1v_2, v_2v_3, \ldots, v_{n-1}v_n, v_nv_1\}$.

Figure 1.6 shows graphs C_3, C_4, C_5 and C_6.

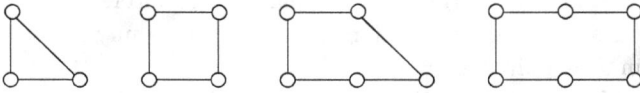

Fig. 1.6. Graphs C_3, C_4, C_5 and C_6.

For every integer $n \geq 1$, the *n-complete graph* is the graph K_n with

vertex set $V(K_n) = \{v_1, v_2, \ldots, v_n\}$ and

edge set $E(K_n) = \{v_iv_j : 1 \leq i < j \leq n\}$.

Figure 1.7 shows graphs K_1, K_2, K_3, K_4 and K_5.

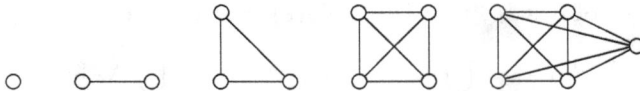

Fig. 1.7. Graphs K_1, K_2, K_3, K_4 and K_5.

Note that $K_1 = P_1$, $K_2 = P_2$ and $K_3 = C_3$.

A graph G is an *r-partite graph* if its vertex set $V(G)$ is the disjoint union of r nonempty sets $V_1, V_2, \ldots, V_r$ such that every edge of the graph has one vertex in some V_i and the other vertex in another V_j. In the definition, $\{V_1, V_2, \ldots, V_r\}$ is called the *r-partition* of the r-partite graph. A 2-partite graph is also called a *bipartite graph*. The *complete r-partite graph* $K_{n_1, n_2, \ldots, n_r}$ is the r-partite graph with the r-partition $\{V_1, V_2, \ldots, V_r\}$ satisfying $|V_i| = n_i$ for $1 \leq i \leq r$ and every vertex in V_i is adjacent to every vertex in V_j for $1 \leq i < j \leq r$. A complete 2-partite graph is also called a *complete bipartite graph*. Figure 1.8 shows graphs $K_{5,3}$ and $K_{2,2,2}$.

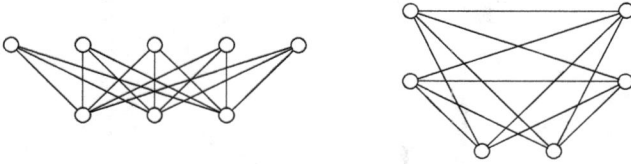

Fig. 1.8. Graphs $K_{5,3}$ and $K_{2,2,2}$.

Suppose $\mathcal{F}$ is a family of sets. The *intersection graph* of $\mathcal{F}$ is the graph obtained by assigning each set in $\mathcal{F}$ to a vertex and joining two distinct vertices with an edge if their corresponding sets intersect. It is the case that every graph is the intersection graph of some family $\mathcal{F}$ satisfying that every set in $\mathcal{F}$ is finite, see Exercise 1.6.

The problem of characterizing the intersection graphs of families of sets having some specific topological property or other pattern is often interesting and frequently has applications in the real world. Interval graphs, chordal graphs, permutation graphs and trapezoid graphs are examples of such graphs.

Relations between vertices

In a graph G, a vertex u is a *neighbor* of v if u is adjacent to v; and u is a *closed neighbor* of v, denoted by $u \sim v$,[1] if $u = v$ or u is adjacent to v. The *neighborhood* v is the set $N_G(v)$ of all neighbors of v. The *closed neighborhood* of v is $N_G[v] = \{v\} \cup N_G(v)$. For a set $A \subseteq V(G)$, let

$$N_G(A) = \bigcup_{v \in A} N_G(v) \ \ \text{and} \ \ N_G[A] = \bigcup_{v \in A} N_G[v].$$

If there is no ambiguity on the graph, $N_G(v), N_G[v], N_G(A)$ and $N_G[A]$ can be written as $N(v), N[v], N(A)$ and $N[A]$ for short, respectively.

The *degree* $\deg(v)$ of a vertex v is the size of $N(v)$. (In a multigraph, $\deg(v)$ is the number of edges incident to v. In a pseudograph, a loop vv contributes 2 to $\deg(v)$.) An *isolated vertex* is a vertex of degree zero. A *leaf* is a vertex of degree one. The minimum degree of a graph G is denoted by $\delta(G)$ and the maximum degree by $\Delta(G)$. A graph G is *r-regular* if $\delta(G) = \Delta(G) = r$, or equivalently all vertices of G are of degree r. A 3-regular graph is also called a *cubic graph*.

---

[1]The notion $u \sim v$ is used for "u is adjacent to v" by some people.

Lemma 1.1 (Handshaking Lemma). *In a graph G, the sum of degrees of all vertices is equal to twice the number of edges. In other words,*

$$\sum_{v \in V(G)} \deg(v) = 2|E(G)|.$$

Proof. By the definition, the degree of a vertex is the number of edges incident to it. When counting the degrees of all vertices, every edge of the graph is counted exactly twice. $\square$

Graph operations

A graph G' is a *subgraph* of another graph G if $V(G') \subseteq V(G)$ and $E(G') \subseteq E(G)$. In the case of $V(G') = V(G)$, graph G' is called a *spanning subgraph* of G. For any nonempty subset S of $V(G)$, the *(vertex-)induced subgraph* $G[S]$ of G is the graph with vertex set S and edge set

$$E(G[S]) = \{uv : u \in S, v \in S \text{ and } uv \in E(G)\}.$$

A graph is H-*free* if it does not contain an induced subgraph isomorphic to H. More generally, for a family $\mathcal{F}$ of graphs, a graph is $\mathcal{F}$-*free* if it does not contain an induced subgraph isomorphic to some graph H in $\mathcal{F}$.

The *deletion* of a proper subset S of $V(G)$ from G, denoted by $G - S$, is the graph $G[V(G)\backslash S]$. For a vertex v of G, graph $G - v$ stands for $G - \{v\}$. The *deletion* of a subset F of $E(G)$ from G is the graph $G - F$ with vertex set $V(G)$ and edge set $E(G)\backslash F$. For an edge e of G, graph $G - e$ stands for $G - \{e\}$.

The *contraction* of a graph G at an edge uv is the graph obtained from G by deleting the edge uv and then identify u and v as a single vertex. If the graph has parallel edges between u and v, loops are created after a contraction.

The *complement* of a graph G is the graph $\overline{G}$ with vertex $V(G)$ and edge set

$$E(\overline{G}) = \{uv : u, v \in V(G), u \neq v \text{ and } uv \notin E(G)\}.$$

Suppose G and H are two graphs with $V(G) \cap V(H) = \emptyset$. The *union* of G and H is the graph $G \cup H$ with vertex set $V(G) \cup V(H)$ and edge set $E(G) \cup E(H)$. The *join* of G and H is the graph $G + H$ with vertex set $V(G) \cup V(H)$ and edge set

$$E(G + H) = E(G) \cup E(H) \cup \{uv : u \in V(G) \text{ and } v \in V(H)\}.$$

The *Cartesian product* of G and H is the graph $G \square H$ with vertex set

$$V(G) \times V(H) = \{(u, v) : u \in V(G) \text{ and } v \in V(H)\},$$

and edge set

$$E(G \square H) = \{(u, v)(u', v') : (u = u', vv' \in E(H)) \text{ or } (uu' \in E(G), v = v')\}.$$

The graph in Figure 1.4(b) is $P_3 \square P_6$.

The *strong product* of G and H is the graph $G \boxtimes H$ with vertex set $V(G) \times V(H)$ and edge set

$$E(G \boxtimes H) = E(G \square H) \cup \{(u, v)(u', v') : uu' \in E(G) \text{ and } vv' \in E(H)\}.$$

The graph in Figure 1.4(a) is $P_3 \boxtimes P_6$.

A *subdivision* of a graph G is the graph obtained from G by a finite number of the following *edge-division operations*: choose an edge uv, add a new vertex x, and replace the edge uv by two new edges ux and xv. Topologically, a subdivision of a graph is the same as the original graph. The inverse operation of an edge-division is as follows: choose a vertex x with $N(x) = \{u, v\}$, delete x, and replace the edges ux and xv by a new edge uv. Note that in case when u is already adjacent to v, a parallel edge is created.

A *minor* of a graph G is the graph obtained from G by a finite number of vertex deletions, edge deletions and edge contractions.

Graph parameters

In a graph G, a *clique* is a set of pairwise adjacent vertices in $V(G)$. An *i-clique* is a clique of size i. A 2-clique is just an edge. A 3-clique is called a *triangle*. The *clique number* $\omega(G)$ of a graph G is the maximum size of a clique of G.

A *clique cover* of a graph G is a family $\mathcal{Q}$ of cliques whose union is $V(G)$. The *clique cover number* $\theta(G)$ of a graph is the minimum size of a clique cover of G. There is no essential difference if the cliques in a clique cover are pairwise disjoint for the definition of $\theta(G)$.

A *stable set* (or *independent set*) is a set of pairwise nonadjacent vertices in $V(G)$. The *stability number* (or *independence number*) $\alpha(G)$ of a graph G is the maximum size of a stable set of G. Note that $\alpha(G) = \omega(\overline{G})$ for every graph G.

A *proper k-coloring* of a graph G is a mapping $f : V(G) \to \{1, 2, \ldots, k\}$ such that $f(u) \neq f(v)$ for $uv \in E(G)$. For $1 \leq i \leq k$, the set $\{v \in V(G) : f(v) = i\}$ is called a *color class*. Note that every color class is stable. So a proper k-coloring is equivalent to a partition $V(G)$ into stable sets. The *chromatic number* $\chi(G)$ of a graph G is the minimum k for which G has a proper k-coloring, or equivalently $V(G)$ is the union of k pairwise disjoint stable sets. Note that $\chi(G) = \theta(\overline{G})$ for every graph G.

The four parameters $\alpha(G), \theta(G), \omega(G), \chi(G)$ play important roles in the perfect graph theory, see Section 1.5 for more introduction on the perfect graph theory.

A *matching* of a graph is a set of edges in which every two edges are not adjacent. A matching is *perfect* if every vertex of the graph is incident to some edge of the matching. The maximum size of a matching of a graph G is denoted by $\alpha'(G)$.

Connection between vertices

For two vertices u and v of a graph G, a *u–v walk* of *length r* is a sequence $(v_0, v_1, \ldots, v_r)$ of vertices, where $v_0 = u$ and $v_r = v$, such that $v_{i-1}v_i \in E(G)$ for $1 \leq i \leq r$. A *trail* (*path*) is a walk in which all edges (vertices) are distinct. A *cycle* is an v_0–v_r walk in which all vertices are distinct except the first vertex v_0 is equal to the last vertex v_r. In a walk $(v_0, v_1, \ldots, v_r)$, a *chord* is an edge v_iv_j with $|i - j| \geq 2$. A chordless path (or chordless cycle) is also called an *induced path* (or *induced cycle*). A graph is *acyclic* if it does not contain any cycle.

A graph is *connected* if for every two vertices u and v, there exists a u–v walk. A graph is *disconnected* if it is not connected. A (*connected*) *component* of a graph is a maximal subgraph which is connected. There is an alternative way to describe the components of a graph. For a graph G, consider a binary relation $\approx$ on $V(G)$ defined by: $u \approx v$ if and only if there is a u–v walk. It is easy to check that $\approx$ is an equivalence relation on $V(G)$, i.e., the following properties hold for all vertices $u, v, w \in V(G)$.

(Reflexivity) $u \approx u$.
(Symmetry) If $u \approx v$, then $v \approx u$.
(Transitivity) If $u \approx v$ and $v \approx w$, then $u \approx w$.

The vertex set $V(G)$ then can be partitioned into equivalence classes, where each class consisting all vertices equivalent to a certain vertex. The equivalence classes induce components of the graph. Note that the vertex set

$V(G)$ (respectively, edge set $E(G)$) is the disjoint union of the vertices sets (respectively, edge sets) of the components.

A *separating set* of a graph G is a set $S \subseteq V(G)$ such that $G - S$ is disconnected. An *a–b separating set* of G is a set $S \subseteq V(G)$ such that vertices a and b are in two different components of $G - S$. A separating set is an *a–b* separating set for some a and b, and the converse is also true.

A separating set or an *a–b* separating set is *minimal* if any proper subset of it is not a separating set or an *a–b* separating set. A minimal separating set is a minimal *a–b* separating set for some a and b, but the converse is not true as shown by the following example. For $n \geq 3$, let G_n be the graph with vertex set $V(G_n) = \{u_i, v_i : 0 \leq i \leq n - 1\}$ and edge set $E(G_n) = \{u_i u_{i+1}, u_i v_i : 0 \leq i \leq n - 1\}$, where $u_n = u_0$. If $n \geq 4$, then $\{u_1, u_3\}$ is a minimal u_0–u_2 separating set, but is not a minimal separating set.

The *connectivity* $\kappa(G)$ of G is the minimum size of a vertex set S such that $G - S$ is disconnected or has only one vertex. A graph is *k-connected* if its connectivity is at least k. A graph being 1-connected is the same as it is connected.

The connectivity of a graph different from a complete graph is the minimum size of a separating set. The complete graph K_n has no separating sets, and its connectivity is $n - 1$, while $\kappa(G) \leq |V(G)| - 2$ when G is not a complete graph.

A *cut-vertex* of G is a vertex v such that $G - v$ has more components than G. In a connected graph G, a vertex v is a cut-vertex if and only if $\{v\}$ is a separating set. Note that in a disconnected graph, if v is a cut-vertex then $\{v\}$ is a separating set, but the converse is not true. A connected graph without cut-vertices may not be 2-connected, since it is possible to be K_1 or K_2. A *block* of a graph G is a maximal connected subgraph of G that has no cut-vertices.

Similar to components of a graph, blocks provide a useful decomposition of a graph. In particular, blocks and cut-vertices have the following properties.

Proposition 1.2. *The following statements hold for every graph G.*

(1) *Two blocks share at most one vertex, and so the set of edge sets of all blocks form a partition of $E(G)$.*

(2) *If two blocks share a vertex, then this vertex is a cut-vertex.*

(3) *A vertex is a cut-vertex if and only if it is shared by two or more blocks.*

The *block-cut-vertex structure* of a graph G is the graph G^* with

$$V(G^*) = \{\text{cut-vertices of } G\} \cup \{\text{blocks of } G\} \text{ and}$$

$$E(G^*) = \{cB : c \text{ is a cut-vertex and } B \text{ is a block containing } c\},$$

see Figure 1.9 for an example. The block-cut-vertex structure G^* of a graph is acyclic. In case when G is connected, G^* is a tree. Every leaf of G^* is a block of G, called a *leaf-block*, which must contain exactly one cut-vertex of G.

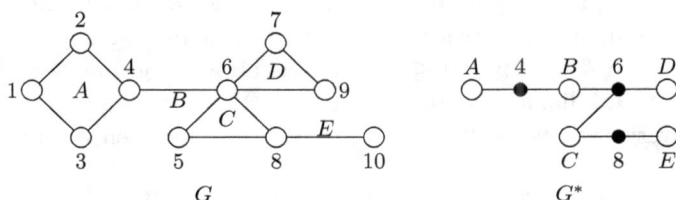

Fig. 1.9. The block-cut-vertex structure G^* of a graph G.

In a graph G, the *distance* from a vertex u to another vertex v, denoted by $d_G(u, v)$ or $d(u, v)$ (if the graph is clear from the context), is the minimum length of a u-v path. If no such path exists, then $d(u, v) = \infty$. The function $d(\cdot, \cdot)$ is a *metric* on $V(G)$. In other words, it satisfies the following properties for all vertices $u, v, w \in V(G)$.

(Positivity)	$d(u, v) \geq 0$; and $d(u, v) = 0$ iff $u = v$.
(Symmetry)	$d(u, v) = d(v, u)$.
(Triangle inequality)	$d(u, w) \leq d(u, v) + d(v, w)$.

In a connected graph G, the *eccentricity* of a vertex v is

$$e(v) = \max\{d(v, u) : u \in V(G)\}.$$

The *radius* rad(G) of G is the minimum eccentricity of a vertex, i.e.,

$$\text{rad}(G) = \min\{e(v) : v \in V(G)\}.$$

A vertex v with $e(v) = \text{rad}(G)$ is called a *center* of G. The set of all centers is denoted by $C(G) = \{v : e(v) = \text{rad}(G)\}$. The *diameter* diam$(G)$ of G is the maximum eccentricity of a vertex, i.e.,

$$\text{diam}(G) = \max\{e(v) : v \in V(G)\}$$
$$= \max\{d(u, v) : u, v \in V(G)\}.$$

It is not hard to see that

$$\text{rad}(G) \leq \text{diam}(G) \leq 2\text{rad}(G)$$

for every connected graph G. The relation between the radius and the diameter of a graph is quite different from that for a circle in the plane; for a circle, the diameter is twice the radius. In general, for any nonnegative integers r and d with $r \leq d \leq 2r$, there is always a graph G with radius $\text{rad}(G) = r$ and diameter $\text{diam}(G) = d$.

For every two sets $A, B \subseteq V(G)$, the distance between A and B is

$$d(A, B) = \min\{d(a, b) : a \in A, b \in B\}.$$

For the case when $A = \{v\}$, we also use $d(v, B)$ for $d(\{v\}, B)$.

For a positive integer k, the kth *power* of a graph G is the graph G^k with vertex set $V(G)$ and edge set $E(G^k) = \{uv : 1 \leq d_G(u, v) \leq k\}$. The second power G^2 is also called the *square* of G and the third power G^3 the *cube* of G.

Terminology of G for simple graphs may properly modified into those for variations of graphs such as infinite graphs, multigraphs, pseudographs and directed graph.

1.3   Variations of domination

Due to different requirements in the applications, many variations of the domination problem have been studied. For instance, in the queen domination problem, one may ask the additional property that two queens don't dominate each other or any two queens must dominate each other. The following are most commonly studied variations of domination, while some others are not mentioned.

Recall that a dominating set of a graph G is a subset D of $V(G)$ such that every vertex in the graph is dominated by some vertex in D. This is equivalent to that $N[v] \cap D \neq \emptyset$ for all $v \in V(G)$ or equivalently

$\cup_{u \in D} N[u] = V(G)$. Different additional conditions lead to variations of domination. Some examples are as follows.

A dominating set D of a graph G is a/an

total dominating set	if $G[D]$ has no isolated vertex,
independent dominating set	if $G[D]$ has no edge,
connected dominating set	if $G[D]$ is connected,
paired-dominating set	if $G[D]$ has a perfect matching,
perfect dominating set	if $|N[v] \cap D| = 1$ for all $v \in V(G) \backslash D$,
efficient dominating set	if $|N[v] \cap D| = 1$ for all $v \in V(G)$,
dominating clique	if D is a clique,
dominating cycle	if D is the vertex set of a cycle.

For these variations of domination, the corresponding *domination numbers* are as follows.

$\gamma_t(G)$	is the total domination number of G.
$\gamma_i(G)$	is the independent domination number of G.
$\gamma_c(G)$	is the connected domination number of G.
$\gamma_{pr}(G)$	is the paired-domination number of G.
$\gamma_{per}(G)$	is the perfect domination number of G.
$\gamma_{eff}(G)$	is the efficient domination number of G.
$\gamma_{cl}(G)$	is the clique domination number of G.
$\gamma_{cy}(G)$	is the cycle domination number of G.

The many-neighbor variations of domination are also considered. For a positive integer k, a *k-dominating set* (respectively, *k-tuple dominating set*) of a graph G is a subset D of $V(G)$ such that every vertex in $V(G) \backslash D$ (respectively, $V(G)$) is dominated by at least k vertices in D. The *k-domination number* $\gamma_k(G)$ (respectively, *k-tuple domination number* $\gamma_{\times k}(G)$) of a graph G is the minimum size of a k-dominating set (respectively, k-tuple dominating set) of G.

The distance variation of domination is also considered. For a positive integer k, a *distance k-dominating set* of a graph G is a subset D of $V(G)$ such that every vertex v in $V(G)$ there exists some vertex u in D with $d(u, v) \le k$. The *distance k-domination number* $\gamma_{disk}(G)$ of a graph G is the minimum size of a distance k-dominating set of G.

The edge variation of domination is also considered. In a graph, an edge is said to dominate itself and all edges adjacent to it. An *edge dominating set* of G is a subset D of $E(G)$ such that every edge in the graph is dominated by some edge in D. The *edge domination number* $\gamma_e(G)$ of a graph G is the minimum size of an edge dominating set of G. Similar to domination, edge domination also has the variations of being total, independent, connected, perfect and efficient.

The concept of domination may also be considered in terms of integer-valued or set-valued functions as follows.

A dominating set D of a graph G corresponds to a *dominating function* which is a function $f : V(G) \to \{0,1\}$ such that $\sum_{u \in N[v]} f(u) \geq 1$ for every $v \in V(G)$. The *value* of f is $f(V(G)) = \sum_{v \in V(G)} f(v)$. Then $\gamma(G)$ is equal to the minimum value of a dominating function of G.

A *signed dominating function* is a function $f : V(G) \to \{1,-1\}$ such that $\sum_{u \in N[v]} f(u) \geq 1$ for every $v \in V(G)$. The *signed domination number* $\gamma_s(G)$ of G is the minimum value of a signed dominating function.

A *Roman dominating function* is a function $f : V(G) \to \{0,1,2\}$ such that every vertex v with $f(v) = 0$ is adjacent to some vertex u with $f(u) = 2$. The *Roman domination number* $\gamma_{Rom}(G)$ of G is the minimum value of a Roman dominating function.

For a positive integer k, a *k-rainbow dominating function* is a function $f : V(G) \to 2^{\{1,2,\dots,k\}}$ such that $f(v) = \emptyset$ implies $\cup_{u \in N(v)} f(u) = \{1,2,\dots,k\}$. The *value* of f is $|f|(V(G)) = \sum_{v \in V(G)} |f(v)|$. The *$k$-rainbow domination number* $\gamma_{krain}(G)$ of G is the minimum value of a k-rainbow dominating function. Note that 1-rainbow domination is the ordinary domination.

A rather different variation motivated from the power system monitoring is as follows. For a positive integer k, suppose D is a vertex subset of a graph G. The following two observation rules are applied iteratively.

(OR1) A vertex in D observes itself and all of its neighbors.

(OR2) If an observed vertex is adjacent to at most k unobserved vertices, then these unobserved vertices become observed as well.

The set D is a *k-power dominating set* of G if all vertices of the graph are observed after repeatedly applying the above two observation rules. Alternatively, let $\mathcal{S}_0(D) = N[D]$ and

$$\mathcal{S}_{i+1}(D) = \cup\{N[v] : v \in \mathcal{S}_i(D), |N(v) \backslash \mathcal{S}_i(D)| \leq k\}$$

for $i \geq 0$. Note that $\mathcal{S}_0(D) \subseteq \mathcal{S}_1(D) \subseteq \mathcal{S}_2(D) \subseteq \cdots$ and there is a non-negative integer t such that $\mathcal{S}_i(D) = \mathcal{S}_t(D)$ for $i \geq t$. Using this notation, D is a k-power dominating set if and only if $\mathcal{S}_t(D) = V(G)$. The *k-power domination number* $\gamma_{kpow}(G)$ of G is the minimum size of a k-power dominating set. For the case of $k = 1$, 1-power domination is called power domination. For more detail of power domination, see Section 3.5.

Figure 1.10 shows a graph G of 19 vertices and 28 edges, the values of $\gamma_\pi(G)$ for various types π of domination and the corresponding optimal sets/functions are given as follows.

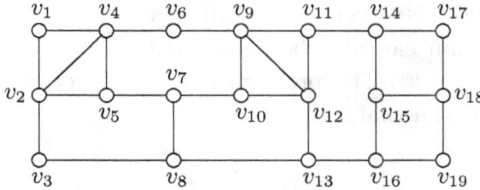

Fig. 1.10. A graph G of 19 vertices and 28 edges.

$$
\begin{aligned}
\gamma(G) &= 5, \quad D^* = \{v_2, v_8, v_9, v_{14}, v_{19}\}. \\
\gamma_t(G) &= 7, \quad D^* = \{v_2, v_4, v_7, v_{10}, v_{12}, v_{15}, v_{18}\}. \\
\gamma_i(G) &= 5, \quad D^* = \{v_2, v_8, v_9, v_{14}, v_{19}\}. \\
\gamma_c(G) &= 9, \quad D^* = \{v_2, v_5, v_7, v_9, v_{10}, v_{11}, v_{14}, v_{15}, v_{16}\}. \\
\gamma_{pr}(G) &= 8, \quad D^* = \{v_2, v_4, v_7, v_{10}, v_{12}, v_{13}, v_{15}, v_{18}\}. \\
\gamma_{per}(G) &= 6, \quad D^* = \{v_4, v_8, v_{12}, v_{13}, v_{16}, v_{17}\}. \\
\gamma_{eff}(G) &= \infty, \text{ there is no efficient dominating set.} \\
\gamma_{cy}(G) &= 11, \quad D^* = \{v_4, v_6, v_9, v_{11}, v_{14}, v_{15}, v_{16}, v_{13}, v_8, v_7, v_5\}. \\
\gamma_{dis2}(G) &= 3, \quad D^* = \{v_6, v_{13}, v_{17}\}. \\
\gamma_{dis3}(G) &= 2, \quad D^* = \{v_7, v_{17}\}. \\
\gamma_{disk}(G) &= 1, \quad D^* = \{v_{11}\} \text{ when } k \geq 4. \\
\gamma_e(G) &= 6, \quad D^* = \{v_2 v_4, v_7 v_8, v_9 v_{12}, v_{14} v_{17}, v_{15} v_{18}, v_{16} v_{19}\}. \\
\gamma_s(G) &= 9, \quad f^*(v_i) = -1 \text{ if } i = 1, 3, 6, 12, 18 \text{ and } f^*(v_i) = 1 \text{ otherwise.} \\
\gamma_{Rom}(G) &= 10, \quad f^*(v_i) = 2 \text{ if } i = 2, 8, 9, 14, 19 \text{ and } f^*(v_i) = 0 \text{ otherwise.} \\
\gamma_{1rain}(G) &= 5, \quad f^*(v_i) = \{1\} \text{ if } i = 2, 8, 9, 14, 19 \text{ and } f^*(v_i) = \emptyset \text{ otherwise.} \\
\gamma_{2rain}(G) &= 9, \quad f^*(v_i) = \{1\} \text{ if } i = 2, 9, 10, 16, 17, \\
&\qquad f^*(v_i) = \{2\} \text{ if } i = 4, 8, 11, 18 \text{ and } f^*(v_i) = \emptyset \text{ otherwise.} \\
\gamma_{3rain}(G) &= 11, \quad f^*(v_i) = \{1\} \text{ if } i = 3, 6, 8, 13, 17, \ f^*(v_i) = \{2\} \text{ if } i = 1, 10, 15, \\
&\qquad f^*(v_i) = \{3\} \text{ if } i = 5, 11, 19 \text{ and } f^*(v_i) = \emptyset \text{ otherwise.} \\
\gamma_{4rain}(G) &= 14, \quad f^*(v_i) = \{1\} \text{ if } i = 3, 6, 8, 13, 17, \ f^*(v_1) = \{2\}, \\
&\qquad f^*(v_i) = \{3\} \text{ if } i = 11, 19, \ f^*(v_i) = \{2, 4\} \text{ for } i = 10, 15, \\
&\qquad f^*(v_5) = \{3, 4\} \text{ and } f^*(v_i) = \emptyset \text{ otherwise.}
\end{aligned}
$$

$\gamma_{5\mathrm{rain}}(G) = 17$, $f^*(v_i) = \{1\}$ if $i = 3, 6, 13, 17$, $f^*(v_i) = \{2\}$ if $i = 1, 8, 11, 19$,
$\qquad f^*(v_i) = \{3, 4, 5\}$ if $i = 5, 10, 15$ and $f^*(v_i) = \emptyset$ otherwise.
$\gamma_{k\mathrm{rain}}(G) = 19$, $f^*(v_i) = \{1\}$ for all i when $k \geq 6$.
$\gamma_{1\mathrm{pow}}(G) = 2$, $\quad D^* = \{v_2, v_{18}\}$.
$\gamma_{k\mathrm{pow}}(G) = 1$, $\quad D^* = \{v_2\}$ when $k \geq 2$.

The (*vertex-*)*weighted* versions of all of the above vertex-subset variations of domination can also be considered. Now, every vertex v has a real-valued weight $w(v)$. The problem is to find a dominating set D in a suitable variation such that

$$w(D) = \sum_{v \in D} w(v)$$

is as small as possible. Denote this minimum value by $\gamma_\pi(G, w)$, where π stands for a variation of the domination problem. When $w(v) = 1$ for all vertices v, the weighted cases become the cardinality cases.

For some variations of domination, the vertex weights may be assumed to be nonnegative as the following lemma shows.

Lemma 1.3. *Suppose G is a graph in which every vertex is associated with a real-valued weight $w(v)$. If $w'(v) = \max\{w(v), 0\}$ for all vertices $v \in V(G)$, then for every $\pi \in \{\emptyset, \mathrm{t}, \mathrm{c}, k, \times k, \mathrm{disk}, \mathrm{kpow}\}$ it is the case that*

$$\gamma_\pi(G, w) = \gamma_\pi(G, w') + \sum_{w(v)<0} w(v).$$

Proof. Denote by A the set of all vertices v with $w(v) < 0$. Suppose D is a π-dominating set of G with $\sum_{v \in D} w(v) = \gamma_\pi(G, w)$. Then

$$\gamma_\pi(G, w') \leq \sum_{v \in D} w'(v)$$

$$= \sum_{v \in D} w(v) - \sum_{v \in D \cap A} w(v)$$

$$\leq \gamma_\pi(G, w) - \sum_{w(v)<0} w(v).$$

On the other hand, for any π-dominating set D of G with $\sum_{v \in D} w'(v) = \gamma_\pi(G, w')$, $D \cup A$ is also a π-dominating set of G and so

$$\gamma_\pi(G, w) \leq \sum_{v \in D \cup A} w(v)$$

$$= \sum_{v \in D} w'(v) + \sum_{v \in A} w(v)$$

$$= \gamma_\pi(G, w') + \sum_{w(v) < 0} w(v).$$

The lemma then follows. $\qquad\square$

More generally, one may consider *vertex-edge-weighted* cases of the domination problems as follows. Besides the weights of vertices, each edge e has a weight $\overline{w}(e)$. The object is then to find a dominating set D in a suitable variation such that

$$\overline{w}(D) = \sum_{v \in D} w(v) + \sum_{u \in V(G) \setminus D} \overline{w}(uu')$$

is as small as possible, where u' is a vertex in D that is adjacent to u. Note that there are many choices of u' except for the perfect domination and its three variations. Denote this minimum value by $\gamma_\pi(G, w, \overline{w})$, where π stands for a variation of domination. When $\overline{w}(e) = 0$ for all edges e, the vertex-edge-weighted cases become the vertex-weighted cases.

Another parameter related to domination is as follows. The *domatic number* $d(G)$ of a graph G is the maximum number r such that G has r pairwise disjoint dominating sets $D_1, D_2, \ldots, D_r$. One can also define *total, independent, connected, paired-, perfect, efficient, clique, cycle, k-, k-tuple, distance k-, edge* and *k-power domatic numbers* $d_t(G)$, $d_i(G)$, $d_c(G)$, $d_{pr}(G)$, $d_{per}(G)$, $d_{eff}(G)$, $d_{cl}(G)$, $d_{cy}(G)$, $d_k(G)$, $d_{\times k}(G)$, $d_{disk}(G)$, $d_e(G)$ and $d_{kpow}(G)$, respectively, according to above variations of domination in similar ways. In some, but not all cases, one may also require that $V(G)$ is equal to the union of $D_1, D_2, \ldots, D_r$.

1.4 Mathematical induction and recurrence

In this book, mathematical induction and the primal-dual method are two basic but powerful approaches for the proofs of many theorems. The proofs using these approaches contain spirits of algorithms. On the other hand, algorithms very often also provide proofs of theorems. These proofs and algorithms are, in the sense, two sides of the same coin.

This section discusses the first approach: mathematical induction.

Mathematical induction is a useful method to prove a property $P(n)$ having a positive integer n as a variable. To prove infinitely many cases $P(1), P(2), P(3), \ldots$, it is done by first proving a simple case, then showing that if the property is true for a given case, then the next case is also true. The following informal metaphor of falling dominoes, given by Diaz (2013), helps to explain the method.

> Imagine a bunch of dominoes on a table. They are set up in a straight line, and you are about to push the first piece to set off the chain reaction that will bring all the dominoes down. For this chain reaction to knock off every piece, you have to make sure that every piece is close enough to the next one. Otherwise, the reaction will stop. First, you have to be absolutely sure to have put all the domino pieces in such a way that each of them, while falling, will knock the next one down. Next, you just need to knock down the first piece. You can then be sure that, eventually, every piece is definitely going to fall.

Formal description of the method is as follows.

Mathematical Induction. Let $P(n)$ be a property having a positive integer n as a variable. If the following two steps are true, then $P(n)$ is true for every positive integer n.

Base Step. $P(1)$ is true.
Inductive Step. For every $n > 1$, if $P(n-1)$ is true, then so is $P(n)$.

The statement "if $P(n-1)$ is true" is called the *induction hypothesis.*

Example 1.1. For every positive integer n, the following identity holds.

$$1^3 + 2^3 + \cdots + n^3 = \frac{n^2(n+1)^2}{4}.$$

Proof.   A proof by induction is as follows.

The identity holds for $n = 1$, since $1^3 = \frac{1^2(1+1)^2}{4}$. For every $n > 1$, assume

$$1^3 + 2^3 + \cdots + (n-1)^3 = \frac{(n-1)^2 n^2}{4}.$$

Then

$$1^3 + 2^3 + \cdots + (n-1)^3 + n^3 = \frac{(n-1)^2 n^2}{4} + n^3 = \frac{n^2(n+1)^2}{4}.$$

Thus, by the induction principle, $1^3 + 2^3 + \cdots + n^3 = \frac{n^2(n+1)^2}{4}$ for every positive integer n. $\square$

It is curious where the identity comes from. If the problem asks "find a formula in term of n for $1^3 + 2^3 + \cdots + n^3$," then how does one get the solution $\frac{n^2(n+1)^2}{4}$. In general, it takes more efforts to get the solution before an induction proof for its correctness. It could be discovered through experiments with small values of n or through analytical methods. The induction proof is more or less a formalism.

The induction principle has many variations. For instance, for a fixed integer n_0, the revised form is as below.

Base Step. $P(n_0)$ is true.
Inductive Step. For every $n > n_0$, if $P(n-1)$ is true, then so is $P(n)$.

As an example, the initial n_0 is chosen to be 6, in order to prove that $2^n \geq 10n$ for every integer $n \geq 6$.

Another variation is a two-step argument with the following revision.

Base Step. $P(1)$ and $P(2)$ are true.
Inductive Step. For every $n > 2$, if $P(n-2)$ and $P(n-1)$ are true, then so is $P(n)$.

As an example, the *Fibonacci numbers* are defined by: $F_1 = F_2 = 1$ and

$$F_n = F_{n-2} + F_{n-1}$$

for $n > 2$. This variation of induction is useful to prove that $1.5^n/3 < F_n < 2^n/1.5$ for every positive integer n.

Another variation is the *strong induction* with the following revision.

Base Step. $P(1)$ is true.
Inductive Step. For every $n > 1$, if $P(n')$ is true for every $n' < n$, then so is $P(n)$.

Note that the inequalities for the Fibonacci numbers can also be proved by the strong induction with two-step initial conditions. The following is another example proved by the strong induction. In this example, we will also demonstrate that a loose property is not easy to be proved by induction.

Example 1.2. The Knuth numbers are defined as: $K_0 = 1$ and

$$K_n = 1 + \min\{2K_{\lfloor(n-1)/2\rfloor}, 3K_{\lfloor(n-1)/3\rfloor}\}$$

for $n > 0$. Then $K_n \geq n$ for every integer $n \geq 0$.

A failed proof. A proof by induction is as follows.
   The inequality holds for $n = 0$, since $K_0 = 1 \geq 0$.
   Assume $n > 0$ and $K_{n'} \geq n'$ for every $n' < n$. Then

$$
\begin{aligned}
K_n &= 1 + \min\{2K_{\lfloor(n-1)/2\rfloor}, 3K_{\lfloor(n-1)/3\rfloor}\} \\
&\geq 1 + \min\{2\lfloor(n-1)/2\rfloor, 3\lfloor(n-1)/3\rfloor\} \\
&\geq 1 + \min\{2(n-1)/2, 3(n-1)/3\} \\
&= n.
\end{aligned}
$$

Thus, by the induction principle, $K_n \geq n$ for every integer $n \geq 0$.   □

The proof is not correct, since $\lfloor(n-1)/2\rfloor < (n-1)/2$ when $n-1$ is not a multiple of 2 and $\lfloor(n-2)/3\rfloor < (n-2)/3$ when $n-2$ is not a multiple of 3. Correct inequalities are

$$\lfloor(n-1)/2\rfloor \geq (n-2)/2 \text{ and } \lfloor(n-1)/3\rfloor \geq (n-3)/3.$$

Therefore, to prove $K_n \geq n$, a useful approach is to prove a stronger statement $K_n \geq n+1$ as follows. This works because we have a stronger induction hypothesis.

A correct proof. We in fact prove a stronger inequality that $K_n \geq n+1$ for every integer $n \geq 0$ by induction as follows.
   The inequality holds for $n = 0$, since $K_0 = 1 \geq 0 + 1$.
   Assume $n > 0$ and $K_{n'} \geq n' + 1$ for every $n' < n$. Then

$$
\begin{aligned}
K_n &= 1 + \min\{2K_{\lfloor(n-1)/2\rfloor}, 3K_{\lfloor(n-1)/3\rfloor}\} \\
&\geq 1 + \min\{2(\lfloor(n-1)/2\rfloor + 1), 3(\lfloor(n-1)/3\rfloor + 1)\} \\
&\geq 1 + \min\{2n/2, 3n/3\} \\
&= n + 1.
\end{aligned}
$$

Thus, by the induction principle, $K_n \geq n + 1$ for every integer $n \geq 0$.
   □

Next is another example to show that in order to prove a property it is better to prove a stronger one. Originally, we want the property that every connected graph of at least two vertices has at least one vertex which is not a cut-vertex. The following gives a stronger property, which is easier to prove than the original one.

Example 1.3. Every connected graph G of at least two vertices has at least two vertices which are not cut-vertices.

Proof. The proof is by induction on $n := |V(G)|$. If $n = 2$, then $G = K_2$ and so the two vertices are not cut-vertices.

Suppose $n \geq 3$ and the assertion is true for every connected graph of $n' < n$ vertices. If G has no cut-vertices, then the assertion holds. Now suppose G has a cut-vertex x. Let $H_1, H_2, \ldots, H_m$ be components of $G - x$, where $m \geq 2$. Then each H_i' obtained from H_i by adding x and all edges between x and its neighbors in $V(H_i)$ is also connected. As $2 \leq |V(H_i')| < n$, by the induction hypothesis, H_i' has at least two non-cut-vertices (note that one is not enough), one of them is not x. This noncut-vertex v_i of H_i' is also a noncut-vertex of G. Hence, G has at least $m \geq 2$ noncut-vertices.

The assertion then follows from the induction principle. $\square$

The following is also an example to illustrate that for solving a problem it is better to solve a more general one. The idea is used as the labeling approach discussed in Chapter 3.

Example 1.4. If x is a leaf adjacent to y in G, then

$$\gamma(G) = \gamma'(G - x),$$

where $\gamma'(G - x)$ is the minimum size of a dominating set D' of $G - x$ such that $y \in D'$.

Proof. Suppose D is a minimum dominating set of G. Since x is of degree one, either x or y is in D. Let $D' = (D \backslash \{x\}) \cup \{y\}$. Then D' is a dominating set of $G - x$ such that $y \in D'$. Hence, $\gamma'(G - x) \leq |D'| \leq |D| = \gamma(G)$.

On the other hand, suppose D' is a dominating set of $G - x$ such that $y \in D'$. Since x is adjacent to y, D' is also a dominating set of G. Hence, $\gamma(G) \leq |D'| = \gamma'(G - x)$.

In conclusion, $\gamma(G) = \gamma'(G - x)$. $\square$

The following is again an example to illustrate that for solving a problem it is better to solve a more general one. The idea is used by the dynamic programming approach discussed in Chapter 3. Recall that the n-path P_n has

$$\text{vertex set } V(P_n) = \{v_1, v_2, \ldots, v_n\} \text{ and}$$
$$\text{edge set } E(P_n) = \{v_1 v_2, v_2 v_3, \ldots, v_{n-1} v_n\}.$$

Example 1.5. Count the number $g(n)$ of functions $f : V(P_n) \to \{2, 3, 4\}$ such that for $i = 2, 3, 4$ there are no i consecutive vertices in $V(P_n)$ all have i as f-values.

Solution. For $i = 2, 3, 4$, let $g_i(n)$ count the number of functions $f : V(P_n) \to \{2, 3, 4\}$ such that $f(v_n) = i$ and for $i = 2, 3, 4$ there exist no i consecutive vertices in $V(P_n)$ all have i as f-values. Then

$$g(n) = g_2(n) + g_3(n) + g_4(n).$$

Some initial values are

$$g_2(1) = g_3(1) = g_4(1) = 1, \ \ g_3(2) = g_4(2) = 3, \ \ g_4(3) = 8.$$

For $n \geq 2$, since $f(n) = 2$ implies $f(n-1) \neq 2$ or equivalently $f(n-1) = 3$ or 4, one has

$$g_2(n) = g_3(n-1) + g_4(n-1).$$

For $n \geq 3$, since $f(n) = 3$ implies either $f(n-1) \neq 3$ or $f(n-1) = 3$ but $f(n-2) \neq 3$, one has

$$g_3(n) = g_2(n-1) + g_4(n-1) + g_2(n-2) + g_4(n-2).$$

For $n \geq 4$, since $f(n) = 4$ implies either $f(n-1) \neq 4$ or $f(n-1) = 4$ but $f(n-2) \neq 4$ or $f(n-1) = f(n-2) = 4$ but $f(n-3) \neq 4$, one has

$$g_4(n) = g_2(n-1) + g_3(n-1) + g_2(n-2) + g_3(n-2) + g_2(n-3) + g_3(n-3).$$

While it is not easy to find formulas in terms of n for $g_2(n), g_3(n), g_4(n)$ and $g(n)$, the values can be computed for $i = 1, 2, \ldots, n$ recursively. □

Another equivalent statement for the induction principle is the *well ordering principle* for positive integers.

Well ordering of positive integers. *Every nonempty subset of the set of all positive integers has a least element.*

In this book, we very often use this version of induction to prove theorems. Here is a simple example.

Example 1.6. For every two vertices u and v in a graph G, there is a u-v walk if and only if there is a u–v path.

Proof. ($\Leftarrow$) This follows from that a path is a walk.

($\Rightarrow$) Suppose there is a u-v walk. Choose a u-v walk $W = (v_0, v_1, \ldots, v_r)$ such that r is as small as possible. Then W is a path. Otherwise, $v_i = v_j$ for some $i < j$. In this case, $(v_0, v_1, \ldots, v_i, v_{j+1}, v_{j+2}, \ldots, v_r)$ is a u-v walk shorter than W, contradicting the minimality of r. Hence, W is a u–v path of G as desired. $\qquad\qquad\square$

1.5 Perfect graph theory and the primal-dual approach

This section discusses the primal-dual approach, in particular the perfect graph theory, related to the domination problem.

Here comes the story.

In 1956, Shannon presented a pioneer paper on the *zero-error capacity* of a noisy channel in communication theory. In terms of graphs, for a graph G it is to determine

$$\Psi(G) = \lim_{n \to \infty} \alpha(G^n)^{1/n},$$

where $G^1 = G$ and $G^n = G^{n-1} \boxtimes G$ for $n \geq 2$.

It is not hard to prove by induction that $\alpha(G)^n \leq \alpha(G^n) \leq \theta(G)^n$, or equivalently $\alpha(G) \leq \alpha(G^n)^{1/n} \leq \theta(G)$. Hence,

$$\alpha(G) \leq \Psi(G) \leq \theta(G).$$

It is then the case that if $\alpha(G) = \theta(G)$, then $\alpha(G) = \Psi(G) = \theta(G)$. Shannon posed the following two questions.

- What are the minimal graphs G for which $\alpha(G) < \Psi(G)$? (He knew that C_5 is the smallest one.)
- What is the zero capacity $\Psi(C_5)$ of C_5?

Lovász (1979) determined that $\Psi(C_5) = \sqrt{5}$. This is one of the reasons in the citation when he won the Wolf Prize in 1999.

According to Berge's recollection (see the papers by Berge (1997) and Berge and Ramírez Alfonsń (2001)), the concept of perfect graphs starts at late 1950s. In 1957, Berge was preparing to write a book on graph theory. His friend Schutzenberger told him that Shannon's work is probably a work missed by mathematicians working on algebra or combinatorics. Berge then started to study the graph parameters $\alpha(G), \theta(G)$ and $\Psi(G)$. He was interested in the conditions for $\alpha(G) = \theta(G)$, or equivalently $\omega(\overline{G}) = \chi(\overline{G})$. He then announced the definition of perfect graphs in a 1961 conference, and formally introduced the concept in a 1963 paper. A graph G is *perfect* if every induced subgraph H of G satisfies $\alpha(H) = \theta(H)$.

Note that to study the "beautiful" property of $\alpha(G) = \theta(G)$, it is better to investigate whether G is perfect. This can be seen from the example that no matter how "ugly" a graph H of n vertices is, the graph G obtain from H by adding n new vertices each connecting to a different vertex in H always has $\alpha(G) = \theta(G) = n$.

Most importantly, Berge raised the following two influential conjectures.

The weak perfect graph conjecture: G is perfect if and only if its complement $\overline{G}$ is perfect.

The strong perfect graph conjecture: G is perfect if and only if it is C_n-free and $\overline{C}_n$-free for every odd $n \geq 5$.

Since then, there are two lines working on perfect graphs. The first one is to prove the above two conjectures. Chordal graphs and comparability graphs were two classical classes of perfect graphs indicating the weak perfect graph conjecture is likely true at very beginning, see Chapters 4 and 7 and the book by Golumbic (1980) for more detail. Note that the strong conjecture implies the weak one. The second line is to design algorithms to compute $\alpha(G), \theta(G), \omega(G)$ and $\chi(G)$ for subclasses of perfect graphs.

The weak perfect graph conjecture was proved by Lovász (1972, 1972a) 10 years later, and the strong perfect graph conjecture was proved by Chudnovsky *et al.* (2006) after 30 more years.

For the spirit of the primal-dual approach, first note that $\alpha(G)$ and $\theta(G)$ form a primal-dual pair. Since every vertex in a stable set S is covered by at least one clique in a clique cover but no two vertices in S can be covered by a same clique, it is the case that the following inequality holds.

Weak duality inequality (clique cover): $\alpha(G) \leq \theta(G)$ for any graph G.

The inequality may be strict as shown by C_{2k+1} and $\overline{C_{2k+1}}$ with $k \geq 2$ that

$$\alpha(C_{2k+1}) = k < k+1 = \theta(C_{2k+1}) \quad \text{and}$$
$$\alpha(\overline{C_{2k+1}}) = 2 < 3 = \theta(\overline{C_{2k+1}}).$$

It is also the case that $\omega(G)$ and $\chi(G)$ form a primal-dual pair. For a proper coloring f of G, since no two vertices in a clique C have the same color, the restriction of f on C is a one-to-one mapping. Then the following inequality holds.

Weak duality inequality (coloring): $\omega(G) \leq \chi(G)$ for any graph G.

The inequality may be strict as shown by C_{2k+1} and $\overline{C_{2k+1}}$ with $k \geq 2$ that

$$\omega(C_{2k+1}) = 2 < 3 = \chi(C_{2k+1}) \quad \text{and}$$
$$\omega(\overline{C_{2k+1}}) = k < k+1 = \chi(\overline{C_{2k+1}}).$$

Similar to the clique cover problem, the domination problem in graphs is a covering-type problem, as the condition $\cup_{v \in D} N_G[v] = V(G)$ for the set D to be a dominating set of G can be viewed as that every vertex in $V(G)$ is covered by at least one $N_G[v]$ with $v \in D$.

In fact, the domination problem is closely related to the clique cover problem. More precisely, for every vertex v in G, $N_G[v]$ is a clique in G^2. It is then the case that a dominating set D of G induces a clique cover $\{N_G[v] : v \in D\}$ of G^2. Consequently,

$$\theta(G^2) \leq \gamma(G). \tag{1.1}$$

Note that the inequality may be strict as shown by C_4, C_5 and the 3-sun S_3 in Figure 1.11, since the squares of these graphs are complete graphs. In other words, $\theta(C_4^2) = \theta(C_5^2) = \theta(S_3^2) = 1 < 2 = \gamma(C_4) = \gamma(C_5) = \gamma(S_3)$.

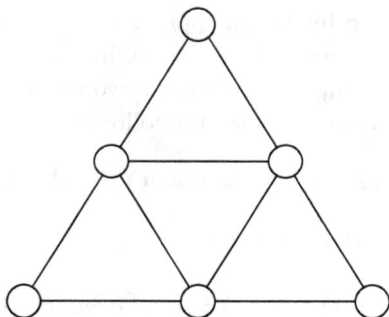

Fig. 1.11. The 3-sun S_3.

On the other hand, there are many graphs G with $\theta(G^2) = \gamma(G)$. The following is an interesting result.

Theorem 1.4. *Suppose G is a $\{C_4, C_5, C_6, S_3\}$-free graph. If C is a nonempty clique in G^2, then $C \subseteq N_G[x]$ for some vertex x, where x may not be in C. Consequently, $\theta(G^2) = \gamma(G)$.*

Proof.   Suppose to the contrary that the theorem is not true. Choose a minimum counterexample $C = \{v_1, v_2, \ldots, v_r\}$ to the theorem, that is, C is a clique in G^2, C is not a subset of any $N_G[x]$ but any proper subset of C is a subset of some $N_G[y]$. Then $r \geq 3$, since if $r \leq 2$ then C being a clique in G^2 clearly implies that C being a subset of some $N_G[x]$.

By the minimality of r, for $1 \leq i \leq r$ there is some u_i such that $C \backslash \{v_i\} \subseteq N_G[u_i]$ but $v_i \notin N_G[u_i]$. And so $|\{u_1, u_2, u_3\}| = 3$. Consider the closed walk $W = (v_1, u_3, v_2, u_1, v_3, u_2, v_1)$, where u_i is ignored if it is the same as some v_j with $j \neq i$. Then W is a cycle of length $\ell \in \{3, 4, 5, 6\}$, depending on how many u_i's are ignored. By the assumption that G is $\{C_4, C_5, C_6\}$-free, $\ell = 6$ and W has at least $\ell - 3 = 3$ chords. The chords can only between u_i's. Then W is S_3, a contradiction. The theorem then follows.   □

If the condition of G being $\{C_4, C_5, C_6, S_3\}$-free is replaced by that G is $\{C_4, C_5, S_3\}$-free, the result "If C is a clique in G^2, then $C \subseteq N_G[x]$ for some vertex x" may not be true. As an example, in C_6 with $V(C_6) = \{v_1, v_2, v_3, v_4, v_5, v_6\}$ and $E(C_6) = \{v_1 v_2, v_2 v_3, v_3 v_4, v_4 v_5, v_5 v_6, v_6 v_1\}$, although $C = \{v_1, v_3, v_5\}$ is a clique in C_6^2, but there is no x with $C \subseteq N_{C_6}[x]$. However, as $\theta(C_6^2) = 2 = \gamma(C_6)$. It is suspected that

"$\theta(G^2) = \gamma(G)$" holds if G is $\{C_4, C_5, S_3\}$-free. If this is true, a more clever proof is necessary.

Question 1. *Is it true that $\theta(G^2) = \gamma(G)$ for every $\{C_4, C_5, S_3\}$-free graph G?*

Similar to the primal-dual relation between $\alpha(G)$ and $\theta(G)$. There is a natural dual of the domination problem as follows. In a graph G, a *2-stable set* is a subset $S \subseteq V(G)$ such that $d(u, v) > 2$ for every two distinct vertices u and v in S. The *2-stability number* $\alpha_2(G)$ of G is the maximum size of a 2-stable set in G.

Since every vertex in a 2-stable set S is dominated by at least one vertex in a dominating set but no two vertices in S can be dominated by the same vertex, it is the case that the following inequality holds.

Weak duality inequality (domination): $\alpha_2(G) \le \gamma(G)$ for any graph G.

Note that the inequality may be strict as shown by C_n that

$$\alpha_2(C_n) = \lfloor n/3 \rfloor \quad \text{and} \quad \lceil n/3 \rceil = \gamma(C_n).$$

Note that $\alpha_2(C_n) = \gamma(C_n)$ only when n is a multiple of 3.

More generally, we have

$$\alpha_2(G) = \alpha(G^2) \le \theta(G^2) \le \gamma(G). \tag{1.2}$$

The equality follows from definitions. The first inequality is the weak duality inequality (clique cover) and the second inequality is (1.1).

For some classes of graphs such as trees (see Chapter 3) and strongly chordal graphs (see Chapter 6), the weak duality inequality (domination) is an equality, called the *strong duality equality* (domination). An interesting approach to prove the equality is to design an algorithm which outputs a dominating set D and a 2-stable set S such that $|D| \le |S|$. Then,

$$|S| \le \alpha_2(G) \le \gamma(G) \le |D| \le |S|$$

and so all inequalities are equalities. Consequently,

(1) S is a maximum 2-stable set of G,
(2) D is a minimum dominating set of G and
(3) the strong duality equality $\alpha_2(G) = \gamma(G)$ holds for the graph G.

This primal-dual approach not only finds a maximum 2-stable set and a minimum dominating set, but also verifies the strongly duality equality (domination), for more details see Chapters 3 and 6.

There is also a primal-dual approach for total domination. Now the dual of total domination is as follows. In a graph G, a *total 2-stable set* is a subset $S \subseteq V(G)$ such that every two distinct vertices $u, v \in S$ are not adjacent to a same vertex of G. The *total 2-stability number* $\alpha_{t2}(G)$ of G is the maximum size of a total 2-stable set in G. Note that a total 2-stable set may possibly contain two adjacent vertices if they are not in a triangle.

The following inequality is similar to that for domination.

Weak duality inequality (total domination): $\alpha_{t2}(G) \leq \gamma_t(G)$ for any graph G.

The inequality may be strict, as shown by the n-cycle C_n that

$$\alpha_{t2}(C_n) = \left\lfloor \frac{n}{4} \right\rfloor + \left\lfloor \frac{n+1}{4} \right\rfloor \quad \text{and} \quad \gamma_t(C_n) = \left\lfloor \frac{n+2}{4} \right\rfloor + \left\lfloor \frac{n+3}{4} \right\rfloor.$$

Note that $\alpha_{t2}(C_n) = \gamma_t(C_n)$ only when n is a multiple of 4.

For some classes of graphs such as trees (see Chapter 3), the weak duality inequality (total domination) is an equality, called the *strong duality equality* (total domination). Again, if one can design an algorithm which outputs a total dominating set D and a total 2-stable set S such that $|D| \leq |S|$. Then,

$$|S| \leq \alpha_{t2}(G) \leq \gamma_t(G) \leq |D| \leq |S|$$

and so all inequalities are equalities. Consequently,

(1) S is a maximum total 2-stable set of G,
(2) D is a minimum total dominating set of G and
(3) the strong duality equality $\alpha_{t2}(G) = \gamma_t(G)$ holds for the graph G.

Again, this primal-dual approach not only finds a maximum total 2-stable set and a minimum total dominating set, but also verifies the strong duality equality (total domination), for more details see Chapter 3.

The primal-dual approach also works for many variations of domination.

1.6 Classes of graphs discussed in this book

In this section, some classes of graphs are introduced. They are not only important in the study of domination, but also fundamental in graph theory.

A *tree* is a connected graph without any cycle. A *directed tree* is an orientation of a tree. A *rooted tree* is a directed tree in which there is a special vertex r, called the *root* of the rooted tree, such that for every vertex v there is a directed r–v path. Trees are probably the simplest structures in graph theory. Problems looking hard in general graphs are often investigated in trees first as a warm up. Domination in trees is introduced in Chapter 3. Many ideas for domination in trees are then generalized to other classes of graphs.

A graph is *chordal* if every cycle of length greater than three has a chord. The class of chordal graphs is one of the classical classes in the *perfect graph theory*, see the book by Golumbic (1980). It turns out to be also very important in the domination theory. Most variations of domination are NP-complete for chordal graphs and even for some of its special subclasses such as split graphs and undirected path graphs. Domination in chordal graphs is investigated in Chapter 4.

It is well-known that a graph is chordal if and only if it is the intersection graph of some subtrees of a certain tree. If these subtrees are paths, this chordal graph is called an *undirected path graph*. If these subtrees are directed paths of a rooted tree, the graph is called a *directed path graph*. If these subtrees are paths in some n-path, the graph is called an *interval graph*. In other words, an interval graph is the intersection graph of intervals in the real line. They play important roles in many applications in the real world. Domination in interval graphs is investigated in Chapter 5.

As an important subclass of chordal graphs, the class of strongly chordal graphs is a star of the domination theory. Strongly chordal graphs include directed path graphs, which in turn include trees and interval graphs. Domination in strongly chordal graphs is studied in Chapter 6.

A *comparability graph* is a graph $G = (V, E)$ that has a *transitive orientation* $G' = (V, E')$, that is, $uv \in E'$ and $vw \in E'$ imply $uw \in E'$. A *cocomparability graph* is the complement of a comparability graph. Cocomparability graphs are generalizations of permutation graphs and interval graphs. Domination in cocomparability graphs is investigated in Chapter 7.

A permutation diagram consists of n points on each of two parallel lines and n line segments matching the points. The intersection graph of the line segments is called a *permutation graph*. Domination in permutation graphs is introduced in Chapter 8.

A graph is *distance-hereditary* if the distance between every two vertices is the same in every connected induced subgraph containing them. Domination in distance-hereditary graphs is studied in Chapter 9.

For more detailed discussions of these classes of graphs, see the remaining chapters of this book.

1.7   Exercises

(1.1) Let $[0..4] = \{0, 1, 2, 3, 4\}$. Suppose graph G has

$$\text{vertex set } V(G) = \{v : v \subseteq [0..4], |v| = 2\} \text{ and}$$
$$\text{edge set } E(G) = \{vv' : v, v' \in V(G), v \cap v' = \emptyset\},$$

and the *Petersen graph* P has

$$\text{vertex set } V(P) = \{u_i, u'_i : i \in [0..4]\} \text{ and}$$
$$\text{edge set } E(P) = \{u_i u_{i+1}, u_i u'_i, u'_i u'_{2i} : i \in [0..4]\},$$

where the indices are taken modulo 5. Prove that $G \cong P$.

(1.2) Verify that the minimum number of kings to dominate all cells of an $m \times n$ chessboard is $\lceil \frac{m}{3} \rceil \lceil \frac{n}{3} \rceil$.

(1.3) Determine the minimum number of kings that dominate all cells of an $2 \times n$ Chinese chessboard. Justify your answer.

(1.4) Verify that Figure 1.3 does show optimal solutions to the chess and the Chinese chess problems on a 3×6 board.

(1.5) Show that removing opposite corner squares from an 8×8 chessboard leaves a sub-board that can not be partitioned into 1×2 and 2×1 rectangles. Using the same argument, make a general statement about all bipartite graphs.

(1.6) Show that every graph is the intersection graph of some family $\mathcal{F}$ satisfying that every set in $\mathcal{F}$ is finite.

(1.7) For every positive integer n, show that there exists a graph G of n vertices such that $G \cong \overline{G}$ if and only if $n \equiv 0, 1 \pmod 4$.

(1.8) Suppose there are n distinct points in the plane. Show that if any pair of points is of distance at least 1, then there are at most $3n$ pairs of points with distance exactly 1.

(1.9) The *degree sequence* of a graph is a list consisting of the degrees of all of its vertices. Let $d = (d_1, d_2, \ldots, d_n)$ be a sequence of integers satisfying $d_1 \geq d_2 \geq \ldots \geq d_n \geq 0$. Show that d is the degree sequence of some graph if and only if $d_1 \leq n - 1$ and

$$d' = (d_2 - 1, d_3 - 1, \ldots, d_{d_1+1} - 1, d_{d_1+2}, d_{d_1+3}, \ldots, d_n)$$

is the degree sequence of some graph.

(1.10) The *girth* of a graph is the minimum length of a cycle in the graph.

(a) Show that a k-regular graph of girth 4 has at least $2k$ vertices. Find all such graphs with exactly $2k$ vertices.

(b) Show that a k-regular graph of girth 5 has at least $k^2 + 1$ vertices. For $k = 2, 3$, find one such graph with exactly $k^2 + 1$ vertices.

(1.11) Show that a $\{P_4, C_3\}$-free connected graph is a complete bipartite graph.

(1.12) (a) Show that for any graph G, at least one of G and $\overline{G}$ is connected.

(b) Show that if G is a P_4-free graph with more than one vertex, then either G or $\overline{G}$ is not connected.

(1.13) Determine $\kappa(K_{m,n})$.

(1.14) For the *n-cube* $Q_n = \underbrace{P_2 \square P_2 \square \cdots \square P_2}_{n}$, determine its connectivity $\kappa(Q_n)$. Justify your answer.

(1.15) Prove Proposition 1.2.

(1.16) Determine the values $\gamma(P_n)$ and $\gamma_t(P_n)$. Justify your answers.

(1.17) Determine the values $\gamma_t(C_n)$ and $\gamma_i(C_n)$. Justify your answers.

(1.18) Determine the values $\gamma_i(K_{m,n})$ and $\gamma_c(K_{m,n})$. Justify your answers.

(1.19) For $Q_n = \underbrace{P_2 \square P_2 \square \cdots \square P_2}_{n}$, determine the values $\gamma_{pr}(Q_n)$, $\gamma_{per}(Q_n)$ and $\gamma_{cy}(Q_n)$. Justify your answers.

(1.20) Determine the value $\gamma_k(C_6 \square C_6)$ for all integers $k \geq 1$. Justify your answer.

(1.21) Determine the value $\gamma_e(C_5 \square C_5)$. Justify your answer.

(1.22) Determine the value $\gamma_s(K_5 \square K_5)$. Justify your answer.

(1.23) Determine the value $\gamma_{Rom}(P_3 \square P_n)$. Justify your answer.

(1.24) Determine the value $\gamma_{rain}(P_4 \square P_n)$. Justify your answer.

(1.25) Determine the value $\gamma_{\text{pow}}(P_5 \,\square\, P_n)$. Justify your answer.

(1.26) For $n_1 \le n_2 \le \cdots \le n_r$, determine the domatic number $d(K_{n_1,n_2,\ldots,n_r})$. Justify your answer.

(1.27) Show that $n^3 + 2n$ is a multiple of 3.

(1.28) Show that the Fibonacci number $F_n = \frac{1}{\sqrt{5}} \left(\left(\frac{1+\sqrt{5}}{2} \right)^n - \left(\frac{1-\sqrt{5}}{2} \right)^n \right)$.

(1.29) Show that the Knuth number $K_n \le 2n + 1$ for every integer $n \ge 0$.

(1.30) Show that if the Knuth number $K_n = n + 1$, then $K_{2n+2} = 2n + 3$ and $K_{3n+3} = 3n + 4$. (Chang, 1990).

(1.31) Determine all connected graphs having exactly two vertices which are not cut-vertices. Justify your answers.

(1.32) Determine the number of regions divided by n lines such that every two of them intersect at exactly one point but no three of them have a common point in the plane. Justify your answer.

(1.33) Show that $\sqrt{5} \le \Psi(C_5)$.

(1.34) Determine all n for which $\theta((P_2 \,\square\, P_n)^2) = \gamma(P_2 \,\square\, P_n)$. Justify your answer.

(1.35) Determine all n for which $\alpha_2((P_2 \,\square\, P_n)^2) = \gamma(P_2 \,\square\, P_n)$. Justify your answer.

(1.36) Determine all n for which $\alpha_{t2}((P_2 \,\square\, P_n)^2) = \gamma_t(P_2 \,\square\, P_n)$. Justify your answer.

(1.37) In a graph G, a *3-stable set* is a subset $S \subseteq V(G)$ such that $d(u, v) > 3$ for distinct $u, v \in S$. The *3-stability number* $\alpha_3(G)$ of G is the maximum size of a 3-stable set in G. Prove the **weak duality inequality (paired-domination):** $2\alpha_3(G) \le \gamma_{\text{pr}}(G)$ for any graph G. Also show that the inequality may be strict, as shown by the n-cycle C_n that for $n \ge 4$,

$$\alpha_3(C_n) = \left\lfloor \frac{n}{4} \right\rfloor \quad \text{and} \quad \gamma_{\text{pr}}(C_n) = 2 \left\lceil \frac{n}{4} \right\rceil.$$

And so $2\alpha_3(C_n) = \gamma_{\text{pr}}(C_n)$ only when n is a multiple of 4.

(1.38) Determine all n for which $2\alpha_3((P_2 \,\square\, P_n)^2) = \gamma_{\text{pr}}(P_2 \,\square\, P_n)$. Justify your answer.

Chapter 2

Design and Analysis of Algorithms

2.1 History of algorithm

Since ancient times, step-by-step procedures, now called algorithms, for solving mathematical problems have been recorded. These include Babylonian mathematics (around 2500 BC), Egyptian mathematics (around 1550 BC), Indian mathematics (around 800 BC and later), the Ifa Oracle (around 500 BC), Greek mathematics (around 240 BC), Chinese mathematics (around 200 BC and later), and Arabic mathematics (around 800 AD). Two well-known examples used by Greek mathematicians are The Sieve of Eratosthenes for finding prime numbers and the Euclidean algorithm for finding the greatest common divisor of two numbers. Arabic mathematicians such as al-Kindi in the ninth century used cryptographic algorithms for code-breaking, based on frequency analysis.

The word algorithm is derived from the name of the Persian scientist and polymath Muhammad ibn Mūsā al-Khwārizmī (romanized, c. 780–850) from Khwarazm. He produced many influential works in mathematics, astronomy and geography. Around 820 CE, he was appointed the astronomer and head of the library of the House of Wisdom in Baghdad.

About 820, al-Khwārizmī wrote the book *On the Calculation with Hindu Numerals*, which spread the Hindu-Arabic numeral system throughout the Middle East and Europe. It was translated into Latin as *Algoritmi de numero Indorum*, in which al-Khwārizmī was translated to (Latin) Algoritmi. In the late medieval period, the corruption of his name came to mean

35

the decimal number system. In the fifteenth century, under the influence of the Greek word $\alpha\rho\iota\theta o\sigma$ (arithmós), number (arithmetic), the Latin word was altered to algorithmus, English algorism. In the seventeenth century, the English term was altered to algorithm. The modern sense of algorithm was introduced in the nineteenth century.

Graph theory is a highly applicable branch of mathematics. In order to solve real world problems by using graph theory, various corresponding algorithms are designed. This is as difficult as the theory itself. An algorithm must not only be correct, but also efficient and memory-saving. By now, there are still many graph-theoretic problems for which sufficiently efficient algorithms are wanted.

Not only the theory of graphs can generate algorithms, algorithmic methods are often used to prove theorems in graph theory. The proof of Euler's theorem for Eulerian trail of a graph is such an example.

This chapter introduces some basic concepts of algorithms, and related issues for graph algorithms. For good references, see the books by Aho *et al.* (1974) and Cormen *et al.* (2022).

2.2   Computational complexity

An *algorithm* is a finite and well-defined list of instructions, which outputs the answer to a problem or the result of a job in finite steps. The study of algorithms mainly concerns two topics:

> one is the *computability* of a program,
> the other is the *computational complexity analysis* of an algorithm.

The first topic, computability, studies whether there exists an algorithm that can solve all problems of a given type.

Mathematicians in the earlier times quickly realized that this is not always possible. The most basic example is the so-called *halting problem*. The problem asks: is there an algorithm that can determine whether any given program ever stops. It can be proved that such an algorithm cannot exist. Please refer to a book on algorithm for details. Another classic example is Matiyasevich's theorem in 1970, which is that no algorithm can determine whether any Diophantus equation (that is, finding integer solutions to an integer coefficient equation) has a solution.

The conclusions for these two examples do not mean that these two questions have no answers. It only means that there are no algorithms to solve them. That is, even if they can be solved, there must be an infinite number of ways to solve them. This is something that algorithms can't do, as an algorithm can only contain a finite number of instructions. To prove that a problem is algorithmically solvable, it is necessary to construct an algorithm for the problem and prove its correctness. Occasionally, though, there are purely existential proofs.

As for the second topic "computational complexity analysis," the focus is on how much time and space (memory) an algorithm needs to solve the problem. In addition to the correctness of the algorithm, it also needs to solve the problem within acceptable time and space. It generally focuses on the time complexity analysis, as the space complexity analysis is relatively easy. The complexity should be discussed from three levels:

> one is the problem itself,
> another is the algorithm, and
> the last is the implementation of the algorithm.

First, a *problem* contains an inquiry to be answered, a requirement to be fulfilled, or a best possible situation or structure to be found. That is the *solution* to the problem. The problem consists of some *parameters* or *variables*. By setting these parameters one gets an *instance* of the problem. As an example, suppose the problem is "to determine whether a program P will stop or not." When the parameter P is "the infinite loop program," the instance then is "determining whether the infinite loop program will stop or not." The corresponding answer to this example of problem is "no." For a given problem, an *algorithm* is a step-by-step procedure that produces a solution based on an instance of the problem. The *implementation* of an algorithm is to convert an algorithm into a program in a computer. Different algorithms and different implementations lead to different levels of efficiency.

Algorithmic graph problems are divided into two categories:

> one is the *optimization problem*, which is to find a best
> > answer satisfying the conditions for the instance;
> the other is the *decision problem*, which answers
> > "yes" or "no" to the problem instance.

At a first look, the optimization problem seems to be harder than the decision problem. But in fact, as long as the latter is solved, the former will also be solved. To see this, consider the following two examples which are related to the domination problem.

The Domination Problem (optimization)
Parameters: A given graph G.
Problem: Determine the domination number $\gamma(G)$.

The Domination Problem (decision)
Parameters: A given graph G and a positive integer k.
Problem: Does G have a dominating set of size at most k?

From an algorithm for the optimization problem, it is easy to derive an algorithm for the decision problem.

On the other hand, suppose an algorithm for the decision problem is found. Then for the optimization problem, set $k = 1, 2, 3, \ldots$ one by one to solve the decision problem. The smallest k for which the answer is "yes" is the answer to the optimization problem.

If G has n vertices, then obviously $\gamma(G) \leq n$. So, the optimization problem can be solved by using the decision problem algorithm at most n times. In fact, using dichotomy,[1] the optimization problem can be solved in at most $\log_2 n$ times. Furthermore, if it is possible to run these n times in parallel, then the time for the optimization problem is the same as the decision problem.

The complexity of an algorithm for a problem is usually expressed as a function of the *input size*, which is the size of the parameters of an instance of the problem. A problem of input size n is said to be solved by an algorithm in $O(f(n))$ time if there is a positive constant $c > 0$ such that the problem is solved in at most $cf(n)$ computational steps. The time complexity of this algorithm is the smallest function f that satisfies the previous description. The notation of the space complexity is similar. The complexity of a problem is the least complexity of an algorithm for the problem.

------

[1]That is the *binary search algorithm*. The concept is simply to divide the current possible range into two from the middle every time. Then try the middle value. Depending on the result, narrow down the possible range to one of the ranges you just split in half and continue. Since the range is reduced to half of the original size each time and so at most $\log_2 n$ steps are needed.

To demonstrate the idea, consider the following example of the *maximum consecutive segment problem:*

> For a given sequence $(a_1, a_2, \ldots, a_n)$ of real numbers, find a segment of consecutive terms $(a_i, a_{i+1}, \ldots, a_j)$, where $1 \leq i \leq j \leq n$, such that $a_i + a_{i+1} + \cdots + a_j$ is maximized; in case all a_i are negative, the answer is the null segment with sum 0.

A naive algorithm is as follows.

```
MAXS ← 0;
for i = 1 to n do
for j = i to n do
{     s_{i,j} ← a_i;
      for k = i + 1 to j do s_{i,j} ← s_{i,j} + a_k;
      if MAXS < s_{i,j} then MAXS ← s_{i,j};
}
```

Since there are n^2 ways to choose i and j and computing the sum $a_i + a_{i+1} + \cdots + a_j$ requires $O(n)$ additions, the total time used in the algorithm is $O(n^3)$. By a slight modification making the computing of $s_{i,j}$ in $O(1)$-time, an $O(n^2)$-time algorithm is possible as follows.

```
MAXS ← 0;
for i = 1 to n do
for j = i to n do
{     if j = i then s_{i,i} ← a_i else s_{i,j} ← s_{i,j-1} + a_j;
      if MAXS < s_{i,j} then MAXS ← s_{i,j};
}
```

The readers are encouraged to design a more efficient algorithm for the problem, hopefully one with $O(n)$ time.

Next consider examples about number comparisons, which are frequently used in computers. Suppose $(a_1, a_2, \ldots, a_n)$ is a sequence of real numbers. The simplest example of number comparison is to find the maximum in the sequence. Here is a straightforward algorithm using $n - 1$ comparisons.

$\text{MAX} \leftarrow a_1;$
for $i = 2$ **to** n **do**
　　　if $\text{MAX} < a_i$ **then** $\text{MAX} \leftarrow a_i;$

Note that $n-1$ comparisons are necessary to find MAX, since using less than $n-1$ comparisons will cause at least one element is not compared, which may or may not be the MAX, and so the MAX can not be found. Similarly, $n-1$ comparisons are necessary and sufficient for finding MIN of the sequence $(a_1, a_2, \ldots, a_n)$.

The above methods ensure that $2n-2$ comparisons are enough for finding both MAX and MIN. However, it is not the least number of comparisons needed. For instance, for $n = 2$, one comparison is enough to find both MAX and MIN at the same time, as seen in the following.

if　　$a_1 > a_2$
then $\text{MAX} \leftarrow a_1$ and $\text{MIN} \leftarrow a_2$
else　$\text{MAX} \leftarrow a_2$ and $\text{MIN} \leftarrow a_1;$

In general, suppose c_n comparisons are necessary and sufficient for finding MAX and MIN on n numbers. To solve the problem on n numbers recursively, first find MAX and MIN for the first $n-2$ numbers by c_{n-2} comparisons. Then use one comparison to find the maximum and the minimum of a_{n-1} and a_n as follows.

if　　$a_{n-1} > a_n$
then $\text{MAX}' \leftarrow a_{n-1}$ and $\text{MIN}' \leftarrow a_n$
else　$\text{MAX}' \leftarrow a_n$ and $\text{MIN}' \leftarrow a_{n-1};$

Hence, we have MAX and MIN for the n numbers by using two more comparisons as follows.

if $\text{MAX}' > \text{MAX}$ **then** $\text{MAX} \leftarrow \text{MAX}';$
if $\text{MIN}' < \text{MIN}$　**then** $\text{MIN} \leftarrow \text{MIN}';$

These conclude that $c_n \leq c_{n-2} + 3$. By solving the following recurrence relation

$$c_1 = 0, c_2 = 1 \quad \text{and} \quad c_n \leq c_{n-2} + 3 \quad \text{for } n \geq 3,$$

one has that $c_n \leq \lceil \frac{3n}{2} \rceil - 2$. The readers are encouraged to prove that the inequality in fact is an equality. Here is the entire algorithm.

$\text{MAX} \leftarrow \text{MIN} \leftarrow a_1;$
if n is even **then** $m \leftarrow 1$ **else** $m \leftarrow 2;$
for $i = m$ **to** n **step by** 2 **do**
$\{ \quad$ **if** $\quad a_i > a_{i+1}$
$\quad\quad$ **then** $\text{MAX}' \leftarrow a_i$ and $\text{MIN}' \leftarrow a_{i+1}$
$\quad\quad$ **else** $\text{MAX}' \leftarrow a_{i+1}$ and $\text{MIN}' \leftarrow a_i;$
$\quad\quad$ **if** $\text{MAX}' > \text{MAX}$ **then** $\text{MAX} \leftarrow \text{MAX}';$
$\quad\quad$ **if** $\text{MIN}' < \text{MIN} \quad$ **then** $\text{MIN} \leftarrow \text{MIN}';$
$\}$

In fact the most frequently used number-comparison problem may be the *sorting* problem:

For a given sequence $(a_1, a_2, \ldots, a_n)$ of real numbers,
permute the items into another sequence $(b_1 \le b_2 \le \ldots \le b_n)$.

A naive algorithm, called the *selection sort*, is as follows.

for $i = 1$ **to** n **do**
$\{ \quad j \leftarrow 1;$
$\quad\quad$ **for** $k = 2$ **to** n **do** $\quad$ **if** $a_k < a_j$ **then** $j \leftarrow k;$
$\quad\quad b_i \leftarrow a_j;$
$\quad\quad a_j \leftarrow \infty;$
$\}$

The time complexity of the selection sort is $O(n^2)$.

Another algorithm, called the *bubble sort*, uses only one array to store the input data as well as the output data.

for $i = n - 1$ **to** 1 **step** -1 **do**
for $j = 1$ **to** i **do**
$\quad$ **if** $a_j > a_{j+1}$ **then** interchange a_j and $a_{j+1};$

The time complexity of the bubble sort is also $O(n^2)$.

There is a more complicated but more efficient algorithm called the *heap sort*, whose time complexity is $O(n \log n)$. Note that theoretically this is best possible, since the solution space has size $n!$ and a comparison divides the solution spaces into two subspaces. Hence, $O(\log_2 n!) = O(n \log n)$ time is needed.

Merge sort is another $O(n \log n)$-time algorithm for sorting n numbers. The idea is as follows. To sort a set S of n numbers, first partition S into two set S' and S'' of size $\lceil \frac{n}{2} \rceil$ and $\lfloor \frac{n}{2} \rfloor$, respectively. Recursively sort S' to get $x_1 \leq x_2 \leq \ldots \leq x_r$ and sort S'' to get $y_1 \leq y_2 \leq \ldots \leq y_s$. Then merge these two sorted sequences into a sequence z by using at most $r + s - 1 = n - 1$ comparisons as follows.

$z \leftarrow$ empty sequence; $i \leftarrow j \leftarrow 1$;
while $i \leq r$ and $j \leq s$ **do**
     if $x_i < y_j$ **then** { put x_i at the end of z; $i \leftarrow i + 1$; }
           else { put y_j at the end of z; $j \leftarrow j + 1$; }
if $i \leq r$ **then** put $x_i, x_{i+1}, \ldots, x_r$ at the end of z;
     else put $y_j, y_{j+1}, \ldots, y_s$ at the end of z;

If the comparisons needed in this algorithm is t_n, then

$$t_n = \begin{cases} 0, & \text{for } n = 1; \\ t_{\lceil \frac{n}{2} \rceil} + t_{\lfloor \frac{n}{2} \rfloor} + n - 1, & \text{for } n \geq 2. \end{cases}$$

If $n = 2^k$ for some integer $k \geq 1$, then $t_n = 2t_{n/2} + n - 1$. Multiply both sides of this equality by 2 and replaced n by $n/2$, and repeat the same process k time to get the equalities as follows:

$$t_n = 2t_{n/2} + n - 1,$$

$$2t_{n/2} = 2^2 t_{n/2^2} + n - 2,$$

$$2^2 t_{n/2^2} = 2^3 t_{n/2^3} + n - 2^2,$$

$$\vdots$$

$$2^{k-2} t_{n/2^{k-2}} = 2^{k-1} t_{n/2^{k-1}} + n - 2^{k-2},$$

$$2^{k-1} t_{n/2^{k-1}} = 2^k t_{n/2^k} + n - 2^{k-1}.$$

Summing up these k equalities and canceling the same terms from both sides of the equality, together with that $t_1 = 0$ and $1 + 2 + 2^2 + \cdots + 2^{k-1} = 2^k - 1 = n - 1$, gives that

$$t_n = n \log_2 n - n + 1.$$

After knowing the equality, it can be verified by induction. First it is clear for $n = 1 = 2^0$. Suppose $t_{\frac{n}{2}} = \frac{n}{2} \log_2 \frac{n}{2} - \frac{n}{2} + 1$, then

$$t_n = 2t_{\frac{n}{2}} + n - 1 = 2\left(\frac{n}{2} \log_2 \frac{n}{2} - \frac{n}{2} + 1\right) + n - 1 = n \log_2 n - n + 1.$$

Hence, $t_n = n \log_2 n - n + 1$ for $n = 2^k$ by induction.

In general when n is not necessarily a power of two, choose a power of two, say m, such that $n \leq m < 2n$. Then

$$t_n \leq t_m = m \log_2 m - m + 1 = m \log_2(m/2) + 1 < 2n \log_2 n + 1,$$

and so $t_n = O(n \log n)$.

2.3 NP-complete problems

To see the difference between $O(n^2)$ and $O(n \log n)$, consider the university entering examination in Taiwan. About $n = 10^5$ students take the exam every year. In order to distribute the students to different departments in universities, it is necessary to sort their scores first. If an efficient sorting algorithm is used, it may take only one day for the sorting job. Note that n^2 is about 1000 times $n \log n$ for $n = 10^5$. So, it might take more than one year in sorting if an $O(n^2)$-time algorithm is used. Consequently, students need to wait for one year to get the result. Of course, this has never happened in Taiwan, as efficient sorting algorithms are used.

An extreme example is the domination problem. For a graph G of n vertices, a naive algorithm is to check all subsets D of $V(G)$. Note that it takes $O(n^2)$ time to determine if a set D is a dominating set or not. After checking all 2^n such sets D, the domination number $\gamma(G)$ is then determined. This algorithm takes $O(n^2 2^n)$ time, which is very large even when n is small.

Table 2.1 shows the comparison between different growing rates. As one can see, even for $n = 100$ the number 2^n is a horribly large number. For example, if a computer performs 10^{20} operations per second, then an algorithm of 2^{100} steps takes about 400 years. No person can see the answer of this program in his lifetime. On the other hand, if the time complexity

Table 2.1. The comparison between different growing rates.

n	$n \log n$	n^2	n^3	n^6	2^n
10	10	10^2	10^3	10^6	$\approx 10^3$
10^2	2×10^2	10^4	10^6	10^{12}	$\approx 10^{30}$
10^3	3×10^3	10^6	10^9	10^{18}	$\approx 10^{301}$
10^4	4×10^4	10^8	10^{12}	10^{24}	$\approx 10^{3010}$
10^5	5×10^5	10^{10}	10^{15}	10^{30}	$\approx 10^{30103}$
10^6	6×10^5	10^{12}	10^{18}	10^{36}	$\approx 10^{301030}$

of the algorithm is $O(n^3)$, then for the problem of input size $n = 100,000$, the answer is obtained in less than a second.

In recent years, efficient algorithms for various problems are studied. Many problems originally only have an *exponential-time algorithm*, which is one taking $O(a^n)$ time for some constant $a > 1$, are improved to have a *polynomial-time algorithm*, which is one taking $O(n^k)$ time for some constant integer $k \geq 1$. It may finally improved to have a *linear-time algorithm*, which is one taking $O(n)$ time. As shown in Table 2.1, the efficiencies of these three levels of algorithms are quite different.

For an exponential-time algorithm, even if the input size n increases slightly, the computation time becomes excessively large. Such algorithms are generally considered to be impractical. On the other hand, polynomial-time algorithms need significantly less computation time and are considered to be more practical. Such algorithms are considered to be more useful than exponential-time algorithms.

In general, an exponential-time algorithm is much less efficient than a polynomial-time algorithm when the input size n is large. However, for some cases, when n is small, an exponential-time algorithm may be more efficient than a polynomial-time algorithm.

In graph theory, there are many problems, such as the domination problem, still have no polynomial-time algorithms. It is desirable to determine whether polynomial-time algorithms are possible for these problems. This involves the concept of so-called P and NP. A decision problem is a *P problem* if there is a polynomial-time algorithm to answer it. The definition of an NP problem is more complicated. A rough explanation is as follows.

A decision problem is an *NP problem* if, when the answer is "yes," it is possible to verify the answer correct.[2]

It is easy to see that all P problems are NP problems, in other words P $\subseteq$ NP. As the answers to a problem is finite, if there are many enough computers to verify all the possible answers at the same time, then an NP problem can be solved in polynomial time. A natural question is whether it is possible to solve any NP problem in polynomial time using only one computer? This is the famous problem "Does P = NP?"[3]

To prove P = NP, it is necessary to verify that every NP problem can be solved in polynomial time. There is an approach for this by the concepts NP-hard and NP-complete. An *NP-hard problem* is one satisfying that if it is in P then any NP problem is also in P. A problem is *NP-complete* if it is in NP and is NP-hard. Therefore, to show P = NP it is only necessary to find an NP-complete problem which is also in P.

The existence of an NP-complete problem was first established by Cook (1971). He proved that the following Satisfiability Problem is NP-complete. In the problem, there are n logical variables $u_1, u_2, \ldots, u_n$, and a logical expression f consisting of these variables, logical and ($\wedge$), logical or ($\vee$), logical not ($\neg$) and parenthesis. After suitable tidy, f can be written as the following *normal form*, in which every $a_{j,i}$ is equal to some u_k or its

---

[2]If the answer to the problem is "no," then it is not necessarily easy to verify the correctness of the answer. If an answer "no" to a problem can be verified in polynomial time, it is called a *co-NP problem*. Some problems belong to both NP and co-NP. For example, the *integer factorization problem*, which is to determine if a given integer m has a factor less than n. Note that it is easy to see that this is an NP problem, but is more difficulty to see that it is co-NP.

In a more complete definition, a P problem (respectively, an NP problem) means a problem that can be solved by a polynomial-time algorithm in a deterministic Turing machine (respectively, nondeterministic Turing machine).

[3]Clay Mathematics Institute (CMI) posted seven Millennium Prize Problems on May 24, 2000. "Does P = NP?" is the first problem. According to the rules by the CMI, solutions to the problems must be published in mathematical journals, and verified by all parties. After two-year verification, one million dollars will be awarded to the solver.

The third problem—the Poincaré conjecture was proved by Russian mathematician Grigori Perelman in 2006. He also won the Fields Medal in the same year for this result. But he didn't show up to accept the award. In 2010, CMI awarded him a million-dollar prize. He also declined to accept the award.

negation $\bar{u}_k := \neg u_k$:

$$(a_{1,1} \vee a_{1,2} \vee \cdots \vee a_{1,n_1}) \wedge (a_{2,1} \vee a_{2,2} \vee \cdots \vee a_{2,n_2}) \wedge$$
$$\cdots \wedge (a_{m,1} \vee a_{3,2} \vee \cdots \vee a_{m,n_m}),$$

where each $a_{j,1} \vee a_{j,2} \vee \cdots \vee a_{j,n_j}$ is called a *clause* and each $a_{j,i}$ a *literal*. An logical expression f is *satisfiable* if it is possible to assign the logical variables true or false so that f is true, or equivalently every clause $a_{j,1} \vee a_{j,2} \vee \cdots \vee a_{j,n_j}$ is satisfiable.

Consider the following decision problem.

The Satisfiability Problem (SAT)
Parameters: A set $U = \{u_1, u_2, \ldots, u_n\}$ of logical variables and a family $\{C_j = a_{j,1} \vee a_{j,2} \vee \cdots \vee a_{j,n_j} : 1 \leq j \leq m\}$ of clauses using the variables in U and their negations.
Problem: Is it possible to assign true/false values to U such that every clause C_j is satisfiable?

By using Turing machines as problems, to be introduced at the end of this section, Cook proved the following theorem.

Theorem 2.1 (Cook, 1971). *SAT is an NP-complete problem.*

Right after Cook's theorem, Karp (1972) obtained many other NP-complete problems, including the 3-satisfiability problem (3-SAT) which is SAT in which every clause has exactly three literals, the vertex cover problem (VC), the independent set problem (IS), the clique problem (CQ), the directed Hamiltonian path problem (DHP), the directed Hamiltonian cycle problem (DHC), the Hamiltonian path problem (HP), the Hamiltonian cycle problem (HC) and the 3-coloring problem (3C) even restricted to planar graphs or 4-regular planar graphs.

Among these NP-complete problems, this book frequently uses two of them to prove the NP-completeness of domination problems. The first problem is (3-SAT).

The 3-Satisfiability Problem (3-SAT)
Parameters: A set $U = \{u_1, u_2, \ldots, u_n\}$ of logical variables and a family $\{C_j = a_{j,1} \vee a_{j,2} \vee a_{j,3} : 1 \leq j \leq m\}$ of clauses using the variables in U and their negations.
Problem: Is it possible to assign true/false values to U such that every clause C_j is satisfiable?

The second problem is (VC). In a graph G, a *vertex cover* of a graph G is a subset C of $V(G)$ such that every edge of the graph contains at least one vertex of C.

The Vertex Cover Problem (VC)
Parameters: A graph G and a positive integer k.
Problem: Does G have a vertex cover of size at most k?

Karp used a reduction-based proof rather than repeating Cook's approach. As Cook already proved that SAT is NP-complete, to prove a problem Γ is NP-complete Karp simply reduced it from a previously known NP-complete problem. Since then a lot of NP-complete problems have been obtained. Garey and Johnson (1979) compiled a comprehensive collection of NP-complete problems, and analyzed their proofs. After the book had published, they still received many NP-complete results from friends. Johnson (1981–2005) then published them as "The NP-completeness column" in 24 papers.

Although hundreds of NP-complete problems have been found, none has been proven to be in P. So, it is still not yet conclude that whether P = NP or not. In fact, people mostly believe that the answer is negative.

At end of this section, the concept of Turing machine is introduced formally as follows. Figure 2.1 shows the picture of a *multitape Turing machine*.

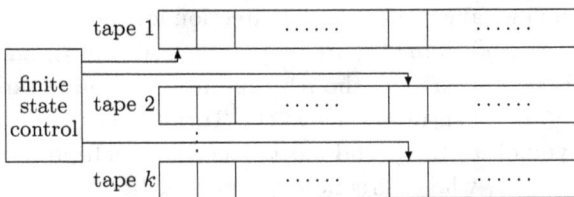

Fig. 2.1. A multitape Turing machine.

It consists of some number k of *tapes*, which is infinite to the right. Each tape is marked off into *cells*, each of which holds one of a finite number of *tape symbols*. One cell on each tape is scanned by a *tape head*, which can read and write. The operation of the Turing machine is determined by a primitive program called a *finite control*. The finite control is always in one of a finite number of *states*, which can be regarded as positions in a program.

One computational step of a Turing machine consists of the following. In accordance with the current state of the finite control and the tape symbols which are under (*scanned by*) each of the tape heads, the Turing machine may do any or all of these operations.

(1) Change the state of the finite control.
(2) Print new tape symbols over the current symbols in any or all of the cells under tape heads.
(3) Move any or all of the tape heads, independently, one cell left (L) or right (R) or keep them stationary (S).

Formally, a *k-tape Turing machine* is a seven-tuple

$$(Q, q_0, q_f, T, I, \mathrm{b}, \delta),$$

where the items are explained as follows.

(1) Q is a finite set of *states* containing an *initial state* q_0 and a *final* (or *accepting*) *state* q_f.
(2) T is a finite set of *tape symbols* and $I \subseteq T$ is the set of *input symbols* and $\mathrm{b} \in T - I$ is the *blank*.
(3) δ, the *next-move function*, is a function from a subset of $Q \times T^k$ to $Q \times (T \times \{\mathrm{L}, \mathrm{R}, \mathrm{S}\})^k$. That is, for some $(k+1)$-tuples consisting of a state and k tape symbols, it gives a new state and k pairs, each pair consisting of a new tape symbol and a direction for the tape head. Suppose $\delta(q, a_1, a_2, \ldots, a_k) = (q', (a'_1, d_1), (a'_2, d_2), \ldots, (a'_k, d_k))$, and the Turing machine is in state q with the ith tape head scanning tape symbol a_i for $1 \leq i \leq k$. Then in one move the Turing machine enters state q', changes symbols a_i to a'_i, and moves the ith tape head in the direction d_i for $1 \leq i \leq k$, where direction d_i stands L (for left) or R (for right) or S (for stationary).

A Turing machine can be made to recognize a language as follows. The tape symbols of the Turing machine include the alphabet of the language, called the input symbols, a special symbol blank, denoted b, and perhaps other symbols. Initially, the first tape holds a string of input symbols, one symbol per cell starting with the leftmost cell. All cells to the right of the cells containing the input string are blank. All other tapes are completely blank. The string of input symbols is *accepted* if and only if the Turing machine, started in the designated initial state, with all tape heads at the left ends of their tapes, makes a sequence of moves in which it eventually enters the accepting state. The *language accepted* by the Turing machine is the set of strings of input symbols so accepted.

Example. The two-tape Turing machine in Figure 2.2 recognizes *palindromes*, a string which reads the same forward and backward, on the alphabet {0, 1} as follows.

(a) Initially input string 0110 on tape 1.
(b) The first cell on tape 2 is marked with a special symbol X, and the content of tape 1 is duplicated onto tape 2.
(c) Then the tape head on tape 2 is moved to the X.
(d) Repeatedly, the head of tape 2 is moved right one cell and the head of tape 1 left one cell, comparing the respective symbols. If all symbols match, the input is a palindrome and the Turing machine enters the accepting state q_5. Otherwise, the Turing machine will at some point have no legal move to make: it will halt without accepting.

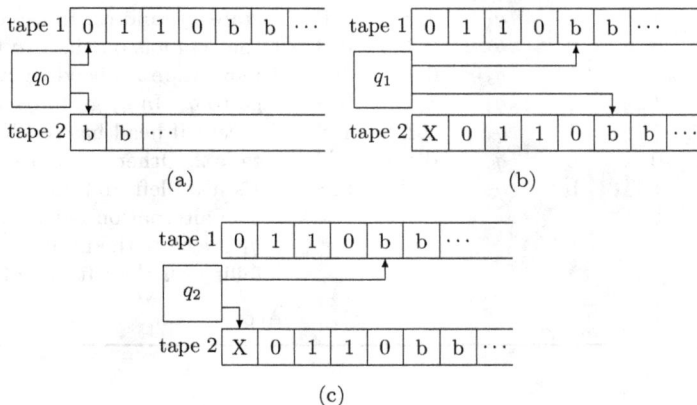

Fig. 2.2. Turing machine processing 0110.

The next-move function of the Turing machine is given by the table of Table 2.2. □

The activity of a Turing machine can be described formally by means of "instantaneous descriptions." An *instantaneous description* (ID) of a k-tape Turing machine M is a k-tuple $(\alpha_1, \alpha_2, \ldots, \alpha_k)$ where each α_i is a string of the form xqy such that xy is the string on the ith tape of M (with trailing

Table 2.2. The next-move function for the Turing machine recognizing palindromes.

Current state	Symbol on: tape 1	Symbol on: tape 2	New state	(New symbol, head move) tape 1	(New symbol, head move) tape 2	Comments
q_0	0	b	q_1	0, S	X, R	If input is nonempty, print X
	1	b	q_1	1, S	X, R	on tape 2 and move head right;
	b	b	q_5	b, S	b, S	go to state q_1. Otherwise, go
						to state q_5.
q_1	0	b	q_1	0, R	0, R	Stay in state q_1 copying tape 1
	1	b	q_1	1, R	1, R	onto tape 2 until b is reached
	b	b	q_2	b, S	b, L	on tape 1. Then go to state q_2.
q_2	b	0	q_2	b, S	0, L	Keep tape 1's head fixed and
	b	1	q_2	b, S	1, L	move tape 2's left until X is
	b	X	q_3	b, L	X, R	reached. Then go to state q_3.
q_3	0	0	q_4	0, S	0, R	Control alternates between
	1	1	q_4	1, S	1, R	states q_3 and q_4. In q_3 compare
q_4	0	0	q_3	0, L	0, S	the symbols on the two tapes,
	0	1	q_3	0, L	1, S	move tape 2's head right, and
	1	0	q_3	1, L	0, S	go to q_4. In q_4 go to q_5 and
	1	1	q_3	1, L	1, S	accept if head has reached b on
	0	b	q_5	0, S	b, S	tape 2. Otherwise move tape
	1	b	q_5	1, S	b, S	1's head left and go back to q_3.
						The alternation between q_3 and
						q_4 prevents the input head from
						falling off the left end of tape 1.
q_5						Accept.

blanks omitted) and q is the current state of M. The symbol immediately to the right of the ith q is the symbol being scanned on the ith tape.

If instantaneous description D becomes instantaneous description D' after one move of the Turing machine M, then we write $D \vdash D'$ (read $\vdash$ as "goes to"). If $D = D_1 \vdash D_2 \vdash \ldots \vdash D_n = D'$ for some $n \geq 1$, then we write $D \vdash^* D'$.

The k-tape Turing machine $M = (Q, q_0, q_f, T, I, \text{b}, \delta)$ *accepts* string $a_1 a_2 \ldots a_n$, where the a_i's are in I, if $(q_0 a_1 a_2 \ldots a_n, q_0, q_0, \ldots, q_0) \vdash^* (\alpha_1, \alpha_2, \ldots, \alpha_k)$ for some α_i's with q_f in them.

As an example, the sequence of instantaneous descriptions entered by the Turing machine of Table 2.2 when presented with the input 010 is shown as follows. Since q_5 is the final state, the Turing machine accepts 010.

$$(q_0010, q_0) \vdash (q_1010, Xq_1) \qquad \vdash (0q_110, X0q_1)$$
$$\vdash (01q_10, X01q_1) \quad \vdash (010q_1, X010q_1)$$
$$\vdash (010q_2, X01q_20) \vdash (010q_2, X0q_210)$$
$$\vdash (010q_2, Xq_2010) \vdash (010q_2, q_2X010)$$
$$\vdash (01q_30, Xq_3010) \vdash (01q_40, X0q_410)$$
$$\vdash (0q_310, X0q_310) \vdash (0q_410, X01q_40)$$
$$\vdash (q_3010, X01q_30) \vdash (q_4010, X010q_4)$$
$$\vdash (q_5010, X010q_5).$$

A *k-tape nondeterministic Turing machine* (NDTM for short) M is a seven-tuple $(Q, q_0, q_f, T, I, \text{b}, \delta)$ where all components have the same meaning as for the ordinary deterministic Turing machine, except that here the next-move function δ is a mapping from $Q \times T^k$ to subsets of $Q \times (T \times \{\text{L}, \text{R}, \text{S}\})^k$. That is, given a state and list of k tape symbols, δ returns a finite set of choices of next move; each choice is a new state, with new tape symbols and k moves of the tape heads. Note that the NDTM M may choose any of these moves, but it cannot choose a next state from one and new tape symbols from another, or make any other combination of moves.

2.4 Data structure

In order to implement an algorithm to a computer program efficiently, it is necessary to organize the input data in a systematic way. This is called the *data structure*.

The simplest data structure is an *array*. In mathematics, this is a variable with indices. A 0-dimensional array is a single variable, a 1-dimensional array is a sequence or a vector, and a 2-dimensional array is a matrix etc. Recursively, a d-dimensional array is a sequence/1-dimensional array of $(d - 1)$-dimensional arrays.

The advantage of using arrays is that it is easy to access various memory locations using indices. In general, every element of an array occupies a same space, say s bytes. For a 1-dimensional array $A_1, A_2, \ldots, A_n$, if the memory address of A_1 is b, then the address of A_i is $b + (i-1)s$ for $1 \le i \le n$.

A higher dimensional array can be viewed as the concatenation of several 1-dimensional arrays. For example, an $n \times m$ matrix can be considered as a 1-dimensional array (by the row-wise ordering)

$$A_{1,1}, A_{1,2}, \ldots, A_{1,m}, A_{2,1}, A_{2,2}, \ldots, A_{2,m}, \ldots \ldots, A_{n,1}, A_{n,2}, \ldots, A_{n,m}.$$

If the address of $A_{1,1}$ is b, then the address of $A_{i,j}$ is $b + (i-1)ms + (j-1)s$. Similar methods work for higher dimensional arrays.

Returning to the algorithms mentioned in the previous paragraph. The $O(n^3)$-time algorithm for the consecutive segment problem can be implemented as follows. Suppose the n real numbers $a_1, a_2, \ldots, a_n$ are stored in the array $A_1, A_2, \ldots, A_n$, or $A[1], A[2], \ldots, A[n]$ or $A[1..n]$ for short. It is not necessary to have an array for $s_{i,j}$. In addition to the original data $A[1..n]$, one more variable TEMP is enough to store $s_{i,j}$ for all time.

```
MAXS ← 0;
for i = 1 to n do
for j = i to n do
{     TEMP ← A[i];
      for k = i + 1 to j do TEMP ← TEMP + A[k];
      if MAXS < TEMP then MAXS ← TEMP;
}
```

As $s_{i,j} = s_{i,j-1} + a_j$, the new value of TEMP can be computed from the old value of TEMP by adding a_j. Hence, the $O(n^2)$-time algorithm for the consecutive segment problem can be implemented as follows.

```
MAXS ← 0;
for i = 1 to n do
{     TEMP ← 0;
      for j = i to n do
      {     TEMP ← TEMP + A[j];
            if MAXS < TEMP then MAXS ← TEMP;
      }
}
```

As said before, a more efficient algorithm for the problem is preferred. The following *dynamic programming approach* provides an $O(n)$-time algorithm. As can be seen, to solve the original problem, more problems needed to be solved. Let m_r be the maximum value of the sum $a_i + a_{i+1} + \cdots + a_j$ among all possible i and j with $1 \leq i \leq j \leq r$. Let p_r be the maximum value of the sum $a_i + a_{i+1} + \cdots + a_r$ among all possible i with $1 \leq i \leq r$. Then the following hold:

$$p_0 = 0 \text{ and } p_r = \max\{p_{r-1}, 0\} + a_r \text{ for } 1 \leq r \leq n;$$

$$m_0 = 0 \text{ and } m_r = \max\{m_{r-1}, p_r\} \text{ for } 1 \leq r \leq n.$$

The value m_n is the final answer. The implementation is as follows. Note that we only need a memory TEMP for p_r, since we may obtain p_r (new TEMP) from p_{r-1} (old TEMP). Similarly, we only need a variable MAXS for m_r.

```
TEMP ← MAXS ← 0;
for r = 1 to n do
{     if TEMP < 0 then TEMP ← A[r]
                  else  TEMP ← TEMP + A[r];
      if MAXS < TEMP then MAXS ← TEMP;
}
```

The selection sort for the sorting problem can be implemented as follows. Suppose the input numbers $a_1, a_2, \ldots, a_n$ are stored in the array $A[1], A[2], \ldots, A[n]$, which is also served as the output array.

```
for i = 1 to n − 1 do
{   m ← i;
    for j = i + 1 to n do if A[j] < A[m] then m ← j;
    TEMP ← A[m];
    for k = m − 1 to i step by −1 do A[k + 1] ← A[k];
    A[i] ← TEMP;
}
```

The bubble sort for the sorting problem can be implemented as follows.

```
for i = n − 1 to 1 step by −1 do
for j = 1 to i do
{   if A[j + 1] < A[j] then
    {   TEMP ← A[j];
        A[j] ← A[j + 1];
        A[j + 1] ← TEMP;
    }
}
```

Various other data structures are derived from arrays. Although these new data structures are essentially arrays, they have additional information to assist the performance of some specific operations. This makes them conceptually different from arrays.

Queue and *stack* are two simplest derived data structures. They are often used in job scheduling. For example, how to arrange the reception order for people who come to the counter for business? A conventional way is to ask them to line up, then first come first serve. This is queuing. Every time a new job comes in, it is put at the end of the queue. The job at the front of the queue is served first, and removed when the service is done. This appears to be the only reasonable approach.

However, in the design of an algorithm, there is another important strategy called stacking, which takes the last-come first-served approach. Every time a new job comes in, it is placed at the top of the stack. The job at the top of the stack is served first, and removed when the service is done.

To implement a queue Q, besides an array DATA, there are variables FRONT and REAR to record the front and the rear of the queue, respectively. Figure 2.3 shows a queue of the five elements in the string 'prime'.

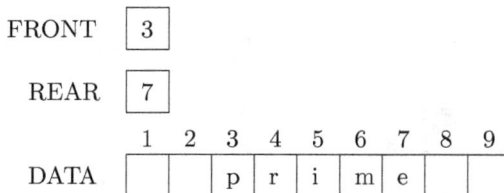

FRONT | 3 |

REAR | 7 |

	1	2	3	4	5	6	7	8	9
DATA			p	r	i	m	e		

Fig. 2.3.   A queue of five elements.

To put an element 'z' into the queue Q, use the function

enqueue(z, Q).

This will cause REAR increasing from 7 to 8, and DATA[8] is altered from its old content to 'z'.

Suppose next is to get an element from the queue. The function

dequeue(Q)

returns the front element 'p' of the queue Q and then increases FRONT from 3 to 4. However, it is not necessary to do anything for the content 'p' in DATA[3].

A careful implementation must also check if FRONT > REAR or REAR is equal to the length of DATA already.

To implement a stack, besides an array DATA, there is a variable TOP to record the top of the stack. Figure 2.4 shows a stack of the six elements in the string 'number'.

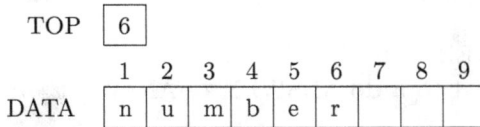

TOP	6								
	1	**2**	**3**	**4**	**5**	**6**	**7**	**8**	**9**
DATA	n	u	m	b	e	r			

Fig. 2.4. A stack of six elements.

To put an element 'z' into the stack, use the function

push(z).

This will cause TOP increasing from 6 to 7, and DATA[7] is altered from its old content to 'z'.

Suppose next is to get an element from the stack. The function

pop

returns the top element 'z' of the stack and then decreases TOP by 1. So, the element returned is 'z' in DATA[7], and TOP then becomes 6. However, it is not necessary to do anything for the content 'z' in DATA[7].

A careful implementation must also check if TOP is 0 or is equal to the length of DATA already.

Although stacks are not as natural as queues for implementing the queuing in the real world, they are useful in computer science. The following is an example for a compiler to deal a recursive algorithm, which is an effortless way of thinking. The factorial of a positive integer is defined as

$$n! = 1 \times 2 \times \cdots \times n.$$

It may also be defined recursively as

$$0! = 1 \text{ and } n! = (n-1)! \times n \text{ when } n \geq 1.$$

The recursive definition is implemented as follows.

```
F(n);
if n = 0 then return 1
        else   return F(n − 1) * n;
```

In fact, a nonrecursive program for computing $n!$ is possible as below. It looks even simpler than the recursive one.

```
ANSWER ← 1;
for i = 1 to n do ANSWER ← ANSWER * i;
```

However, in general, a recursive program is not so easy to be transformed to a nonrecursive one. This can be seen by the famous problem of *Towers of Hanoi*. The problem consists of three rods and a number of disks of various diameters, which can slide onto any rod. The problem begins with the disks stacked on one rod in order of decreasing size, the smallest disk at the top, thus approximating a conical shape. The objective of the problem is to move the entire stack of disks to the last rod, obeying the following three rules.

(1) Only one disk may be moved at a time.
(2) Each move consists of taking the upper disk from one of the stacks and placing it on top of another stack or on an empty rod.
(3) No disk may be placed on top of a disk that is smaller than it.

Suppose there are n disks on rod A initially, and will be moved into rod C, while rod B is a relay rod. The following is a recursive procedure for the puzzle.

HANOI(A, B, C, n);
if $n > 0$ **then**;
{   HANOI($A, C, B, n-1$);
                print "move disk n from rod A to C";
                HANOI($B, A, C, n-1$);
}

Let $t(n)$ be the number of moves using this procedure. Then $t(1) = 1$ and $t(n) = 2t(n-1) + 1$ for $n \geq 2$. Let $t'(n) = t(n) + 1$ for $n \geq 1$. Then $t'(1) = 2$ and

$$t'(n) = t(n) + 1 = 2t(n-1) + 1 + 1 = 2t'(n-1)$$

for $n \geq 2$. In other words, $(t'(1), t'(2), \ldots, t'(n))$ is a geometric sequence with initial term 2 and common ratio 2 and so $t'(n) = 2^n$ implying $t(n) = 2^n - 1$ for $n \geq 1$. In fact, $2^n - 1$ is the minimum number of moves required to solve the Hanoi puzzle with n disks.

Note that the procedure HANOI is not as easy as the function F to transform into a nonrecursive one.

Although recursion is not necessary for computing $n!$, the following uses the recursive function F to explain how a compiler implements recursion by using stacks. A compiler reserves a segment of memory to store the values of all variables, viewed as a stack called DATA. Suppose $A = F(2)$ is to be performed, that is, 2! is computed and then stored in A. Figure 2.5 shows the changing of the memory contents during the performance of $F(2)$. More detailed explanations are as follows.

L1: Before performing $A = F(2)$, the stack for the data has TOP $= 3$. Location 3 is reserved for the content for A.

L2: When entering F with $n = 2$, locations $4 \sim 6$ are reserved for $F(n)$ (unknown so far), n ($= 2$), old TOP ($= 3$). New TOP is 6.

L3: When entering F with $n = 1$, locations $7 \sim 9$ are reserved for $F(n)$ (unknown so far), n ($= 1$), old TOP ($= 6$). New TOP is 9.

L4: When entering F with $n = 0$, locations $10 \sim 12$ are reserved for $F(n)$ (unknown so far), n ($= 0$), old TOP ($= 9$). New TOP is 12.

L5: At the end of F with $n = 0$, the value $F(0)$ is found as 1 and stored in DATA[10]. TOP is still 12 now.

L6: Return to the previous call, at the end of F with $n = 1$, the value $F(1)$ is found as 1 and stored in DATA[7]. TOP is 9 now.

L7: Return to the previous call, at the end of F with $n = 2$, the value $F(2)$ is found as 2 and stored in DATA[4]. TOP is 6 now.

L8: Return to the previous call, $A = F(2)$ is found to be 2 and stored in DATA[3]. TOP is 3 now.

memory for $F(n)$ | $F(n)$ | n | TOP |

	A											
DATA	1	2	3	4	5	6	7	8	9	10	11	12
1: TOP=3 before performing $A = F(2)$			○									
2: TOP=6 when entering F with $n = 2$					2	③						
3: TOP=9 when entering F with $n = 1$					2	3		1	⑥			
4: TOP=12 when entering F with $n = 0$					2	3		1	6		0	⑨
5: TOP=12 at the end of F with $n = 0$					2	3		1	6	1	0	⑨
6: TOP=9 at the end of F with $n = 1$					2	3	1	1	⑥	1	0	9
7: TOP=6 at the end of F with $n = 2$				2	2	③	1	1	6	1	0	9
8: TOP=3 after performing $A = F(2)$			②	2	2	3	1	1	6	1	0	9

Fig. 2.5.   A circled position is the top of the stack for computing $F(2)$.

2.5   Lists and representations of graphs

Recall that in the selection sort, at every iteration a minimum item is selected from the remaining sequence and then deleted from the sequence. In the implementation, the sequence is stored in the 1-dimensional array $A[1..n]$. At iteration i, where $1 \leq i \leq n - 1$, a minimum $A[m]$ is selected from $A[i..n]$. Before moving $A[m]$ to $A[i]$, it is first stored in TEMP. Then $A[i..m-1]$ is moved to $A[i+1..m]$. After this, restore TEMP into $A[i]$. The drawback of this implementation is that in order to delete an element $A[m]$ from $A[i..n]$, it needs to move a segment of data $A[i..m-1]$ to $A[i+1..m]$. This costs too much. In particular, if the deletion operation is performed frequently, this is very time-consuming.

For many examples, deleting an item from or inserting an item into a sequence are necessary. For this purpose, this section introduces a slightly more complicated derivation of arrays.

A *list* consists of some linearly linked *records* of the same size, each has some *fields* storing *data* or *pointers*. Figure 2.6 shows two *singly linked lists*, where Λ is the special symbol to indicate the end of the list.

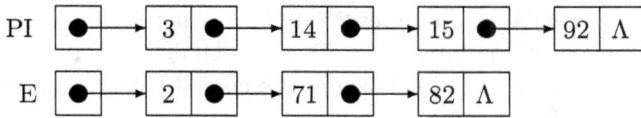

Fig. 2.6. Singly linked lists PI and E.

The difference between an array and a list is that data of an array are stored in order, but data in a list are not necessarily stored in order while are linked by pointers. The lists PI and E are stored in the arrays as shown in Figure 2.7.

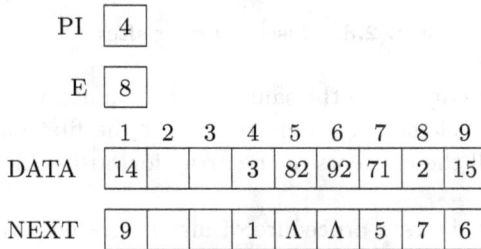

Fig. 2.7. Representations of lists PI and E by arrays.

In the array representations for these two lists, there are two variables PI and E, storing the locations for the first elements of the lists in the array DATA. For example, PI = 4 indicates that the first item 3 is stored in DATA[4]. To get the next item, NEXT[4] = 1 indicating that the next item 14 is stored in DATA[1]. Similarly, NEXT[1] = 9 gives the next item 15 in DATA[9], and NEXT[9] = 6 gives the next item 92 in DATA[6], while NEXT[6] = Λ indicating the end of the list. The list E can be read similarly.

Every data structure has its advantages and disadvantages. The usage of data structures depends on the demands.

A list has an obvious disadvantage in reading a specified item. For example, to read the third item of the list PI, it is necessary to read the elements one by one starting from the first item. On the other hand, it is easy to read any ith item in an array.

The advantage of a list is the convenience for inserting data. As an example, the operation of inserting a new data 99 right after 14 in the list PI can be done as follows. First, find an available location in DATA for storing 99, say DATA[3] in Figures 2.7 and 2.8. Then let 14 point to it, and let 99 use the original pointer of 14. This can be done by altering contents of only three memory cells as shown in the circled locations in Figure 2.8.

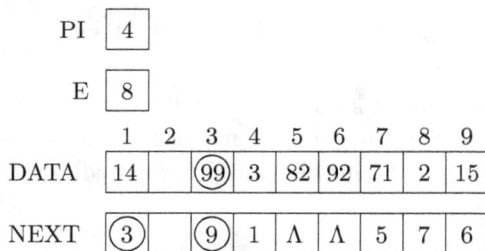

Fig. 2.8.   Insert 99 right after 14.

It is relatively hard to do the same thing in an array. Consider an array with thousands of elements. To insert data at the first position, it is necessary to move all the elements of the array forward, which is significantly inefficient.

The following shows a procedure of inserting a new data x right after an item at position i in a singly linked list, where avl is an available space in DATA.

insert(DATA, NEXT, x, i, avl);
DATA[avl] $\leftarrow x$;
NEXT[avl] $\leftarrow$ NEXT[i];
NEXT[i] $\leftarrow$ avl;

To enable efficient deletion in a list, it is better to have a PREV pointer to reference the previous record. In this way, to delete an item it is enough to suitably relink PREV and NEXT. This is a *doubly linked list.* Figures 2.9 and 2.10 are the revisions of Figures 2.6 and 2.7 from singly linked lists into doubly linked lists, respectively.

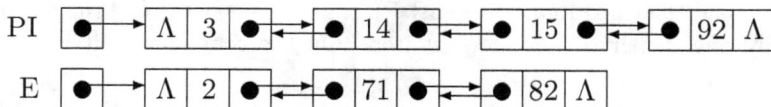

Fig. 2.9.   Doubly linked lists PI and E.

		1	2	3	4	5	6	7	8	9
PI	4									
E	8									
DATA		14			3	82	92	71	2	15
NEXT		9			1	Λ	Λ	5	7	6
PREV		4			Λ	7	9	8	Λ	1

Fig. 2.10. Representations of doubly linked lists PI and E.

Sometimes, depending on specific applications, more complicated data structures may be necessary.

Having the concepts of arrays and lists, it is ready to see the representations of graphs using data structures.

Suppose G is a graph with $V(G) = \{1, 2, \ldots, n\}$. The *adjacency matrix* of G is the $n \times n$ matrix $A = [a_{ij}]$, where

$$a_{ij} = \begin{cases} 1, & \text{if } ij \in E; \\ 0, & \text{if } ij \notin E. \end{cases}$$

By the definition, A is a symmetric (0,1)-matrix whose diagonal elements are all 0. A 2-dimensional array can store all information of a graph. For the case of a directed graph, $a_{ij} = 1$ represents only an edge from i to j. In general, a_{ij} may be not equal to a_{ji}.

The advantage of this representation is that it is easy to determine whether two vertices are adjacent or not. It is also easy to add a new edge to or to delete an old edge from a graph. On the other hand, it takes $O(n)$ time to determine the neighborhood $N(i)$, and $O(n^2)$ time to count the number of edges in $E(G)$. In the case when $\deg(i)$ is relatively smaller than n, or $|E(G)|$ is relatively smaller than $n(n-1)/2$, this becomes inefficient. The $O(n^2)$-space for the matrix is also waste as many useless 0's occupied memories.

When a graph algorithm checks neighborhoods $N(i)$ of vertices i frequently, it is better to use *adjacency lists* to represent a graph. The adjacency lists have a list, which may be singly or doubly linked depending on the situation, for every vertex storing all neighbors of this vertex.

Figure 2.11 shows the adjacency matrix and the adjacency lists for a graph with 5 vertices and 6 edges.

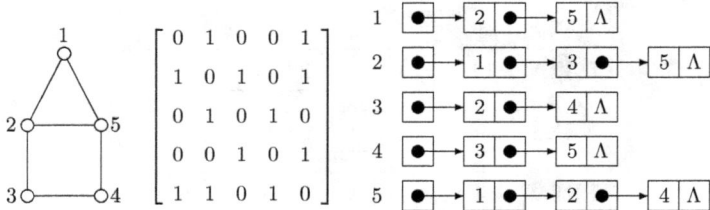

$$\begin{bmatrix} 0 & 1 & 0 & 0 & 1 \\ 1 & 0 & 1 & 0 & 1 \\ 0 & 1 & 0 & 1 & 0 \\ 0 & 0 & 1 & 0 & 1 \\ 1 & 1 & 0 & 1 & 0 \end{bmatrix}$$

Fig. 2.11.   The adjacency matrix and the adjacency lists of a graph.

For a graph of n vertices and m edges, the adjacency lists need only $O(n+m)$ space, which is much better than the $O(n^2)$ space for the adjacency matrix when the graph is sparse. For a vertex i it takes only $\deg(i)$ time to determine the neighborhood $N(i)$. Also, it takes only $O(m)$ time to count the number of edges in $E(G)$. On the other hand, it takes $O(\deg(i))$ or $O(\deg(j))$ time to determine if vertex i is adjacent to vertex j. The deletion of an edge is also not so easy as using the adjacency matrix. These differences need to be considered in the real applications.

Finally, a method different from the above is introduced. Suppose G is a graph with $V(G) = \{1, 2, \ldots, n\}$ and $E(G)$ is represented by m pairs of vertices. The neighborhoods of all vertices are stored in a *sequential space*. In other words, there is an array DATA in which $N(1), N(2), \ldots, N(n)$ are stored in this order. Let $N(i)$ be stored in $\text{DATA}[b_i..b_{i+1} - 1]$, and so $\deg(i) = b_{i+1} - b_i$. There is also an array BEG to store the values $b_1, b_2, \ldots, b_n, b_{n+1}$. As an example, assume $n = 9$ and the edge set

$$E(G) = \{14,\ 15,\ 92,\ 65,\ 35,\ 89,\ 79,\ 32,\ 38,\ 46\}.$$

Figure 2.12 show arrays BEG and DATA, where N_i represents $N(i)$ for the sequential space representation of the graph.

BEG	1	3	5	8	10	13	15	16	18	21											

	1	2	3	4	5	6	7	8	9	10	11	12	13	14	15	16	17	18	19	20	21
DATA	4	5	9	3	5	2	8	1	6	1	6	3	5	4	9	9	3	2	8	7	
	N_1	N_2	N_3		N_4	N_5				N_6			N_7	N_8	N_9						

Fig. 2.12.   The sequential space representation of a graph.

2.6 Case study: Finding components of a graph

This section presents a detail implementation for finding the components of a graph. The sequential space representation is used to store a graph. In the rest of the book, only algorithms are presented without detail implementations.

Figure 2.13 is a graph of 9 vertices, 10 edges and 3 components.

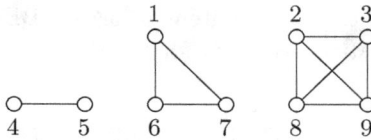

Fig. 2.13. A graph of 9 vertices, 10 edges and 3 components.

In the detail implementation, the input data for the graph are "9" for the number of vertices, and a sequence of pairs of vertices ending at 0 0 for the edges:

$$1\,6 \quad 1\,7 \quad 2\,3 \quad 2\,8 \quad 2\,9 \quad 3\,8 \quad 3\,9 \quad 4\,5 \quad 6\,7 \quad 8\,9 \quad 0\,0.$$

First, it needs a procedure to input the data for the graph and create a sequential space representation.

```
input-graph(V, BEG, DATA);
input V;
for i = 1 to V do DEG[i] ← 0;
input i, j;
while i ≠ 0 do
{       DEG[i] ← DEG[i] +1;
        DEG[j] ← DEG[j] +1;
        input i, j;
}
BEG[1] ← 1
for i = 1 to V do BEG[i + 1] ← BEG[i] + DEG[i];
re-open the input file so that the input data can be read again;
for i = 1 to V do TEMP[i] ← BEG[i];
```

```
input V, i, j;
while i ≠ 0 do
{      DATA[TEMP[i]] ← j;
       DATA[TEMP[j]] ← i;
       TEMP[i] ← TEMP[i] + 1;
       TEMP[j] ← TEMP[j] + 1;
       input i, j;
}
```

Figure 2.14 shows the final content of arrays BEG and DATA, where N_i represents $N(i)$. The array DEG for the degrees of all vertices are also found.

DEG	2	3	3	1	1	2	2	3	3												
BEG	1	3	6	9	10	11	13	15	18	21											
	1	2	3	4	5	6	7	8	9	10	11	12	13	14	15	16	17	18	19	20	21
DATA	6	7	3	8	9	2	8	9	5	4	1	7	1	6	2	3	9	2	3	8	
	N_1		N_2			N_3			N_4	N_5	N_6		N_7		N_8			N_9			
COMP	1	6	7	0	2	3	8	9	0	4	5	0	Λ								

Fig. 2.14.   The sequential space representation of the graph in Figure 2.13.

In the following, a *breadth-first search* (BFS) is used to find all components of a graph. At beginning, all vertices are marked by 0. After a vertex is scanned by the BFS, it is marked by 1. The algorithm finds a vertex CUR marked by 0 and creates a STACK containing CUR, whenever such a vertex CUR exists. This starts forming a component. Whenever the stack is not empty, pop an element i from the stack. Put i into the current component and push all neighbors of i marked by 0 into the stack. The component is completed when the stack is empty. Put 0 at end of the component.

For the final content of array COMP, see Figure 2.14.

```
find-components(V, BEG, DATA, COMP);
for i = 1 to V do MARK[i] ← 0;
POS ← 1;
CUR ← 1;
while CUR ≤ V do
{      TOP ← 1;
       STACK[1] ← CUR;
```

```
    while TOP ≠ 0 do
    {       i ← STACK[TOP];
            TOP ← TOP −1;
            MARK[i] ← 1;
            COMP[POS] ← i;
            POS ← POS +1;
            for j = BEG[i] to BEG[i + 1] −1 do
                if MARK[j] = 0 then
                    {           TOP ← TOP +1;
                                STACK[TOP] ← j;
                    }
    }
    COMP[POS] ← 0;
    POS ← POS +1;
    while CUR ≤ V and MARK[CUR] = 1 do
        CUR ← CUR +1;
}
COMP[POS] ← Λ;
```

2.7 Exercises

(2.1) A naive method to determine if a number n being prime is to check if i divides n for i from 2 to $n-1$. This takes $O(n)$ divisions. Design an algorithm to determine if a number n is prime using $O(\sqrt{n})$ divisions.

(2.2) (a) For $m = an + r$, where m, n, a, r are nonnegative integers such that $0 \le r < n$, prove that the greatest common divisor $\gcd(m, n) = \gcd(n, r)$.

(b) Show that (a) can be used to design an $O(\log m + \log n)$-time algorithm for finding $\gcd(m, n)$.

(2.3) (a) Show that for nonnegative integers a and b there exist integers x and y such that $ax + by = \gcd(a, b)$.

(b) Design an algorithm to find such x and y. (Hint: Revise an algorithm for finding $\gcd(a, b)$.)

(2.4) (a) For positive integers a, b and n, show that the equation $ax \equiv b$ (mod n) has integral solutions if and only if $\gcd(a, n)$ is a divisor of b. Moreover, if $ax \equiv b$ (mod n) is solvable, it has exactly $\gcd(a, n)$ solutions modulo n.

(b) Design an algorithm to determine, for positive integers a, b and n, if $ax \equiv b \pmod{n}$ is solvable; in the case of a positive answer, find the $\gcd(a, n)$ solutions modulo n.

(2.5) For a polynomial $f(x) = a_n x^n + a_{n-1} x^{n-1} + \cdots + a_0$ and a real number a, design an $O(n)$-algorithm to compute $f(a)$.

(2.6) Show that $\lceil \frac{3n}{2} \rceil - 2$ comparisons are necessary and sufficient for finding the MAX and the MIN among n real numbers.

(2.7) Recall that the recursion for the time complexity of the merge sort is as follows:

$$
t_n =
\begin{cases}
0, & \text{for } n = 1; \\
t_{\lceil \frac{n}{2} \rceil} + t_{\lfloor \frac{n}{2} \rfloor} + n - 1, & \text{for } n \geq 2.
\end{cases}
$$

Prove that $t_n = n \lceil \log_2 n \rceil - 2^{\lceil \log_2 n \rceil} + 1$.

(2.8) **(Bucket sort or radix sort)**
Suppose $(a_1, a_2, \ldots, a_n)$ is a sequence of integers in the range 0 to $m - 1$, where m is not too large. Design an $O(n)$-time algorithm to sort the sequence.

(2.9) Design an efficient algorithm to determine if a sequence of integers $a_1 < a_2 < \ldots < a_n$ has an item $a_i = i$. What is the time complexity of your algorithm?

(2.10) **(Monge arrays)**
An $m \times n$ array $M = (a_{i,j})_{m \times n}$ of real numbers is a *Monge array* if for all i, j, k and ℓ with $1 \leq i < k \leq m$ and $1 \leq j < \ell \leq n$,

$$a_{i,j} + a_{k,\ell} \leq a_{i,\ell} + a_{k,j}.$$

In other words, whenever two rows and two columns of the array are picked, the four elements at the intersections of the rows and the columns satisfy the property that the sum the upper-left and the lower-right elements is less than or equal to the sum of the upper-right and the lower-left elements.

(a) Show that an array is Monge if and only if for $1 \leq i \leq m - 1$ and $1 \leq j \leq n - 1$,

$$a_{i,j} + a_{i+1,j+1} \leq a_{i,j+1} + a_{i+1,j}.$$

(b) Let $f(i)$ be the least index j such that $a_{i,j}$ is minimum in row i in an $m \times n$ Monge array. Show that $f(1) \leq f(2) \leq \ldots \leq f(m)$.

(c) Here is a divide-and-conquer algorithm that computes $f(i)$'s of an $m \times n$ Monge array.

> Construct a submatrix M' of M consisting of the even-numbered rows of M. Recursively determine $f'(i)$'s for M'. Then compute $f(i)$'s for odd-numbered rows of M.

Explain how to compute $f(i)$'s for odd-numbered rows of M in $O(m + n)$ time, providing that $f'(i)$'s for M' are known.

(d) Write the recurrence describing the running time of the algorithm described in (c). Show that its solution is $O(m + n \log_2 m)$.

(2.11) For r matrices $M_1, M_2, \ldots, M_r$, where M_i is an $n_{i-1} \times n_i$ matrix, design an algorithm to compute their product $M_1 M_2 \cdots M_r$ using the minimum number of real number products.

(2.12) Provide a Turing machine which prints 0^{2n} on tape 2 when started with 0^n on tape 1.

(2.13) Provide a Turing machine which prints 0^{n^2} on tape 2 when started with 0^n on tape 1.

(2.14) Provide a Turing machine which accepts inputs of the form $0^n 10^{n^2}$.

(2.15) A *context-free grammar* in Chomsky normal form G is a four-tuple (N, S, Σ, P), where

1. N is a finite set of *nonterminal symbols* containing a special symbol S,
2. Σ is a finite set of *terminal symbols*,
3. P is a finite set of pairs, called *productions*, of the form $A \rightarrow BC$ or $A \rightarrow a$ where A, B, C are in N and a is in Σ.

We write $\alpha A \gamma \Longrightarrow \alpha \beta \gamma$ if α, β, γ are strings of nonterminals and terminals and $A \rightarrow \beta$ is in P. The *language generated by* G is the set

$$L(G) = \{w : S \overset{*}{\Longrightarrow} w\},$$

where $\overset{*}{\Longrightarrow}$ is the reflexive and transitive closure of $\Longrightarrow$. As an example, a context-free grammar G with $N = \{S\}$, $\Sigma = \{a, b\}$ and the following two productions has $L(G) = \{a^n b^n : n \geq 1\}$:

$$S \rightarrow aSb, \quad S \rightarrow ab.$$

While the following productions generate the language $\{a^n b^n : n \geq 0\}$:

$$S \rightarrow aSb, \ S \rightarrow \varepsilon.$$

(a) Define a context-free grammar which generates the following language:

$$\{a^m b^n a^{2m} : m \geq 0, n \geq 0\}.$$

(b) Write an $O(n^3)$-time algorithm to determine whether a given string $w = a_1 a_2 \ldots a_n$ is in $L(G)$, where $G = (N, S, \Sigma, P)$ is a context-free grammar in Chomsky normal form.
[Hint: Let $m_{i,j} = \{A : A \in N \text{ and } A \overset{*}{\Longrightarrow} a_i a_{i+i} \ldots a_j\}$. Note that $w \in L(G)$ if and only if $S \in m_{1,n}$. Use a dynamic programming approach to compute the $m_{i,j}$'s.]

(2.16) For an $m_1 \times m_2 \times \cdots \times m_d$ array A, where each element occupies s bytes and the memory address of $A_{1,1,\ldots,1}$ is b, find the memory address of $A_{i_1, i_2, \ldots, i_d}$.

(2.17) Write an algorithm to reverse the items of a singly linked list.

(2.18) Suppose A and B are singly linked lists whose data are stored in the array DATA and the index array is NEXT. The following program makes the union of these two lists and stores the result in A.

$$t \leftarrow A;$$
$$\textbf{while } \text{NEXT}[t] \neq \Lambda \textbf{ do } \{ \ tt \leftarrow t; \quad t \leftarrow \text{NEXT}[tt]; \ \}$$
$$\text{NEXT}[tt] \leftarrow B;$$

Modify the data structure of singly linked lists, keeping them still singly linked, so that the union of two lists takes only $O(1)$ time.

(2.19) Suppose a doubly linked list A stores data in DATA and uses index arrays NEXT and PREV.

(a) Write a program that adds a new item x after an item i in A.
(b) Write a program that deletes an item i.

(2.20) Show that the necessary and sufficient moves for the problem of Towers of Hanoi on n disks is $2^n - 1$.

(2.21) Design a nonrecursive algorithm for the problem of Towers of Hanoi.

(2.22) For a graph G with vertex set $V(G) = \{v_1, v_2, \ldots, v_n\}$ and edge set $E(G) = \{e_1, e_2, \ldots, e_m\}$, the *incidence matrix* of G is the $n \times m$ matrix $M = [b_{i,j}]$, where

$$
b_{i,j} = \begin{cases} 1, & \text{if } v_i \text{ and } e_j \text{ are incident}; \\ 0, & \text{if } v_i \text{ and } e_j \text{ are not incident}. \end{cases}
$$

Show that $MM^{\mathrm{T}} = A + D$, where A is the adjacency matrix of G and $D = [d_{i,j}]$ with $d_{i,i} = \deg(v_i)$ and $d_{i,j} = 0$ when $i \neq j$.

(2.23) For a directed graph G with vertex set $V(G) = \{v_1, v_2, \ldots, v_n\}$ and edge set $E(G) = \{e_1, e_2, \ldots, e_m\}$, the *incidence matrix* of G is the $n \times m$ matrix $M = [b_{i,j}]$, where

$$
b_{i,j} = \begin{cases} 1, & \text{if } v_i v_{i'} = e_j; \\ -1, & \text{if } v_{i'} v_i = e_j; \\ 0, & \text{if } v_i \text{ and } e_j \text{ are not incident}. \end{cases}
$$

Show that M is *totally unimodular*, that is, any square submatrix of M has determinant 0, 1 or -1.

(2.24) (a) Prove that the incidence matrix of a bipartite graph is totally unimodular.

(b) Find infinitely many graphs whose incidence matrices are not totally unimodular.

(2.25) Suppose A is the adjacency matrix of a graph G with vertex set $V(G) = \{1, 2, \ldots, n\}$. For an integer $k \geq 0$, let $[c_{i,j}] := A^k$ and $[d_{i,j}] := (A + I)^k$. Explain the meaning of $c_{i,j}$ and $d_{i,j}$.

(2.26) **(Topological sort)**

(a) Show that every acyclic digraph G has at least one vertex of indegree zero.

(b) Write an $O(n + m)$-time algorithm to order the vertices of an acyclic digraph G of n vertices and m edges into $[v_1, v_2, \ldots, v_n]$ such that if $v_i v_j \in E(G)$ then $i < j$.

(2.27) Write an $O(n+m)$-time algorithm, for two specified vertices s and t in an acyclic digraph G of n vertices and m edges, to find a maximum number of internally vertex-disjoint s–t paths.

(2.28) **(Stable marriage problem)**

Let B be a set of n boys and G a set of n girls. Every boy ranks the girls from 1 to n and every girl ranks the boys from 1 to n. A *matching* is a one-to-one correspondence of boys and girls. A matching is *stable* if for every two boys b_1 and b_2 and their matched girls g_1 and g_2, the following two conditions hold:

(1) either b_1 ranks g_1 higher than g_2, or g_2 ranks b_2 higher than b_1;
(2) either b_2 ranks g_2 higher than g_1, or g_1 ranks b_1 higher than b_2.

Show that a stable matching always exists and write an algorithm to find such a matching.

Chapter 3

Trees

3.1 Trees are graphs with the simplest structures

A *forest* (or *acyclic graph*) is a graph without cycles. A *tree* is a connected forest. In other words, a tree is a connected graph without cycles. Figure 3.1 shows all six trees with six vertices, where isomorphic trees are counted as the same.

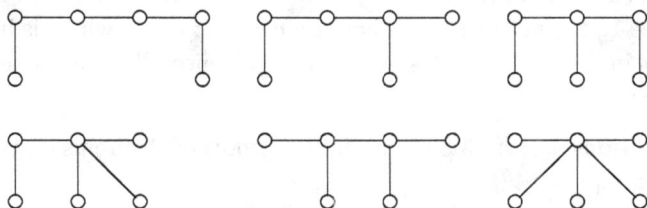

Fig. 3.1. All six trees with six vertices.

Note that cycles of a graph make its structure complicated, since between two vertices there are various paths going through different cycles. In this sense, trees/forests are graphs with the simplest structures, as they have no cycles. People very often study graph problems starting from trees/forests as the first attempt when they have no idea how to approach the problem in general graphs.

Cockayne *et al.* (1975) gave the first and a linear-time algorithm for the domination problem in trees by using the labeling approach. Various algorithmic results have since been investigated by many researchers. Before discussing their result, basic properties of trees are introduced in this section.

Proposition 3.1. *For every vertex u in a tree T of at least two vertices, a vertex v at maximum distance from u is a leaf.*

Proof.   Let $P = (u = v_0, v_1, \ldots, v_r = v)$ be a u–v path of largest possible length. As T has at least two vertices, $r \geq 1$. Then v is a leaf, for otherwise it has a neighbor v' different from v_{r-1}. If v' is in P, then a cycle is formed which is a contradiction to that T is acyclic, otherwise (P, v') is a u–v' path of length larger than the length of P, a contradiction to the assumption on v.   □

Consequently, we have the following proposition.

Proposition 3.2. *Every tree of at least two vertices has at least two leaves.*

Proof.   Choose a vertex x. By Proposition 3.1, there exists a leaf y at maximum distance from x. Again, there exists another leaf z at maximum distance from y.   □

Proposition 3.3. *If v is a leaf in a tree T, then $T - v$ is also a tree.*

Proof.   First, since T is acyclic, so is $T - v$. For every two vertices x and y in $T - v$, since they are also in T, there exists an x–y path P in T. Since v is of degree 1, it is not in P for otherwise $v = x$ or $v = y$ which is impossible as v is not in $T - v$. Then P is also in $T - v$. Hence, $T - v$ is connected and so is a tree.   □

Propositions 3.2 and 3.3 are useful for proofs of theorems on trees. Here are two examples.

Proposition 3.4. *A tree T of n vertices has exactly $n - 1$ edges.*

Proof.   The proposition is obviously true for $n = 1$. Suppose $n \geq 2$ and the proposition is true for $n - 1$. By Proposition 3.2, T has a leaf v. By Proposition 3.3, $T - v$ is a tree. Also, $T - v$ has one vertex and one edge less than T. As $T - v$ has exactly $n - 1$ vertices, by the induction hypothesis it has exactly $n - 2$ edges. Hence, T has exactly $n - 1$ edges as desired.   □

Proposition 3.5. *If T is a tree of k edges and G is a graph with minimum degree $\delta(G) \geq k$, then G has a subgraph isomorphic to T.*

Proof.   The proposition is obviously true for $k = 0$. Suppose $k \geq 1$ and the proposition is true for $k - 1$. Then by Proposition 3.2, T has a leaf v.

Let u be the only neighbor of v in T. By Propositions 3.3 and 3.4, $T' := T - v$ is a tree with $k - 1$ edges. By the induction hypothesis, G has a subgraph G' isomorphic to $T - v$. Suppose vertex u in T' corresponds to vertex u' in G'. Since $\deg_G(u') \geq \delta(G) \geq k = |V(G')|$, vertex u' in G has a neighbor v' not in $V(G')$. Adding vertex v' and edge $u'v'$ to G' yields a subgraph of G isomorphic to T. $\qquad\square$

Erdős and Sós (1963) (see Erdős (1964) and Problem 31 in the book by Bondy and Murty (2008)) proposed the following conjecture that is stronger than Proposition 3.5. Note that $\delta(G) \geq k$ implies $m \geq nk/2$ and so $m > n(k-1)/2$, but $m > n(k-1)/2$ does not imply $\delta(G) \geq k$.

Conjecture 3.6 (Erdő-Sós Conjecture, 1963). *Suppose T is a tree of k edges and G is a graph with n vertices and m edges. If $m > n(k-1)/2$, then G has a subgraph isomorphic to T.*

Also, $G = K_k$ shows that the condition $m \geq n(k-1)/2$ does not guarantee that G has a subgraph isomorphic to T.

Since then many investigations on this conjecture have been done, see the paper by Rozhoň (2019) and its references.

Propositions 3.2 and 3.3 are also useful for the design of algorithms for problems in trees, for instance, as shown in the next section for the labeling algorithm of the domination problem in trees. Suppose v is a leaf of a tree T. Since only v and its only neighbor may dominate v, it is relatively easy to derive a minimum dominating set of T from a minimum dominating set of $T - v$; see Theorem 3.11 for a detailed description. In the algorithm side, the theorem is used repeatedly.

Suppose

$$T_1 \leftarrow T \text{ and } v_1 \text{ is a leaf of } T_1,$$

$$T_2 \leftarrow T_1 - v_1 \text{ and } v_2 \text{ is a leaf of } T_2,$$

$$T_3 \leftarrow T_2 - v_2 \text{ and } v_3 \text{ is a leaf of } T_3,$$

$$\vdots$$

$$T_{n-1} \leftarrow T_{n-2} - v_{n-2} \text{ and } v_{n-1} \text{ is a leaf of } T_{n-1},$$

$$T_n \leftarrow T_{n-1} - v_{n-1} \text{ and } v_n \text{ is the only vertex in } T_n.$$

The sequence $[v_1, v_2, \ldots, v_n]$ makes the implementation of Theorem 3.11 as well as many other algorithms for trees easier. Such a sequence is called a *tree ordering* of T. In other words, a tree ordering is a permutation $[v_1, v_2, \ldots, v_n]$ of the vertices of T satisfying the following property.

For $1 \leq i < n$, vertex v_i is adjacent to exactly one v_j with $j > i$.   (TO)

In the above description, the only neighbor v_j of v_i with $j > i$ or, more precisely, v_{j_i} with $j_i > i$, is called the *parent* of v_i in this tree ordering. Sometimes it is convenient to consider v_n as the parent of v_n itself, that is $j_n = n$. The list $[v_{j_1}, v_{j_2}, \ldots, v_{j_n}]$ is called the *parent list* of T corresponding to the tree ordering $[v_1, v_2, \ldots, v_n]$.

A tree may have many different tree orderings. Figure 3.2 shows a tree ordering of a tree of 15 vertices, and its corresponding parent list $[v_4, v_4, v_7, v_7, v_8, v_8, v_9, v_9, v_{12}, v_{13}, v_{13}, v_{15}, v_{15}, v_{15}, v_{15}]$.

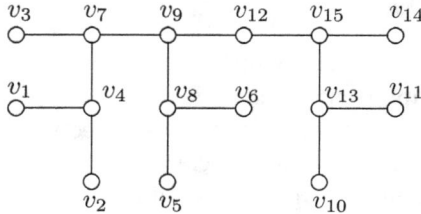

Fig. 3.2.   An example of the tree ordering $[v_1, v_2, \ldots, v_{15}]$.

Not only a tree has a tree ordering, the converse is also true. So, in addition to their use in algorithms, tree orderings also provide a characterization of trees.

Proposition 3.7. *A graph has a tree ordering if and only if it is a tree.*

Proof.   From the arguments above, a tree has a tree ordering.

On the other hand, suppose a graph G has a tree ordering $[v_1, v_2, \ldots, v_n]$. As every vertex v_i with $1 \leq i < n$ has a neighbor v_j with $i < j \leq n$, every vertex has a path to v_n. This shows that G is connected. If G has a cycle C, then the vertex v_i in C with the minimum index has two neighbors with larger indices, a contradiction. Hence, G is a tree. $\square$

Two more propositions for cycles in connected graphs are needed before characterizing trees.

Proposition 3.8. *If e is an edge in the cycle C of a connected graph G, then $G - e$ is still connected.*

Proof. For every two vertices u and v in $G-e$, as they are also vertices of the connected graph G, there is a u–v walk W in G. For every appearance of e in W, replace it by $C - e$ to get a u–v walk in $G - e$. This shows that $G - e$ is also connected. □

A *spanning tree* of a graph is a spanning subgraph of the graph that is also a tree.

Proposition 3.9. *A graph has a spanning tree if and only if it is connected.*

Proof. If graph G has a spanning tree, then for every two vertices u and v there is a u–v path in the spanning tree which is also in G. Hence, G is connected.

On the other hand, suppose G is connected. Choose a spanning subgraph H of G with the minimum number of edges among all spanning subgraphs. Suppose H has a cycle C, then by Proposition 3.8 for any edge e in C, the spanning subgraph $H - e$ of G is connected but has fewer edges than H, a contradiction. Hence, H is a tree as desired. □

Next, we give several characterizations of trees. Only the proofs for equivalences between (1), (2), (3) and (6) are given. Note that (1) $\Leftrightarrow$ (7) is Proposition 3.7.

Theorem 3.10. *The following statements are equivalent for any graph G with n vertices.*

(1) *G is acyclic and connected, i.e., G is a tree.*
(2) *G is acyclic and has exactly $n - 1$ edges.*
(3) *G is connected and has exactly $n - 1$ edges.*
(4) *G is acyclic, but adding a new edge to G results a graph having cycles.*
(5) *G is connected, but deleting an edge from G results a disconnected graph.*
(6) *For any two vertices u and v, there exists a unique u–v path in G.*
(7) *G has a tree ordering.*

Proof. (1) $\Rightarrow$ (2). This follows from Propositions 3.4.

(2) $\Rightarrow$ (3). Suppose G has r components $G_1, G_2, \ldots, G_r$, with $n_1, n_2, \ldots, n_r$ vertices, respectively. Then $n = \sum_{i=1}^{r} n_i$. For $1 \le i \le r$, since each G_i is a tree, by Propositions 3.4, it has exactly $n_i - 1$ edges. Therefore, $n - 1 = \sum_{i=1}^{r} (n_i - 1) = n - r$, and so $r = 1$. Consequently, $G = G_1$ is connected as desired.

(3) $\Rightarrow$ (1). By Proposition 3.9, G has a spanning tree T. By Proposition 3.4, T has $n-1$ edges, which is the same as G. Hence, $G = T$ is a tree.

(1) $\Rightarrow$ (6). Suppose G is a tree. For any two vertices u and v, there exists a u–v path in G since G is connected. Suppose there exist two distinct u–v paths $u = u_0, u_1, \ldots, u_p = v$ and $u = v_0, v_1, \ldots, v_q = v$. Choose the index r such that $u_i = v_i$ for $1 \le i \le r$ but $u_{r+1} \ne v_{r+1}$. Next choose the first $s > r+1$ and the first $t > r+1$ such that $u_s = v_t$. Then $u_r, u_{r+1}, \ldots, u_s = v_t, v_{t-1}, \ldots, v_r = u_r$ is a cycle, a contradiction. Hence, the u–v path is unique.

(6) $\Rightarrow$ (1). Suppose for any two vertices u and v, there exists a unique u–v path in G. First, G is connected. If G has a cycle C, then there exist two paths between every two vertices in the cycles, a contradiction. Hence, G is acyclic and so is a tree. $\qquad\square$

To design graph algorithms, it is often to traverse all vertices of a graph according to certain vertex search rules. If the graph is connected, a traversal search produces a spanning tree and some related information. A traversal search on a general graph is just the traversal searches on all of its components.

There are two typical graph traversal searches, called the *depth-first search* (DFS for short) and the *breadth-first search* (BFS for short). The usages of spanning trees generated by these two traversal searches often make graph algorithms more efficient.

In various graph traversal searches, for a given connected graph G, it first chooses a vertex r as the *root* of a tree. The traversal search starts from the root r. During the traversal search, it goes from a vertex to the next vertex through an edge, called a *tree edge*. All tree edges form a spanning tree. In the tree, there is a unique path between every vertex v and the root r. All vertices in this path, including v, are called the *ancestors* of v, where the one adjacent to v is the *parent* of v. On the other hand, those having v as an ancestor are the *descendants* of v, where adjacent descendants are the *children*. The edges that are not tree edges are divided into two classes. An edge contains a vertex and an ancestor is called a *back edge*, the others are *cross edges*. At the end of the traversal search, all vertices of the graph are visited, and every edge of the graph is a tree edge, a back edge or a cross edge.

If the graph is disconnected, a traversal search is just the same as performing a traversal search for every component. In fact, this can be viewed conversely that a traversal search of a general graph identifies all components of the graph.

Figure 3.3 shows a traversal spanning tree of a connected graph, in which r is the root; vertex v_3 has four ancestors v_3, v_2, v_1, r; vertex v_2 has five descendants v_2, v_3, v_4, v_5, v_6; vertex v_1 is the parent of v_2; vertex v_4 has two children v_5 and v_6; e_1 is a cross edge; e_2 is a back edge.

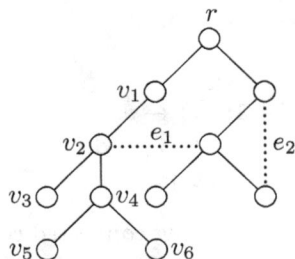

Fig. 3.3. A traversal spanning tree.

A version of depth-first search was investigated in the nineteenth century by French mathematician Charles Pierre Trémaux as a strategy for solving mazes.

In terms of graphs, a DFS for a connected graph starts by choosing a vertex a as the root. It then goes from a to a neighbor b and select ab as a tree edge, where a is the parent of b. Next goes from b to an unvisited neighbor c and choose bc as a tree edge, where b is the parent of c. Continuing this process, each time, it goes from the current vertex to an unvisited neighbor and choose the edge between them as a tree edge, where the old vertex is the parent of the new vertex. Eventually, it arrives a vertex whose neighbors all have been visited. It then goes back to the parent of this vertex and continue the process.

Note that a DFS only produces tree edges and back edges, but does not produce cross edges. Suppose there exists a cross edge xy for which x is not an ancestor of y and y is not an ancestor of x. Assume that x is visited before y. At that time, y is an unvisited neighbor of x and so may be chosen as a next vertex to visit, a contradiction.

A DFS may also be performed for directed graphs. In this case, a traversal from a to b is allowed only when ab is a (directed) edge. Now cross edges are possibly produced.

A recursive procedure for DFS is as follows. It may also be implemented by using a stack.

Depth-First Search. Find a depth-first ordering $[v_1, v_2, \ldots, v_n]$ of a connected graph G.

for all vertices v of G **do** label v unscanned;
$i \leftarrow 0$; $T \leftarrow \emptyset$;   //* T is a DFS spanning tree. *//
choose an arbitrary vertex v;   DFS(v);

procedure DFS(v);
label v scanned; $i \leftarrow i + 1$; $v_i \leftarrow v$;
for every unscanned vertex $u \in N(v)$ **do**
     $\{ T \leftarrow T \cup \{vu\}$;   DFS($u$); $\}$

BFS and its application in finding connected components of graphs (see Section 2.6 for an illustration) were invented in 1945 by Konrad Zuse, in his (rejected) Ph.D. thesis on the Plankalkül programming language, but this was not published until 1972. BFS was reinvented by Moore (1959), who used it to find a shortest path out of a maze. Later it was developed by Lee (1961) into a wire routing algorithm.

In terms of graphs, a BFS for a connected graph first chooses a vertex a as the root. It then goes from a to all neighbors $b, c, \ldots$ and choose $ab, ac, \ldots$ as tree edges, where a is the parent of $b, c, \ldots$. Next goes from the next visited vertex b to all its un-visited neighbors. And then the next earliest visited vertex c. Continue the same process.

On contrary to DFS, BFS only produces tree edges and cross edges, but no back edges. The arguments are similar to those for DFS.

Breadth-First Search. Find a breadth-first ordering $[v_1, v_2, \ldots, v_n]$ of a connected graph G.

for all vertices v of G **do** label v unscanned;
create an empty queue Q;
$i \leftarrow 0$; $T \leftarrow \emptyset$;   //* T is a BFS spanning tree. *//
choose an arbitrary vertex v;   BFS(v);

procedure BFS(v);
label v scanned;   enqueue(v, Q); $i \leftarrow i + 1$; $v_i \leftarrow v$;

while Q is not empty **do**
{ $u \leftarrow$ dequeue(Q);
      for all unscanned vertex $w \in N(u)$ **do**
        $\{ T \leftarrow T \cup \{uw\}$; enqueue$(w, Q)$; $i \leftarrow i + 1$; $v_i \leftarrow w$; $\}$
}

At the end of this section, a linear-time algorithm is shown for recognizing if a graph is a tree by using a *reverse BFS*. In the case of a positive answer, a tree ordering and its corresponding parent list are also given. Assume that the structure of the graph is stored in its adjacency lists or a sequential space representation.

Algorithm TreeRecognition: Test if a graph G is a tree. If it is, then a tree ordering $[v_1, v_2, \ldots, v_n]$ and the parent list $[p_1, p_2, \ldots, p_n]$ are given.

for all vertices v of G **do** label v unscanned;
choose a vertex r; $v_n \leftarrow r$; label r scanned; $p_n \leftarrow r$; $j \leftarrow n$;
for $i = n$ **to** 1 **step** -1 **do**
{   **if** $i < j$ **then** report "G is not connected" and **stop**;
    for every $u \in N(v_i)$ with $u \neq p_i$ **do**
       if u is scanned **then** report "G has a cycle" and **stop**
       else $\{ j \leftarrow j - 1$; $v_j \leftarrow u$; label u scanned; $p_j \leftarrow v_i$; $\}$
}

Note that Figure 3.2 is a tree ordering produced by Algorithm TreeRecognition. This reverse BFS ordering is a *strong tree ordering*, as besides being a tree ordering it has more useful properties. The first property is:

all children of a vertex are consecutive in the ordering.   (STO1)

This property enables an algorithm in trees to process simultaneously a group of leaves having the same parent, see Algorithm kRainbowDomTreeL in Section 3.2 for an example. Another property is

if v_i is farther to the root than v_j, then $i < j$.   (STO2)

3.2   Labeling approach for domination in trees

Cockayne *et al.* (1975) gave the first and a linear-time algorithm for the domination problem in trees by the labeling approach, which is a naive but useful method. They took the philosophy that to prove a statement by induction it is better to prove a stronger one. This yields a labeling algorithm described as follows.

The algorithm starts processing a leaf v of a tree T, which is adjacent to a unique vertex u. To dominate v, a minimum dominating set D of T must contain u or v. However, since $N[v] \subseteq N[u]$, it is better to have u in D rather than v in D. So, one can keep the information "required" in u and delete v from T. At some iteration, if a "required" leaf v adjacent to u is processed, it is necessary to include v in D and remove it from T. As now there is a vertex in D that dominates u, so u is labeled by "free" unless it was labeled "required" already. For convenience, all vertices are labeled by "bound" initially.

More precisely, suppose the vertex set of a graph G is partitioned into three sets B, F and R, where B consists of *bound* vertices, F consists of *free* vertices and R consists of *required* vertices. A *mixed dominating set* of G (*with respect to B, F and R*) is a subset $D \subseteq V(G)$ such that the following conditions hold.

(MD1) $R \subseteq D$.
(MD2) Every vertex in B is dominated by some vertex in D.

Free vertices need not be dominated by D but may be included in D in order to dominate bound vertices. The *mixed domination number* $\gamma^{\mathrm{m}}(G)$ of the graph G is the minimum size of a mixed dominating set in G. Note that mixed domination is the ordinary domination for the case when $B = V(G)$ and $F = R = \emptyset$.

The algorithm by Cockayne *et al.* (1975) for mixed domination in trees is based on the following theorem.

Theorem 3.11. *Suppose T is a tree with bound, free and required vertices. If v is a leaf adjacent to u, then the following statements hold.*

(1) *If $v \in B$ and u is relabeled as required in $T - v$, then $\gamma^{\mathrm{m}}(T) = \gamma^{\mathrm{m}}(T - v)$.*
(2) *If $v \in F$, then $\gamma^{\mathrm{m}}(T) = \gamma^{\mathrm{m}}(T - v)$.*

(3) *If $v \in R$ and $u \in R$, then $\gamma^m(T) = \gamma^m(T-v) + 1$.*

(4) *If $v \in R$ with $u \notin R$ and u is relabeled as free in $T-v$, then $\gamma^m(T) = \gamma^m(T-v) + 1$.*

In terms of mixed dominating sets, if D' is a minimum mixed dominating set of $T-v$, then D' (respectively, $D' \cup \{v\}$) is a minimum mixed dominating set of T when $v \in B \cup F$ (respectively, $v \in R$).

Proof. (1) Since u is required in $T-v$, any minimum mixed dominating set D' of $T-v$ always contains u and hence is also a mixed dominating set of T. Thus, $\gamma^m(T) \leq |D'| = \gamma^m(T-v)$. On the other hand, suppose D is a minimum mixed dominating set of T. Since v is bound in T, either u or v is in D. In any case, $D' = (D\backslash\{v\}) \cup \{u\}$ is a mixed dominating set of $T-v$, in which u is considered as a required vertex. So, $\gamma^m(T-v) \leq |D'| \leq |D| = \gamma^m(T)$. Hence, $\gamma^m(T) = \gamma^m(T-v)$.

(2) Since v is free, any minimum mixed dominating set D' of $T-v$ is also a mixed dominating set of T. Thus, $\gamma^m(T) \leq |D'| = \gamma^m(T-v)$. On the other hand, suppose D is a minimum mixed dominating set of T. If $v \notin D$, then D is also a mixed dominating set of $T-v$. If $v \in D$, then $(D\backslash\{v\}) \cup \{u\}$ is a mixed dominating set of $T-v$, whose size is at most $|D|$. In either case, $\gamma^m(T-v) \leq |D| = \gamma^m(T)$. Hence, $\gamma^m(T) = \gamma^m(T-v)$.

(3) If D' is a minimum mixed dominating set of $T-v$, then $D' \cup \{v\}$ is a mixed dominating set of T. Thus, $\gamma^m(T) \leq |D' \cup \{v\}| = \gamma^m(T-v) + 1$. On the other hand, any minimum mixed dominating set D of T contains both u and v. Then $D\backslash\{v\}$ is a mixed dominating set of $T-v$. So, $\gamma^m(T-v) \leq |D\backslash\{v\}| = \gamma^m(T) - 1$. Hence, $\gamma^m(T) = \gamma^m(T-v) + 1$.

(4) If D' is a minimum mixed dominating set of $T-v$, then $D' \cup \{v\}$ is a mixed dominating set of T. Thus, $\gamma^m(T) \leq |D' \cup \{v\}| = \gamma^m(T-v) + 1$. On the other hand, any minimum mixed dominating set D of T contains v. Since u is free in $T-v$, $D\backslash\{v\}$ is a mixed dominating set in $T-v$. So, $\gamma^m(T-v) \leq |D\backslash\{v\}| = \gamma^m(T) - 1$. Hence, $\gamma^m(T) = \gamma^m(T-v) + 1$. $\square$

Note that the theorem works for a general graph. The only essential condition is that v is a leaf adjacent to a unique vertex u.

The theorem then gives the following linear-time algorithm.

Algorithm DomTreeL: For a tree T with bound, free and required vertices, find a minimum mixed dominating set of T, providing a tree ordering $[v_1, v_2, \ldots, v_n]$ and the parent list $[p_1, p_2, \ldots, p_n]$ are given.

$D \leftarrow \emptyset$;
for $i = 1$ **to** $n-1$ **do**
    if v_i is bound **then** relabel p_i as required
    else if v_i is required **then**
        $\{ D \leftarrow D \cup \{v_i\}$; **if** p_i is bound **then** relabel p_i as free; $\}$
if v_n is not free **then** $D \leftarrow D \cup \{v_n\}$;

Soon, Slater (1976) generalized the above idea to solve the distance k-domination problem in trees. In fact he solved a slightly more general problem called the *R-domination problem*. Now, each vertex v is associated with an ordered pair $R_v = (a_v, b_v)$, where $a_v \geq 0$ and $b_v \geq 1$ are integers. The dominating set D is chosen so that each vertex v is within distance a_v from some vertex in D. The integer b_v indicates that there is already a vertex in the current D that is at distance b_v from v. More precisely, an *R-dominating set* of a graph G is a vertex subset D such that for any vertex v in G, at least one of the following two conditions holds.

(RD1) $d(u,v) \leq a_v$ for some $u \in D$.
(RD2) $b_u + d(u,v) \leq a_v$ for some $u \in V(G)$.

The *R-domination number* $\gamma^R(G)$ of the graph G is the minimum size of an R-dominating set in G. Note that R-domination with each $R_v = (k, k+1)$ is the distance k-domination.

The algorithm by Slater for R-domination in trees is based on the following theorem whose proof is omitted.

Theorem 3.12. *Suppose T is a tree in which each vertex v has a label $R_v = (a_v, b_v)$, where a_v is a nonnegative integer and b_v a positive integer. If v is a leaf adjacent to u, then the following statements hold.*

(1) *If $a_v = 0$ and b_u is reset by 1 in $T - v$, then $\gamma^R(T) = \gamma^R(T - v) + 1$.*
(2) *If $a_v \geq b_v$ and b_u is reset by $\min\{b_u, b_v + 1\}$ in $T - v$, then $\gamma^R(T) = \gamma^R(T - v)$.*
(3) *If $0 < a_v < b_v$ and a_u is reset by $\min\{a_u, a_v - 1\}$ in $T - v$, then $\gamma^R(T) = \gamma^R(T - v)$.*

In terms of mixed R-dominating sets, if D' is a minimum R-dominating set of $T - v$, then $D' \cup \{v\}$ (respectively, D') is a minimum R-dominating set of T when $a_v = 0$ (respectively, $a_v > 0$).

It is also the case that the theorem works for a general graph.

The theorem then gives the following linear-time algorithm. The final step is based on the fact that if T has exactly one vertex v, then v is in the minimum R-dominating set if and only if $a_v < b_v$.

Algorithm RDomTreeL: For a tree T in which every vertex v has a label $R_v = (a_v, b_v)$ where $a_v \geq 0$ and $b_v \geq 1$ are integers, find a minimum R-dominating set of T, providing a tree ordering $[v_1, v_2, \ldots, v_n]$ and the parent list $[p_1, p_2, \ldots, p_n]$ are given.

$D \leftarrow \emptyset$;
for $i = 1$ **to** $n - 1$ **do**
    if $a_{v_i} = 0$ **then** $\{ D \leftarrow D \cup \{v_i\}; \quad b_{p_i} \leftarrow 1; \}$
    else if $a_{v_i} \geq b_{v_i}$ **then** $b_{p_i} \leftarrow \min\{b_{p_i}, b_{v_i} + 1\}$
                else $a_{p_i} \leftarrow \min\{a_{p_i}, a_{v_i} - 1\}$;
if $a_{v_n} < b_{v_n}$ **then** $D \leftarrow D \cup \{v_n\}$;

The labeling approach was also used for a linear-time algorithm for total domination in trees by Laskar *et al.* (1984). Recall that a total dominating set of a graph G is a subset D of $V(G)$ such that every vertex v in V is adjacent to some vertex u in D. Note that a vertex v no longer dominates itself in this variation of domination. Hence, a graph with isolated vertices has no total dominating set.

Now the vertex set of a graph G is partitioned into four sets B, F, R_1 and R_2, each consisting of vertices labeled by B, F, R_1 and R_2, respectively. A *mixed total dominating set* of G (*with respect to B, F, R_1 and R_2*) is a subset $D \subseteq V(G)$ such that the following conditions hold.

(MTD1) $R_1 \cup R_2 \subseteq D$.
(MTD2) Every vertex in $B \cup R_1$ is adjacent to some vertex in D.

The *mixed total domination number* $\gamma_t^m(G)$ of the graph G is the minimum size of a mixed total dominating set in G. Note that mixed total domination is the ordinary total domination when $B = V(G)$ and $F = R_1 = R_2 = \emptyset$.

Note that K_1 has no total dominating set, but it has a mixed total dominating set if and only if the only vertex is labeled by F or R_2.

The algorithm by Laskar *et al.* (1984) for mixed total domination in trees is based on the following theorem.

Theorem 3.13. *Suppose T is a tree with B, F, R_1 and R_2 vertices. If v is a leaf adjacent to u, then the following statements hold when $|V(T)| \geq 2$.*

(1) *If $v \in F$, then $\gamma_t^m(T) = \gamma_t^m(T - v)$.*

(2) *If $v \in B$ with $u \in F \cup R_2$ and u is relabeled as R_2 in $T - v$, then*
$\gamma_t^m(T) = \gamma_t^m(T - v)$.

(3) *If $v \in B$ with $u \in B \cup R_1$ and u is relabeled as R_1 in $T - v$, then*
$\gamma_t^m(T) = \gamma_t^m(T - v)$.

(4) *If $v \in R_1$ and u is relabeled as R_2 in $T-v$, then $\gamma_t^m(T) = \gamma_t^m(T-v)+1$.*

(5) *If $v \in R_2$ with $u \in R_1 \cup R_2$ and u is relabeled as R_2 in $T - v$, then*
$\gamma_t^m(T) = \gamma_t^m(T - v) + 1$.

(6) *If $v \in R_2$ with $u \in F \cup B$ and u is relabeled as F in $T - v$, then*
$\gamma_t^m(T) = \gamma_t^m(T - v) + 1$.

In terms of mixed total dominating sets, if D' is a minimum mixed total dominating set of $T-v$, then D' (respectively, $D'\cup\{v\}$) is a minimum mixed total dominating set of T when $v \in F \cup B$ (respectively, $v \in R_1 \cup R_2$).

If $V(T) = \{v\}$, then the minimum mixed total dominating set is $\emptyset$ (respectively, $\{v\}$) when v is labeled by F (respectively, R_2).

The above method for mixed total domination is slightly more complicated than that for mixed domination. In fact, it is possible to redesign the labels to make it easier. The idea in the new labeling scheme comes from the facts that a vertex is "needed to be dominated" and "required to be chosen to dominate others" are two independent events. Therefore, they can be handled separately.

Now, every vertex v has a label (a_v, b_v), where $a_v = B$ if v is needed to be dominated and $a_v = F$ otherwise, $b_v = R$ if v is required to be chosen to dominate others and $b_v = \emptyset$ otherwise. In this way, $(B, \emptyset), (F, \emptyset), (B, R)$ and (F, R) correspond to B, F, R_1 and R_2 in the original setting, respectively. In the new version, a mixed total dominating set is a vertex subset D such that the following conditions hold.

(new-MTD1) If $b_v = R$, then $v \in D$.

(new-MTD2) If $a_v = B$, the v is adjacent to some vertex in D.

Note that if T has just one vertex v, then $\gamma_t^m(T) = \infty, 1$ or 0 depending on $a_v = B$, $(a_v, b_v) = (F, R)$ or $(a_v, b_v) = (F, \emptyset)$.

The supporting theorem now is simplified as follows.

Theorem 3.14. *Suppose T is a tree of at least two vertices with $(B, \emptyset)$, $(F, \emptyset)$, (B, R) and (F, R) vertices, in which v be a leaf adjacent to u.*

(1) *If $a_v = B$, then $\gamma_t^m(T)$ remains the same after relabeling b_u by R.*
(2) *If $a_u = B$ and $|V(T)| = 2$, then $\gamma_t^m(T)$ remains the same after relabeling b_v by R.*
(3) *If $b_v = R$, then $\gamma_t^m(T)$ remains the same after relabeling a_u by F.*

After these relabeling, the following hold.

(4) *If $b_v = \emptyset$, then $\gamma_t^m(T) = \gamma_t^m(T - v)$.*
(5) *If $b_v = R$, then $\gamma_t^m(T) = \gamma_t^m(T - v) + 1$.*

In terms of mixed total dominating sets, if D' is a minimum mixed total dominating set of $T - v$, then D' (respectively, $D' \cup \{v\}$) is a minimum mixed total dominating set of T when $b_v = \emptyset$ (respectively, $b_v = R$).

For the case, where T has at least two vertices, after a sequence of applications of Theorem 3.14 a final tree containing just one vertex v reaches. Because the reductions (2) and (3) are applied, it must be the case that $a_v = F$; and so the problem is feasible.

As can be seen, this theorem gives a much simpler algorithm for the total domination problem in trees. The readers are encouraged to design one.

Next is an example of labeling algorithm using *strong* tree orderings. Chang *et al.* (2010) investigated the k-rainbow domination problem in trees. For technical reasons, they in fact considered a more general problem. A *k-assignment* L of a graph G assigns every vertex v of G a pair $L(v) = (a_v, b_v)$ of integers in $\{0, 1, \ldots, k\}$. A *k-L-rainbow dominating function* is a function $f : V(G) \to 2^{\{1,2,\ldots,k\}}$ such that for every vertex v in G the following conditions hold.

  (L1) $|f(v)| \geq a_v$.
  (L2) $|\bigcup_{u \in N(v)} f(u)| \geq b_v$ whenever $f(v) = \emptyset$.

The *k-L-rainbow domination number* $\gamma_{kL\text{rain}}(G)$ of the graph G is the minimum weight of a k-L-rainbow dominating function of G. A k-L-rainbow dominating function f is *optimal* if $w(f) = \gamma_{kL\text{rain}}(G)$. Note that k-rainbow domination is k-L-rainbow domination with $L(v) = (0, k)$ for each $v \in V(G)$.

Theorem 3.15. *Suppose L is a k-assignment of a graph G, in which x is a leaf adjacent to y. Let $G' = G - x$ and L' be the restriction of L on $V(G')$, except that when $a_x > 0$ let $b'_y = \max\{0, b_y - a_x\}$. Then the following statements hold.*

(1) If $a_x > 0$, then $\gamma_{kL\mathrm{rain}}(G) = \gamma_{kL'\mathrm{rain}}(G') + a_x$.

(2) If $a_x = 0$ and $a_y \geq b_x$, then $\gamma_{kL\mathrm{rain}}(G) = \gamma_{kL'\mathrm{rain}}(G')$.

Note that after applying Theorem 3.15, for any leaf x adjacent to y one may assume that $a_x = 0$ and $b_x > a_y$. The following theorem then takes care of the remaining case.

Theorem 3.16. *Suppose L is a k-assignment of a graph G, in which $N(y)$ consists of s leaves $x_1, x_2, \ldots, x_s$ and at most one other vertex z. Assume $a_{x_i} = 0$ for $1 \leq i \leq s$ and $b_{x_1} \geq b_{x_2} \geq \ldots \geq b_{x_s} > a_y$. Let i^* be the minimum index such that $b_{x_{i^*}} + i^* = \min\{b_{x_i} + i : 1 \leq i \leq s+1\}$, where $b_{x_{s+1}} = a_y$. If $G' = G - \{x_1, x_2, \ldots, x_s\}$ and L' is the restriction of L on $V(G')$ with modifications that $a'_y = b_{x_{i^*}}$ and $b'_y = \max\{0, b_y - i^* + 1\}$, then $\gamma_{kL\mathrm{rain}}(G) = \gamma_{kL'\mathrm{rain}}(G') + i^* - 1$.*

The theorems above give the following linear-time algorithm for the k-L-rainbow domination problem in trees.

Algorithm kRainbowDomTreeL: For a tree T in which every vertex v has a label which is a pair $L(v) = (a_v, b_v)$ of integers in $\{0, 1, \ldots, k\}$, find $r := \gamma_{kL\mathrm{rain}}(T)$, providing a *strong* tree ordering $[v_1, v_2, \ldots, v_n]$ and the parent list $[p_1, p_2, \ldots, p_n]$ are given.

$r \leftarrow 0; \quad \ell \leftarrow 1;$
while $\ell < n$ **do**
$\{ \qquad$ let m be the largest index such that $\ell \leq m < n$
$\qquad\qquad$ and all vertices $v_\ell, v_{\ell+1}, \ldots, v_m$ have the
$\qquad\qquad$ same parent $y = p_\ell = p_{\ell+1} = \ldots = p_m;$
$\qquad$ **for** $\ell \leq i \leq m$ with $a_{v_i} > 0$ **do**
$\qquad\qquad \{ r \leftarrow r + a_{v_i}; \quad b_y \leftarrow \max\{0, b_y - a_{v_i}\}; \}$
$\qquad$ let $A = \{v_i : \ell \leq i \leq m, a_{v_i} = 0, b_{v_i} > a_y\};$
$\qquad$ use a bucket sort to sort b_{v_i} for all v_i in A
$\qquad\qquad$ and assume that $A = \{x_1, x_2, \ldots, x_s\}$
$\qquad\qquad$ with $b_{x_1} \geq b_{x_2} \geq \cdots \geq b_{x_s} > b_{x_{s+1}} := a_y;$
$\qquad$ let i^* be the minimum index such that
$\qquad\qquad b_{x_{i^*}} + i^* = \min\{b_{x_i} + i : 1 \leq i \leq s+1\};$
$\qquad r \leftarrow r + i^* - 1; \quad a_y \leftarrow b_{x_{i^*}}; \quad b_y \leftarrow \max\{0, b_y - i^* + 1\};$
$\qquad \ell \leftarrow m + 1;$
$\}$
if $a_{v_n} > 0$ **then** $r \leftarrow r + a_{v_n}$ **else if** $b_{v_n} > 0$ **then** $r \leftarrow r + 1;$

For more papers of algorithms on the variations of domination in trees using the labeling approach, see the references at the end of the book marked by (3.2).

As these labeling algorithms suggest, they only work for problems whose solutions have "local properties." For instance, consider the independent domination problem in the tree T of Figure 3.4. The unique minimum independent dominating set of T is $\{v_1, v_3\}$. If the tree ordering $[v_1, v_2, v_4, v_3, v_5]$ is given, the algorithm must be clever enough to put the leaf v_1 into the solution at the first iteration. If another tree ordering $[v_5, v_4, v_3, v_2, v_1]$ is given, the algorithm must be clever enough not to put the leaf v_5 into the solution at the first iteration. So, the algorithm must be one that not only looks at a leaf and its only neighbor, but also has some idea about the global structure of the tree. This is the meaning that the solution does not have a "local property." Labeling algorithms usually do not work for these kind of problems, including the independent domination problem in trees.

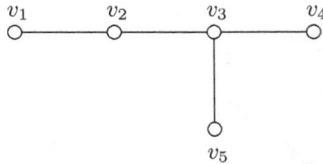

Fig. 3.4.   A tree T with a unique minimum independent dominating set.

3.3   Dynamic programming approach for domination in trees

Dynamic programming is a powerful approach for solving many discrete optimization problems, see the books by Bellman and Dreyfus (1962), Nemhauser (1966) and Dreyfus and Law (1977) for detail discussions of this technique.

The main idea of the dynamic programming approach for domination is to transform the "bottom-up" labeling approach into a "top-down" one. This approach is, once again, an implementation of mathematical induction.

Now a specific vertex v is chosen from G. One may consider (G, v) as a graph rooted at the chosen vertex v. A minimum dominating set D of (G, v) either contains v or does not. So, it is useful to consider the following two domination problems which are the ordinary domination problem with boundary conditions.

$$\gamma^1(G, v) = \min\{|D| : D \text{ is a dominating set of } G \text{ and } v \in D\}.$$

$$\gamma^0(G, v) = \min\{|D| : D \text{ is a dominating set of } G \text{ and } v \notin D\}.$$

It is clear to have the following equality, since a dominating set of (G, v) either contain v or does not.

Lemma 3.17. $\gamma(G, v) = \min\{\gamma^1(G, v), \gamma^0(G, v)\}$ *for every rooted graph* (G, v).

The dynamic programming approach for the domination problem in trees needs the following graph operation. The *composition* of two rooted graphs (G, v) and (H, u) is the rooted graph (I, v) obtained from the disjoint union of (G, v) and (H, u) by joining a new edge vu, see Figure 3.5 for an example.

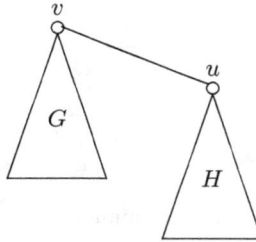

Fig. 3.5.   The composition of two rooted graphs (G, v) and (H, u).

To solve these variant domination problems, it is useful to use $\gamma^1(G, v), \gamma^0(G, v), \gamma^1(H, u)$ and $\gamma^0(H, u)$ to compute $\gamma^1(I, v)$ and $\gamma^0(I, v)$. During the derivation, there is a problem as follows.

Suppose D is a dominating set of (I, v) with $v \in D$. Then $D = D' \cup D''$, where $D' = D \cap V(G)$ is a dominating set of G with $v \in D'$ and $D'' = D \cap V(H)$ is a subset of $V(H)$ which dominates $V(H) - \{v\}$. There are two cases. In the case of $u \in D''$, D'' is a dominating set of H. On the other hand, if $u \notin D''$, then D'' is a dominating set of $H - v$ but may not be a dominating set of H. In order to resolve this difficulty in the latter case, the following new problem is introduced.

$$\gamma^{00}(G, v) = \min\{|D| : D \text{ is a dominating set of } G - v\}.$$

It is the case that $\gamma^{00}(G, v) \le \gamma^0(G, v)$, since a dominating set D of G with $v \notin D$ is also a dominating set of $G - v$. Note that a dominating set D of $G - v$ does not contains v, and it may or may not dominate v in G.

The dynamic programming solution for the domination problem in trees is based on the following theorem.

Theorem 3.18. *For the composition (I, v) of (G, v) and (H, u), the following statements hold.*

(1) $\gamma^1(I, v) = \min\{\gamma^1(G, v) + \gamma^1(H, u), \gamma^1(G, v) + \gamma^{00}(H, u)\}$.

(2) $\gamma^0(I, v) = \min\{\gamma^0(G, v) + \gamma^0(H, u), \gamma^{00}(G, v) + \gamma^1(H, u)\}$.

(3) $\gamma^{00}(I, v) = \min\{\gamma^{00}(G, v) + \gamma^1(H, u), \gamma^{00}(G, v) + \gamma^0(H, u)\}$.

Proof. (1) The formula follows from the fact that D is a dominating set of (I, v) with $v \in D$ if and only if $D = D' \cup D''$, where $D' = D \cap V(G)$ is a dominating set of (G, v) with $v \in D'$, and either $D'' = D \cap V(H)$ is a dominating set of (H, u) with $u \in D''$ or a dominating set of $H - u$.

(2) The formula follows from the fact that D is a dominating set of (I, v) with $v \notin D$ if and only if $D = D' \cup D''$, where either $D' = D \cap V(G)$ is a dominating set of (G, v) with $v \notin D'$ and $D'' = D \cap V(H)$ is a dominating set of (H, u) with $u \notin D''$, or else D' is a dominating set of $G - v$ and D'' is a dominating set of (H, u) with $u \in D''$.

(3) The formula follows from the fact that D is a dominating set of $I - u$ if and only if $D = D' \cup D''$, where $D' = D \cap V(G)$ is a dominating set of $G - u$ and $D'' = D \cap V(H)$ is a dominating set of (H, u). $\square$

Lemma 3.17 and Theorem 3.18 then give the following linear-time algorithm for the domination problem in trees. It starts with n isolated vertices, and then applies $n - 1$ composition operations to merge them into a tree.

Algorithm DomTreeD: Determine $\gamma(T)$ for a tree T providing a tree ordering $[v_1, v_2, \ldots, v_n]$ and the parent list $[p_1, p_2, \ldots, p_n]$ are given.

for $i = 1$ **to** n **do**
    $\{ \gamma^1(v_i) \leftarrow 1; \quad \gamma^0(v_i) \leftarrow \infty; \quad \gamma^{00}(v_i) \leftarrow 0; \}$
for $i = 1$ **to** $n - 1$ **do**
$\{ \quad \gamma^1(p_i) \leftarrow \min\{\gamma^1(p_i) + \gamma^1(v_i), \gamma^1(p_i) + \gamma^{00}(v_i)\};$
    $\gamma^0(p_i) \leftarrow \min\{\gamma^0(p_i) + \gamma^0(v_i), \gamma^{00}(p_i) + \gamma^1(v_i)\};$
    $\gamma^{00}(p_i) \leftarrow \min\{\gamma^{00}(p_i) + \gamma^1(v_i), \gamma^{00}(p_i) + \gamma^0(v_i)\};$
$\}$
$\gamma(T) \leftarrow \min\{\gamma^1(v_n), \gamma^0(v_n)\};$

The advantage of the dynamic programming approach is that it also works for problems whose solutions have no local property. As an example, Beyer *et al.* (1977) solved the independent domination problem by this approach.

Similar to the domination problem, there are also three modified problems for independent domination. For a rooted graph (G, v), consider the following problems with boundary conditions.

$\gamma_i^1(G, v) = \min\{|D| : D$ is an independent dominating set of G and $v \in D\}.$
$\gamma_i^0(G, v) = \min\{|D| : D$ is an independent dominating set of G and $v \notin D\}.$
$\gamma_i^{00}(G, v) = \min\{|D| : D$ is an independent dominating set of $G - v\}.$

The dynamic programming algorithm for independent domination in trees is based on the following results. All formulas are similar to those for domination, except that $\gamma_i^1(I, v)$ differs from that for $\gamma^1(I, v)$, since any independent dominating set containing v does not contain u. The proofs are similar to those for domination and thus omitted.

Lemma 3.19. $\gamma_i(G, v) = \min\{\gamma_i^1(G, v), \gamma_i^0(G, v)\}$ *for every rooted graph* (G, v).

Theorem 3.20. *For the composition* (I, v) *of* (G, v) *and* (H, u)*, the following statements hold.*

(1) $\gamma_i^1(I, v) = \gamma_i^1(G, v) + \gamma_i^{00}(H, u).$
(2) $\gamma_i^0(I, v) = \min\{\gamma_i^0(G, v) + \gamma_i^0(H, u), \gamma_i^{00}(G, v) + \gamma_i^1(H, u)\}.$
(3) $\gamma_i^{00}(I, v) = \min\{\gamma_i^{00}(G, v) + \gamma_i^1(H, u), \gamma_i^{00}(G, v) + \gamma_i^0(H, u)\}.$

Lemma 3.19 and Theorem 3.20 then give the following linear-time algorithm for the independent domination problem in trees.

Algorithm IndDomTreeD: Determine $\gamma_i(T)$ for a tree T providing a tree ordering $[v_1, v_2, \ldots, v_n]$ and the parent list $[p_1, p_2, \ldots, p_n]$ are given.

for $i = 1$ **to** n **do**
$\quad \{ \ \gamma_i^1(v_i) \leftarrow 1; \quad \gamma_i^0(v_i) \leftarrow \infty; \quad \gamma_i^{00}(v_i) \leftarrow 0; \ \}$
for $i = 1$ **to** $n - 1$ **do**
$\{ \quad \gamma_i^1(p_i) \leftarrow \gamma_i^1(p_i) + \gamma_i^{00}(v_i);$
$\quad\quad \gamma_i^0(p_i) \leftarrow \min\{\gamma_i^0(p_i) + \gamma_i^0(v_i), \gamma_i^{00}(p_i) + \gamma_i^1(v_i)\};$
$\quad\quad \gamma_i^{00}(p_i) \leftarrow \min\{\gamma_i^{00}(p_i) + \gamma_i^1(v_i), \gamma_i^{00}(p_i) + \gamma_i^0(v_i)\};$
$\}$
$\gamma_i(T) \leftarrow \min\{\gamma_i^1(v_n), \gamma_i^0(v_n)\};$

Dynamic programming was used to solve the vertex-edge-weighted case of the domination problem in trees by Natarajan and White (1978). Recall that in a graph G every vertex v has a weight $w(v)$ and every edge e has a weight $\overline{w}(e)$. The *vertex-edge-weighted domination number* $\gamma(G, w, \overline{w})$ is the minimum value of a dominating set D of G:

$$\widehat{w}(D) = \sum_{v \in D} w(v) + \sum_{u \in V(G) \setminus D} \overline{w}(uu'),$$

where D is a dominating set of G, and u' is a vertex in D that is used to dominate u not in D.

The following derivation keeps the spirit but is slightly different from that given by Natarajan and White.

$\gamma^1(G, v, w, \overline{w}) = \min\{\widehat{w}(D) : D \text{ is a dominating set of } G \text{ and } v \in D\}.$

$\gamma^0(G, v, w, \overline{w}) = \min\{\widehat{w}(D) : D \text{ is a dominating set of } G \text{ and } v \notin D\}.$

$\gamma^{00}(G, v, w, \overline{w}) = \min\{\widehat{w}(D) : D \text{ is a dominating set of } G - v\}.$

The dynamic programming algorithm for vertex-edge-weighted domination in trees is based on the following two results.

All formulas are similar to those for domination, except two places. Suppose D is an optimal solution. For $\gamma^1(I, v, w, \overline{w})$, there are two cases. In the case, where $u \in D$ or $u \notin D$ but u is dominated by some $z \in V(H)$, the answer is $\gamma^1(G, v, w, \overline{w}) + \gamma(H, u, w, \overline{w})$. In case of $u \notin D$ and u is dominated by v, the answer is $\gamma^1(G, v, w, \overline{w}) + \overline{w}(vu) + \gamma^{00}(H, u, w, \overline{w})$.

The formula for $\gamma^0(I, v, w, \overline{w})$ has a similar modification.

Lemma 3.21. $\gamma(G, w, \overline{w}) = \min\{\gamma^1(G, v, w, \overline{w}), \gamma^0(G, v, w, \overline{w})\}$ *for every rooted graph* (G, v).

Theorem 3.22. *For the composition* (I, v) *of* (G, v) *and* (H, u), *the following statements hold.*

(1) $\gamma^1(I, v, w, \overline{w}) = \min\{\gamma^1(G, v, w, \overline{w}) + \gamma(H, u, w, \overline{w}), \gamma^1(G, v, w, \overline{w}) + \overline{w}(vu) + \gamma^{00}(H, u, w, \overline{w})\}.$

(2) $\gamma^0(I, v, w, \overline{w}) = \min\{\gamma^0(G, v, w, \overline{w}) + \gamma(H, u, w, \overline{w}), \gamma^{00}(G, v, w, \overline{w}) + \overline{w}(vu) + \gamma^1(H, u, w, \overline{w})\}.$

(3) $\gamma^{00}(I, v, w, \overline{w}) = \gamma^{00}(G, v, w, \overline{w}) + \gamma(H, u, w, \overline{w}).$

The lemma and the theorem above then give the following linear-time algorithm for the vertex-edge-weighted domination problem in trees.

Algorithm VEWDomTreeD: Determine the vertex-edge-weighted domination number of a tree T, providing a tree ordering $[v_1, v_2, \ldots, v_n]$ and the parent list $[p_1, p_2, \ldots, p_n]$ are given.

for $i = 1$ **to** n **do**
    $\{\ \gamma(v_i) \leftarrow \gamma^1(v_i) \leftarrow w(v_i);\quad \gamma^0(v_i) \leftarrow \infty;\quad \gamma^{00}(v_i) \leftarrow 0;\ \}$
for $i = 1$ **to** $n - 1$ **do**
$\{\quad \gamma^1(p_i) \leftarrow \gamma^1(p_i) + \min\{\gamma(v_i), \overline{w}(p_i v_i) + \gamma^{00}(v_i)\};$
    $\gamma^0(p_i) \leftarrow \min\{\gamma^0(p_i) + \gamma(v_i), \gamma^{00}(p_i) + \overline{w}(p_i v_i) + \gamma^1(v_i)\};$
    $\gamma^{00}(p_i) \leftarrow \gamma^{00}(p_i) + \gamma(v_i);$
    $\gamma(p_i) \leftarrow \min\{\gamma^1(p_i), \gamma^0(p_i)\};$
$\}$
$\gamma(T, w, \overline{w}) \leftarrow \gamma(v_n);$

For more papers on variations of domination in trees using the dynamic programming approach, see the references at the end of the book marked by (3.3).

3.4 Primal-dual approach for domination in trees

The primal-dual approach is a powerful method in many combinatorial optimization problems, including covering-type problems, e.g., the set cover problem, the hitting set problem and the domination problem.

Farber (1981) gave a linear-time algorithm for the weighted domination problem by means of linear programming, in which primal-dual concept plays an important role. We present the idea only for the cardinality case in terms of graphs as introduced in Section 1.5. For the weighted case, see Chapter 6 for strongly chordal graphs which include trees.

Recall that the weak duality inequality (domination) says that

$$\alpha_2(G) \leq \gamma(G) \text{ for any graph } G.$$

And if there exists a dominating set D and a 2-stable set S such that $|D| \leq |S|$, then

$$|S| \leq \alpha_2(G) \leq \gamma(G) \leq |D| \leq |S|,$$

and so all inequalities are equalities. Consequently, S is a maximum 2-stable set of G, D is a minimum dominating set of G and the following equality holds.

Strong duality equality (domination): $\alpha_2(G) = \gamma(G)$.

The following presents an algorithm to find such sets D and S for a tree. The algorithm starts from a leaf v adjacent to u. It also uses the same idea in the labeling algorithm that u is more powerful than v, as $N[v] \subseteq N[u]$. Instead of choosing v, vertex u is put into D. Furthermore, v is included into S. More precisely, the algorithm can be written in two ways, as follows.

Algorithm DomTreePD: Find a minimum dominating set D and a maximum 2-stable set S of a tree T, providing a tree ordering $[v_1, v_2, \ldots, v_n]$ and the parent list $[p_1, p_2, \ldots, p_n]$ are given.

$D \leftarrow \emptyset$; $S \leftarrow \emptyset$; $D \leftarrow \emptyset$; $S \leftarrow \emptyset$;
for $i = 1$ **to** n **do**   **for** $i = 1$ **to** n **do** label v_i unscanned;
  if $N[v_i] \cap D = \emptyset$ **then**   **for** $i = 1$ **to** n **do**
  $\{$ $D \leftarrow D \cup \{p_i\}$;   **if** v_i is unscanned **then**
      $S \leftarrow S \cup \{v_i\}$; $\{$ $D \leftarrow D \cup \{p_i\}$;
  $\}$ $S \leftarrow S \cup \{v_i\}$;
                        label all vertices in $N[p_i]$ scanned;
                      $\}$

To verify the correctness of the algorithm, it is sufficient to prove that D is a dominating set, S is a 2-stable set and $|D| \leq |S|$.

First, D is a dominating set, as the if-then statement forces that a previous undominated vertex v_i is dominated by its parent p_i at the end of iteration i.

Suppose S is not a 2-stable set, that is, there exist v_i and $v_{i'}$ in S such that $i < i'$ but $d_T(v_i, v_{i'}) \leq 2$. Let $T_i = T[\{v_i, v_{i+1}, \ldots, v_n\}]$. Then T_i contains v_i, p_i and $v_{i'}$. Since $d(v_i, v_{i'}) \leq 2$, the unique v_i–$v_{i'}$ path in T is either $(v_i, p_i = v_{i'})$ or $(v_i, p_i, v_{i'})$. In either case, $p_i \in N[v_{i'}]$. Thus, at the end of iteration i, D contains p_i. When the algorithm processes $v_{i'}$ later, $N[v_{i'}] \cap D \neq \emptyset$ which implies that S cannot contain $v_{i'}$, a contradiction. This proves that S is a 2-stable set.

The inequality $|D| \leq |S|$ follows from that when p_i is added into D, which may or may not already be in D, a new vertex v_i is always added into S.

Consequently, the following theorem holds.

Theorem 3.23. *Algorithm DomTreePD gives a minimum dominating set D and a maximum 2-stable set S of a tree T with $|S| = \alpha_2(T) = \gamma(T) = |D|$ in linear time.*

The primal-dual approach also works for total domination in trees.

Recall, as introduced in Section 1.5, that the weak duality inequality (total domination) says that

$$\alpha_{t2}(G) \le \gamma_t(G) \text{ for any graph } G.$$

And if there exist a total dominating set D and a total 2-stable set S such that $|D| \le |S|$, then

$$|S| \le \alpha_{t2}(T) \le \gamma_t(T) \le |D| \le |S|,$$

and so all inequalities are equalities. Consequently, D is a minimum total dominating set, S is a maximum total 2-stable and the following equality holds.

Strong duality equality (total domination): $\alpha_{t2}(G) = \gamma_t(G)$.

The following presents an algorithm to find such D and S. The algorithm starts from a leaf v adjacent to u. It is the case that u is the only vertex adjacent to v. So, u must be put into D. Additionally, v is added to S. More precisely, the algorithm written in two ways is as follows.

Algorithm TotalDomTreePD: Find a minimum total dominating set D and a maximum total 2-stable set S of a tree T, providing a tree ordering $[v_1, v_2, \dots, v_n]$ and the parent list $[p_1, p_2, \dots, p_n]$ are given.

$p_n \leftarrow$ a neighbor of v_n; $p_n \leftarrow$ a neighbor of v_n;

$p_n \leftarrow$ a neighbor of v_n; $p_n \leftarrow$ a neighbor of v_n;
$D \leftarrow \emptyset$; $S \leftarrow \emptyset$; $D \leftarrow \emptyset$; $S \leftarrow \emptyset$;
for $i = 1$ **to** n **do**   **for** $i = 1$ **to** n **do** label v_i unscanned;
  if $N(v_i) \cap D = \emptyset$ **then**   **for** $i = 1$ **to** n **do**
  { $D \leftarrow D \cup \{p_i\}$;   **if** v_i is unscanned **then**
    $S \leftarrow S \cup \{v_i\}$;   { $D \leftarrow D \cup \{p_i\}$;
  } $S \leftarrow S \cup \{v_i\}$;
                                label all neighbors of p_i scanned;
                                }

To verify the algorithm, it is sufficient to prove that D is a total dominating set, S is a total 2-stable set and $|D| \le |S|$.

First, D is a total dominating set, as the if-then statement forces that a previous undominated vertex v_i is dominated by its parent p_i at the end of iteration i.

Suppose to the contrary that S is not a total 2-stable set, that is, there exist two vertices v_i and $v_{i'}$ in S such that $i < i'$ and v_i and $v_{i'}$ are adjacent to a same vertex, that is, $d_T(v_i, v_{i'}) = 2$. Let $T_i = T[\{v_i, v_{i+1}, \ldots, v_n\}]$. Then T_i contains v_i, p_i and $v_{i'}$. Since $d(v_i, v_{i'}) = 2$, the unique v_i-$v_{i'}$ path in T is $(v_i, p_i, v_{i'})$. In this case, $p_i \in N(v_{i'})$. Thus, at the end of iteration i, D contains p_i. When the algorithm processes $v_{i'}$ later, $N(v_{i'}) \cap D \neq \emptyset$ which causes that S does not contain $v_{i'}$, a contradiction. This proves that S is a total 2-stable set.

The inequality $|D| \leq |S|$ follows from that when p_i is added into D, which may or may not already be in D, a new vertex v_i is always added into S.

Consequently, the following theorem holds.

Theorem 3.24. *Algorithm TotalDomTreePD gives a minimum total dominating set D and a maximum total 2-stable set S of a tree T with $|S| = \alpha_{t2}(T) = \gamma_t(T) = |D|$ in linear time.*

At the end of this section, we discuss an application of the primal-dual approach for the upper domination and the fractional upper domination problems in trees.

A *minimal dominating set* of a graph G is a dominating set none of its proper subsets is a dominating set. For every vertex v in a dominating set D, a *D-private neighbor* of v is a vertex dominated by v but not dominated by any other vertex in $D - \{v\}$. The following proposition characterizes minimal dominating sets.

Proposition 3.25. *A dominating set D of G is minimal if and only if every vertex in D has at least one D-private neighbor.*

Note that the domination number $\gamma(G)$ of a graph G is the minimum size of a minimal dominating set. The *upper domination number* $\Gamma(G)$ of a graph G is the maximum size of a minimal dominating set. It is clear that

$$\gamma(G) \leq \Gamma(G) \quad \text{for every graph } G.$$

The inequality may be strict, as shown by the n-cycle C_n that $\gamma(C_n) = \lceil \frac{n}{3} \rceil$ but $\Gamma(C_n) = \lfloor \frac{n}{2} \rfloor$.

In a graph G, a maximum stable set S is a dominating set, since for every vertex v not in S, the set $S \cup \{v\}$ is no longer a stable set and so v is adjacent to some vertex in S. Also, S is a minimal dominating set, as any vertex in S can only be dominated by itself. Hence, the following inequality holds.

$$\alpha(G) \leq \Gamma(G) \quad \text{for every graph } G. \tag{3.1}$$

The inequality may be strict, as shown by $K_n \square K_2$ that $\alpha(K_n \square K_2) = 2$ but $\Gamma(K_n \square K_2) = n$ for $n \geq 2$.

Cheston *et al.* (1990) introduced the concept of *fractional upper domination* as follows. Recall that a dominating set D of G can be viewed as a *dominating function*, which is a function $f : V(G) \to \{0, 1\}$ such that

$$f(N[v]) \geq 1 \quad \text{for all } v \in V(G). \tag{3.2}$$

A dominating function is *minimal* if there is no dominating function $g \leq f$ (that is, $g(v) \leq f(v)$ for all $v \in V(G)$), but $g \neq f$. Proposition 3.25 then is transformed as follows.

Proposition 3.26. *A dominating function f of G is minimal if and only if for every v with $f(v) > 0$ there is some $u \in N[v]$ such that $f(N[u]) = 1$.*

Now consider the above concepts for fractional versions. A *fractional dominating function* of a graph G is a function $f : V(G) \to [0, 1]$ such that (3.2) holds. A fractional dominating function is *minimal* if there is no fractional dominating function $g \leq f$ but $g \neq f$. The following proposition is clear.

Proposition 3.27. *A fractional dominating function f of G is minimal if and only if for every v with $f(v) > 0$ there is some $u \in N[v]$ such that $f(N[u]) = 1$.*

The *upper fractional domination number* $\Gamma_f(G)$ of a graph G is the maximum value $f(V(G))$ of a minimal fractional dominating function f of G. Cheston *et al.* (1990) proved the existence of $\Gamma_f(G)$ by the fact that a continuous real-valued function over a compact domain in the

n-dimensional space attains a maximum. By the definition, it is the case that

$$\Gamma(G) \leq \Gamma_f(G) \quad \text{for every graph } G. \tag{3.3}$$

An example G with $\Gamma(G) < \Gamma_f(G)$ was given by Cheston *et al.* (1990).

By inequalities (3.1) and (3.3), $\alpha(G) \leq \Gamma(G) \leq \Gamma_f(G)$ for every graph G. Although there exist graphs for which the inequalities are strict, Cheston *et al.* (1990) proved that the inequalities are in fact equalities for trees.

Theorem 3.28. $\alpha(T) = \Gamma(T) = \Gamma_f(T)$ *for every tree* T.

To prove the theorem, we consider the following algorithm which is used to calculate $\alpha(T)$ for a tree T by means of a primal-dual method. Note that a clique in a tree is either a vertex or an edge.

Algorithm AlphaThetaTreePD. Find a maximum stable set S and a minimum clique cover $\mathcal{C}$ of a tree T proving a tree ordering $[v_1, v_2, \ldots, v_n]$ and the parent list $[p_1, p_2, \ldots, p_n]$ are given.

$S \leftarrow \emptyset; \quad \mathcal{C} \leftarrow \emptyset;$
for $i = 1$ **to** n **do** label v_i by s;
for $i = 1$ **to** n **do**
    if v_i is labeled by s **then**
    $\{ \quad S \leftarrow S \cup \{v_i\};$
        if p_i is labeled by z **then** $\mathcal{C} \leftarrow \mathcal{C} \cup \{v_i\};$
        else $\{ \mathcal{C} \leftarrow \mathcal{C} \cup \{v_i p_i\}; \quad$ label p_i by $z; \}$
    $\}$

Based on the algorithm, we now have the following results.

Theorem 3.29. *Algorithm AlphaThetaTreePD produces a maximum stable set and a minimum clique cover of a tree* T *in linear time. Also,* $\alpha(T) = \theta(T) = \Gamma(T) = \Gamma_f(T)$.

Proof. The running time is clearly linear. Consider S and $\mathcal{C}$ at the end of the algorithm. First, S is a stable set, since the parent of a vertex labeled by s is always labeled by z. In fact, S contains exactly those vertices labeled by s. Next, $\mathcal{C}$ is clear a clique cover. Finally, $|S| = |\mathcal{C}|$, since when a vertex is added into S a clique is also added into $\mathcal{C}$. Then

$$|S| \leq \alpha(G) \leq \theta(T) \leq |\mathcal{C}| = |S|.$$

Hence, all inequalities are equalities. Therefore, S is a maximum stable set, $\mathcal{C}$ is a minimum clique cover and $\alpha(T) = \theta(T)$.

For the equalities on $\Gamma(T)$ and $\Gamma_f(T)$, consider the minimal dominating function f corresponding to S which is a minimal dominating set. That is

$$f(v_i) = 1 \text{ if } v_i \text{ is labeled by } s, \quad f(v_i) = 0 \text{ if } v_i \text{ is labeled by } z.$$

Then f is also a minimal fractional dominating function. Choose a minimal fractional dominating function g with value $\Gamma_f(T)$.

We claim that $g = f$ and so value $\Gamma_f(T) = g(V(T)) = f(V(T)) = |S| = \alpha(T)$ implying that $\alpha(T) = \Gamma(T) = \Gamma_f(T)$. Suppose to the contrary that $g \neq f$. Then there exists a least index i such that $g(v_i) \neq f(v_i)$. We may assume that g is chosen so that i is as large as possible. Consider two cases.

Case 1. v_i is labeled by z. In this case, $g(v_i) > 0 = f(v_i)$.

By the algorithm, v_i has a child v_j labeled by s, where $j < i$. Then $g(v_j) = f(v_j) = 1 > 0$. There is some $v_k \in N[v_j]$ such that $g(N[v_k]) = 1$. Since $g(v_i) > 0$ and $g(v_j) = 1$, v_k is a child of v_j and so is labeled by z. Again, v_k has a child v_ℓ labeled by s and so $g(v_\ell) = f(v_\ell) = 1$. Hence, $1 = g(N[v_k]) \geq g(v_j) + g(v_\ell) = 1 + 1$, a contradiction.

Case 2. v_i is labeled by s. In this case, $g(v_i) < 1 = f(v_i)$.

By the algorithm, every child v_j is labeled by z and so $g(v_j) = f(v_j) = 0$. Then $g(v_i) + g(p_i) = g(N[v_i]) \geq 1$. We claim that $g(v_i) + g(p_i) = 1$.

Suppose to the contrary that $g(v_i) + g(p_i) > 1$. Then $g(p_i) \leq 1$ implies $0 < g(v_i) < 1$. Consider the vertex $v_{j'} \in N[v_i]$ for which $g(N[v_{j'}]) = 1$. In this case, $v_{j'}$ is not a child of v_i, for otherwise $1 = g(N[v_{j'}]) = g(v_i) +$ (some integer), a contradiction. Then $v_{j'} = v_i$ or $v_{j'} = p_i$. Since $g(N[p_i]) \geq g(N[v_i]) \geq 1$, it is the case that $g(v_i) + g(p_i) = 1$.

Now, consider g' obtained from g by the following modification:

(1) $g'(v_i) = 1$ and $g'(p_i) = 0$,
(2) for any $u \in N(p_i) \setminus \{v_i\}$, if $g(N[u] \setminus \{p_i\}) < 1$, then $g'(u) = g(u) + 1 - g(N[u] \setminus \{p_i\})$.

It can be verified that g' is a minimal fractional dominating function and $g'(V(T)) \geq g(V(T))$. Hence, the value of g' is also $\Gamma_f(T)$. But $i' > i$, a contradiction to the maximality of i. $\qquad\square$

3.5 Power domination in trees

The concept of power domination was introduced by Haynes *et al.* (2002), inspired by a problem in the electric power system industry. Electric power companies need to continually monitor their system's state as defined by a set of state variables (e.g., the voltage magnitude at loads and the machine phase angle at generators). One method of monitoring these variables is to place phase measurement units, called PMUs, at selected locations in the system. Because of the high cost of a PMU, it is desirable to minimize their number while maintaining the ability to monitor (observe) the entire system. A system is said to be observed if all of the state variables of the system can be determined from a set of measurements (e.g., voltages and currents).

Let G be a graph representing an electric power system, where a vertex represents an electrical node. A PMU measures the state variable for the vertex at which it is placed and its incident edges and their end-vertices; these vertices and edges are considered observed for monitoring purposes. The other observation rules are (P1), (P2) and (P3). Once all vertices and all edges of the graph are observed, the system is observed.

(P1) Any vertex that is incident to an observed edge is observed.

(P2) Any edge joining two observed vertices is observed.

(P3) If all edges, except one edge e, incident to a vertex are observed, then edge e is also observed.

According to (P2), Kneis *et al.* (2006) proved that the description of the observation is only necessary on vertices:

(Q1) A vertex with a PMU can observe itself and all its neighbors;

(Q2) If all neighbors, except u, of a vertex are observed, then u is also observed.

Formally, for a subset $D \subseteq V(G)$, perform the following steps in order to result M.

(M1) Initialize $M = D$. Add to M, one by one, any vertex u not already in M which is adjacent to a vertex v in D, until no such vertex u remains. Say that u is *propagated* from v.

(M2) If M has a vertex v for which all neighbors, except u, are in M, then add u into M. If v is propagated from r, then u is propagated from r. Repeat the propagation until no such vertex v remains.

Note that the observation rules imply some propagating behavior, which is a phenomenon quite different from the standard domination problem. A *power dominating set* of a graph G is a subset $D \subseteq V(G)$ for which the final M is equal to $V(G)$. The *power domination number* $\gamma_{\text{pow}}(G)$ is the minimum size of a power dominating set of G.

Returning to the paper by Haynes *et al.* (2002). They gave a linear-time algorithm for the power domination problem in trees by means of spiders defined below.

A *spider* is a tree with at most one vertex of degree at least 3. A vertex of degree at least 3 is called the *root* of the spider if it exists, otherwise any vertex can be considered as the root. A *spider partition* of a tree T is a partition $\{V_1, V_2, \ldots, V_s\}$ of $V(T)$ such that every induced subgraph $T[V_i]$ is a spider. The *spider number* $\text{sp}(T)$ of a tree T is the minimum size of a spider partition of T.

Here is the relation between the spider number and the power domination number of a tree.

Lemma 3.30. $\text{sp}(T) \leq \gamma_{\text{pow}}(T)$ *for every tree* T.

Proof.   Suppose D is a minimum power dominating set of T. Every $v \in D$ propagates to some of its neighbors by rule (M1), and then each observed vertex only propagates to one of its neighbors by rule (M2). Hence, the set V_v of all vertices propagated from v induces a spider rooted at v. And so $\text{sp}(T) \leq |D| = \gamma_{\text{pow}}(T)$.   □

Lemma 3.31. $\gamma_{\text{pow}}(T) \leq \text{sp}(T)$ *for every tree* T.

Proof.   Suppose $\text{sp}(T) = s$ and $\{V_1, V_2, \ldots, V_s\}$ is a spider partition of $V(T)$, where for each i, r_i is the root of the spider $T[V_i]$. For the case of $s = 1$, $\{r_1\}$ is clearly a power dominating set of T. Hence, $\gamma_{\text{pow}}(T) \leq \text{sp}(T)$.

Suppose $s \geq 2$ and the lemma holds for $s' < s$. Then the graph H with vertex set $\{V_1, V_2, \ldots, V_s\}$ and edge set

$$\{V_i V_j : i \neq j, \text{ there exists an edge } v_i v_j \in E(T) \text{ with } v_i \in V_i \text{ and } v_j \in V_j\}$$

is a tree. Choose a leaf of H, say V_s. There is exactly one edge $vu \in V(T)$ with $v \in V_s$ and $u \in V - V_s$. Also, $T' := T - V_s$ is a tree with a spider partition $\{V_1, V_2, \ldots, V_{s-1}\}$ and so $\text{sp}(T') \leq s-1$. Choose a minimum power dominating set D' of T'. By the induction hypothesis, $|D'| = \gamma_{\text{pow}}(T') \leq \text{sp}(T') \leq s - 1$.

Consider $D := D' \cup \{r_s\}$ and the following propagation starting from D. First, observe vertices of V_s from r_s until v is observed. Next, observe all vertices of T' from D'. Note that this is possible as the vertices in T' adjacent only to v (outside T'), which has already been observed. Finally, observe the remaining unobserved vertices in V_s. These give that $\gamma_{\text{pow}}(T) \leq |D| = |D'| + 1 = (s-1) + 1 = s = \text{sp}(T)$ as desired. $\qquad\square$

The lemmas are combined as the following theorem. Note that the theorem is also true for forests by considering all components.

Theorem 3.32. $\gamma_{\text{pow}}(T) = \text{sp}(T)$ *for every tree T.*

Next, we present an alternative method for finding the spider number of a tree. Now, the tree T is considered to be rooted at r. Recall that the concepts of ancestors, parent, descendants and children are introduced in Section 3.1. In particular, $D(v)$ denotes the set of all descendants of v. A vertex v is of *type i* if it has exactly i descendants each of them having at least two children. Let $\text{ch}[v]$ count the number of children of v and $\text{ch}i[v]$ count the number of type i children of v.

The following is the base of the algorithm for finding $\text{sp}(T)$ of a tree T. It is a revision of Lemma 3.31.

Theorem 3.33. *Suppose v is a vertex in a rooted tree T such that either*

(1) $\text{ch}[v] = 2$ *and* $\text{ch}0[v] = \text{ch}1[v] = 1$, *or*

(2) *v is of type 1 and its parent p has another child v' of type 1 or two children v'' and v''' of type 0.*

If $T' := T - D(v)$, then $\text{sp}(T) = \text{sp}(T') + 1$.

Proof. First, in either case (1) or (2), $T[D(v)]$ is a spider. Hence, a spider partition of T' and $D(v)$ form a spider partition of T, and so $\text{sp}(T) \leq \text{sp}(T') + 1$.

Next, choose a minimum spider partition $\mathcal{S}$ of T.

If $V_i \subseteq D(v)$ for some $V_i \in \mathcal{S}$, then T' is an induced subgraph of $T - V_i$ and $\mathcal{S} \backslash \{V_i\}$ is a spider partition of $T - V_i$. Hence,

$$\text{sp}(T') \leq \text{sp}(T - V_i) \leq \text{sp}(T) - 1. \tag{3.4}$$

Otherwise, $D(v)$ is a proper subset of some $V_i \in \mathcal{S}$. As $D(v)$ has a vertex x with at least two children, V_i is the union of $D(v)$ and the vertices of a p-y path $P = (p, p', \ldots, y)$ in T', where p is the parent of v.

For case (1), V_i has two vertices x and v both of degree at least 3, a contradiction to that V_i is a spider. For case (2), we consider three subcases.

(a) p' does not exist or $p' = v'$. In this subcase, choose $z = v'$.

(b) $p' = v''$ by symmetry. In this subcase, choose $z = v'''$.

(c) p' is the parent of p. In this subcase, choose $z = v'''$.

In either subcase, $D(z)\backslash V(P)$ is partitioned into at least one spiders in $\mathcal{S}$. Then $(\mathcal{S}\backslash\{V_i,$ spider partition of $D(z)\backslash V(P)\}) \cup \{D(z) + P\}$ is a spider partition of T' and so inequality (3.4) holds.

Both $\mathrm{sp}(T) \le \mathrm{sp}(T') + 1$ and $\mathrm{sp}(T') \le \mathrm{sp}(T) - 1$ give the theorem. $\square$

Note that v is of type 0 if and only if $D(v)$ induces a path. Also, v is of type 1 if and only if either $\mathrm{ch}[v] = \mathrm{ch1}[v] = 1$ or $\mathrm{ch}[v] = \mathrm{ch0}[v] \ge 2$.

Here is the implementation of the theorem. Note that during the algorithm, every vertex v_i keeps $\mathrm{ch}[v_i] = \mathrm{ch0}[v_i] + \mathrm{ch1}[v_i]$, $\mathrm{ch1}[v_i] \le 1$ and $\mathrm{ch0}[v_i] \le 1$ when $\mathrm{ch1}[v_i] = 1$.

Algorithm PowerDomTree: Compute $\gamma_{\mathrm{pow}}(T)$ for a tree T, providing a tree ordering $[v_1, v_2, \ldots, v_n]$ and the parent list $[p_1, p_2, \ldots, p_n]$ are given.

for $i = 1$ **to** n **do** $\mathrm{ch}[v_i] \leftarrow \mathrm{ch0}[v_i] \leftarrow \mathrm{ch1}[v_i] \leftarrow 0$;

$\gamma_{\mathrm{pow}} \leftarrow 0$;

for $i = 1$ **to** $n - 1$ **do**

{   **if** $\mathrm{ch}[v_i] = 2$ and $\mathrm{ch0}[v_i] = \mathrm{ch1}[v_i] = 1$

      then $\gamma_{\mathrm{pow}} \leftarrow \gamma_{\mathrm{pow}} + 1$; $//* D(v_i)$ is deleted from the tree. $*//$

      else $//*$ Now $\mathrm{ch}[v_i] = \mathrm{ch0}[v_i]$ or $\mathrm{ch}[v_i] = \mathrm{ch1}[v_i] = 1$. $*//$

            if $\mathrm{ch0}[p_i] \ge 2$ or $\mathrm{ch1}[p_i] = 1$

            then $\gamma_{\mathrm{pow}} \leftarrow \gamma_{\mathrm{pow}} + 1$; $//* D(v_i)$ is deleted from the tree. $*//$

            else { $//*$ Now $\mathrm{ch}[p_i] = \mathrm{ch0}[p_i] \le 1$ $*//$

                  $\mathrm{ch}[p_i] \leftarrow \mathrm{ch}[p_i] + 1$;

                  if $\mathrm{ch}[v_i] = \mathrm{ch0}[v_i] = 1$ **then** $\mathrm{ch0}[p_i] \leftarrow \mathrm{ch0}[p_i] + 1$

                                          else $\mathrm{ch1}[p_i] \leftarrow \mathrm{ch1}[p_i] + 1$;

            }

}

if $\mathrm{ch}[v_n] \ge 1$ **then** $\gamma_{\mathrm{pow}} \leftarrow \gamma_{\mathrm{pow}} + 1$;

Guo *et al.* (2008) provided an improved algorithm using a post-order traversal of the tree.

Lyu (2023) provided a labeling algorithm for power domination in trees. The following gives an alternative labeling algorithm. suppose the vertex set of a graph G is partitioned into three sets B, O and R, where B consists of *bound* vertices, O consists of *observed* vertices and R consists of *required* vertices. For a subset $D \subseteq V(G)$, performing the following steps in order results in M.

(mM1) Initialize $M = D$. Add to M, one by one, any vertex u not already in M which is adjacent to a vertex v in D, until there is no such vertex u. Call u being *propagated* from v.

(mM2) $M = M \cup O$.

(mM3) If M has a vertex v for which all neighbors, except u, are in M, then add u into M. If v is propagated from r, then u is propagated from r. Repeat the propagation until there is no such vertex v.

A *mixed power dominating set* of G (*with respect to* B, O, R) is a subset $D \subseteq V(G)$ with $R \subseteq D$ for which the final M is equal to $V(G)$. The *mixed power domination number* $\gamma_{\text{pow}}^{\text{m}}(G)$ of G is the minimum size of a mixed power dominating set of G. Note that mixed power domination is the ordinary power domination for the case when $B = V(G)$ and $O = R = \emptyset$.

The algorithm is based on the following lemmas, in which G is a graph with bound, observed and required vertices B, O and R, respectively.

Lemma 3.34. *If G' is obtained from G by relabeling the common neighbor u of two bound leaves v and v' by R, then $\gamma_{\text{pow}}^{\text{m}}(G) = \gamma_{\text{pow}}^{\text{m}}(G')$.*

Proof. If D is a minimum mixed power dominating set of G, then $D \cap \{u, v, v'\} \neq \emptyset$, for otherwise v and v' will never be put into M, a contradiction. Hence, $(D \backslash \{u, v, v'\}) \cup \{u\}$ is a mixed power dominating set of G' and so $\gamma_{\text{pow}}^{\text{m}}(G') \leq |(D \backslash \{u, v, v'\}) \cup \{u\}| \leq |D| = \gamma_{\text{pow}}^{\text{m}}(G)$. If D' is a minimum mixed power dominating set of G', then $u \in D'$, and so D' is a mixed power dominating set of G as $v, v' \in M$ after step (mM1). Hence, $\gamma_{\text{pow}}^{\text{m}}(G) \leq |D'| = \gamma_{\text{pow}}^{\text{m}}(G')$. Therefore, $\gamma_{\text{pow}}^{\text{m}}(G) = \gamma_{\text{pow}}^{\text{m}}(G')$. $\qquad \square$

Lemma 3.35. *If G' is obtained from G by relabeling all bound neighbors of a required vertex v by O, then $\gamma_{\text{pow}}^{\text{m}}(G) = \gamma_{\text{pow}}^{\text{m}}(G')$.*

Proof. The lemma follows from the fact that for a set $D = D'$ containing v, the set M for G is equals to the set M' for G' after step (mM2). $\qquad \square$

Lemma 3.36. *If G' is obtained from G by relabeling the only bound neighbor u of an observed vertex v by O, then $\gamma_{\mathrm{pow}}^{\mathrm{m}}(G) = \gamma_{\mathrm{pow}}^{\mathrm{m}}(G')$.*

Proof.   The lemma follows from the fact that for a set $D = D'$, the set M for G is equals to the set M' for G' after step (mM2) and one more propagation from v to u, since v has exactly one bound neighbor u. $\square$

Lemma 3.37. *If G' is obtained from G by deleting a required vertex v with no bound neighbors, then $\gamma_{\mathrm{pow}}^{\mathrm{m}}(G) = \gamma_{\mathrm{pow}}^{\mathrm{m}}(G') + 1$.*

Proof.   Since $N_G[v]$ contains no bound vertex, a sequence of propagations for G starting from D containing v is a sequence of propagations for G' stating from $D\backslash\{v\}$, and vice versa. Hence, $\gamma_{\mathrm{pow}}^{\mathrm{m}}(G) = \gamma_{\mathrm{pow}}^{\mathrm{m}}(G') + 1$. $\square$

Lemma 3.38. *If G' is obtained from G by deleting an observed vertex v with no bound neighbors, then $\gamma_{\mathrm{pow}}^{\mathrm{m}}(G) = \gamma_{\mathrm{pow}}^{\mathrm{m}}(G')$.*

Proof.   First, a minimum mixed power domination set of G does not contain v, since $N_G[v]$ contains no bound vertex. Then, a sequence of propagations for G starting from D not containing v is a sequence of propagations for G' stating from D, and vice verse. Hence, $\gamma_{\mathrm{pow}}^{\mathrm{m}}(G) = \gamma_{\mathrm{pow}}^{\mathrm{m}}(G')$. $\square$

Lemma 3.39. *Suppose v is a bound leaf adjacent to a bound/observed vertex u with $\deg_G(u) \leq 2$. If G' is obtained from G by deleting v and relabeling u by B when $\deg_G(u) = 2$, then $\gamma_{\mathrm{pow}}^{\mathrm{m}}(G) = \gamma_{\mathrm{pow}}^{\mathrm{m}}(G')$.*

Proof.   If $\deg_G(u) = 1$, then $G[\{u, v\}]$ is a component of G. Since v is bound, $\gamma_{\mathrm{pow}}^m(G[\{u, v\}]) = \gamma_{\mathrm{pow}}^m(G[\{u\}])$ and so $\gamma_{\mathrm{pow}}^{\mathrm{m}}(G) = \gamma_{\mathrm{pow}}^{\mathrm{m}}(G')$.

Now suppose u is adjacent to v and w. Then u is bound in G' by the assumption. Choose a minimum mixed power dominating set D. We may assume that $v \notin D$ for otherwise $(D\backslash\{v\}) \cup \{u\}$ is a minimum mixed power dominating set. Since $\deg_G(u) = 2$, the propagations only possible from w to u but not possible from u to w. Hence, D is also a mixed dominating set of G' and so $\gamma_{\mathrm{pow}}^{\mathrm{m}}(G') \leq |D| = \gamma_{\mathrm{pow}}^{\mathrm{m}}(G)$.

On the other hand, a minimum mixed power dominating set D' of G' is also a mixed power dominating set, since after all vertices of $V(G')$ are observed vertex u can propagate to v. Hence, $\gamma_{\mathrm{pow}}^{\mathrm{m}}(G) \leq |D'| = \gamma_{\mathrm{pow}}^{\mathrm{m}}(G')$.

Therefore, $\gamma_{\mathrm{pow}}^{\mathrm{m}}(G) = \gamma_{\mathrm{pow}}^{\mathrm{m}}(G')$. $\square$

The lemmas above then give the following linear-time algorithm for the mixed power domination problem in trees.

Algorithm PowerDomTreeL: For a tree T in which every vertex has a label bound, observed or required, find $s := \gamma_{\text{pow}}^{\text{m}}(T)$, providing a *strong tree ordering* $[v_1, v_2, \ldots, v_n]$ and the parent list $[p_1, p_2, \ldots, p_n]$ are given.

$s \leftarrow 0; \quad \ell \leftarrow 1;$

while $\ell < n$ **do**

{   let m be the largest index such that $\ell \leq m < n$ and all vertices
         in $L := \{v_\ell, v_{\ell+1}, \ldots, v_m\}$ have the same parent u;

    let r be the number of required vertices in L;

    if L has at least two bound vertices **then** relabel u by R;

    if u is required **then** relabel all bound vertices in L by O;

    if $r \geq 1$ and u is bound **then** relabel u by O;

    if L has some observed vertex and u is bound **then** relabel u by O;

    if $L = \{v\}$ with v bound and $u \neq v_n$ **then** relabel u by B;

    $s \leftarrow s + r; \quad \ell \leftarrow m + 1;$

}

if v_n is not observed **then** $s \leftarrow s + 1;$

Theorem 3.40. *Algorithm PowerDomTreeL gives $\gamma_{\text{pow}}^{\text{m}}(T)$ of a tree whose vertices are labeled by bound, observed or required in linear time.*

Proof.   The first if statement uses Lemma 3.34. The second if statement uses Lemma 3.35. By now, if u is required, then L has no bound vertices; if u is not required, then L has at most one bound vertex.

The third if statement uses Lemma 3.35. The forth if statement uses Lemma 3.36. By now, if u is bound, then u has exactly one child which is bound.

The fifth if statement uses Lemma 3.39.

The statement $s \leftarrow s + r$ follows from Lemmas 3.37–3.39.   □

3.6   Paired-domination in trees

The concept of paired-domination was introduced by Haynes and Slater (1998). For a vertex subset D of a graph G, think of every $v \in D$ as the location of a guard capable of protecting every vertex in $N[v]$. Then domination requires every vertex in G is protected; and for total domination, each guard is protected by another. For paired-domination, it requires the guards' locations to be selected as adjacent pairs of vertices so that every guard is assigned one other and they are designated as backups for each other.

More formally, a *paired-dominating set* of a graph G is a dominating set D such that $G[D]$ has a perfect matching. In other words, $D = \{v_1, v_2, v_3, v_4, \ldots, v_{2r-1}, v_{2r}\}$ such that $v_{2i-1}v_{2i} \in E(G)$ for $1 \leq i \leq r$. The *paired-domination number* $\gamma_{\text{pr}}(G)$ of a graph G is the minimum size of a paired-dominating set of G.

Paired-dominating sets and total dominating sets have the same property that a vertex in the sets is dominated by a neighbor. However, the pairing of vertices makes paired-domination quite different from total domination. Qiao *et al.* (2003) gave a linear-time algorithm for the paired-domination problem in trees.

Start the observation with the n-path P_n with

$$\text{vertex set } V(P_n) = \{v_0, v_1, \ldots, v_{n-1}\} \text{ and}$$
$$\text{edge set } E(P_n) = \{v_0v_1, v_1v_2, \ldots, v_{n-2}v_{n-1}\}.$$

Intuitively, pairs of adjacent vertices with two unselected vertices between consecutive pairs will be enough, see the examples as follows (Figure 3.6).

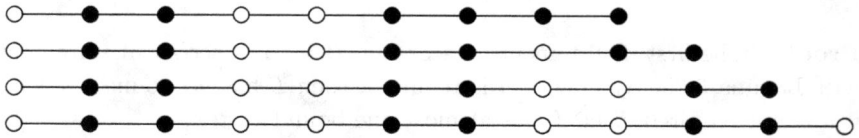

Fig. 3.6. Paired-dominating set in paths.

In general, D_n as below is a paired-dominating set of P_n of size $2\lceil\frac{n}{4}\rceil$:

$$D_n = \begin{cases} \{v_i : i \equiv 1, 2 \ (\text{mod } 4), 0 \leq i \leq n-1\} \\ \cup \{v_{n-2}, v_{n-1}\}, & \text{if } n \equiv 1, 2 \ (\text{mod } 4); \\ \{v_i : i \equiv 1, 2 \ (\text{mod } 4), 0 \leq i \leq n-1\}, & \text{if } n \equiv 3, 0 \ (\text{mod } 4). \end{cases}$$

Hence, $\gamma_{\text{pr}}(P_n) \leq 2\lceil\frac{n}{4}\rceil$ for $n \geq 2$. This is in fact an equality. Suppose a minimum paired-dominating set D of P_n has r pairs. Then $n \leq 2r + 1 + 2(r-1) + 1 = 4r$ and so $r \geq \lceil\frac{n}{4}\rceil$, yielding $\gamma_{\text{pr}}(P_n) \geq 2\lceil\frac{n}{4}\rceil$. This is because, $|D| = 2r$, $V(G)\backslash D$ has at most one vertex on the left (respectively, right) of the first (respectively, last) pair in D, and $V(G)\backslash D$ has at most two vertices between two consecutive pairs in D.

While paths capture the property of depth in trees, stars capture the property of breadth in trees. Next consider spiders. As every four vertices need two vertices to paired-dominate them, only spiders in which each leg is of length at most three are considered, see an example as follows (Figure 3.7).

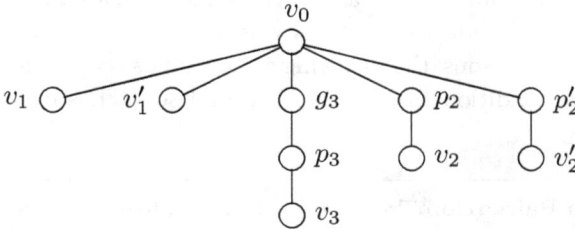

Fig. 3.7. A spider.

To paired-dominate v_2, choosing its parent p_2 and grandparent v_0 is better than choosing v_2 and p_2. After v_0 having chosen, it is no longer possible to choose p_2' and v_0 for paired-dominating v_2'. In the case of the spider only has legs of length 1 and 3, p_3 and g_3 are good for paired-dominating v_3. To paired-dominate v_1, choosing v_0 and its parent p_0 is better. However, if for some reasons, p_0 was chosen before, it is only possible to choose v_1 and v_0.

The two possibilities to choose a pair to paired-dominate an undominated vertex make things complicated. The algorithm for the paired-domination in trees by Qiao *et al.* (2003) is a little complicated. The verification of the correctness is also long.

In the following a strong tree ordering is used to make the algorithm simple. Then a primal-dual approach is to make the verification of the correctness easier.

A *3-stable* set of a graph is a vertex subset S in which $d(u, v) > 3$ for every two distinct vertices u and v in S. The *3-stability number* $\alpha_3(G)$ of a graph G is the maximum size of a 3-stable set in G.

Suppose S is a 3-stable set and D is a paired-dominating set of a graph G. Define a function $f : S \to D$ by $f(v) = u$ for some $u \in D$ which dominates v; if there are several possible u, choose any one. Suppose u and v are two distinct vertices in $V(G)$. Then $f(u) \neq f(v)$, for otherwise $d(u, v) \leq 2$, a contradiction. Also, $\{f(u), f(v)\}$ is not a matching pair in D, for otherwise $d(u,v) \le 3$, again a contradiction. Therefore, the following inequality holds.

Weak duality inequality (paired-domination): $2\alpha_3(G) \le \gamma_{\mathrm{pr}}(G)$ for every graph G having no isolated vertices.

The inequality can be strict as shown by the n-cycles that $\alpha_3(C_n) = \lfloor \frac{n}{4} \rfloor$ and $\gamma_{\mathrm{pr}}(C_n) = 2\lceil \frac{n}{4} \rceil$ for every integer $n \ge 3$.

For technical reasons, the following algorithm uses a strong tree ordering with an extra condition that v_n is a leaf and so v_n has exactly one child which is v_{n-1}.

Algorithm PairedDomTreePD: For a tree T, find a maximum 3-stable set S and a minimum paired-dominating set D, providing a *strong* tree ordering $[v_1, v_2, \ldots, v_n]$ and the parent list $[p_1, p_2, \ldots, p_n]$ are given, where the root v_n has only one child v_{n-1}.

$S \leftarrow D \leftarrow \emptyset$; $\ell \leftarrow 1$;
while $\ell \le n-3$ **do**
$\{$ let m be the largest index such that $\ell \le m \le n-3$ and vertices
         $v_\ell, v_{\ell+1}, \ldots, v_m$ have the same parent $p_\ell = p_{\ell+1} = \ldots = p_m$;
         if $N[v_i] \cap D = \emptyset$ for some i with $\ell \le i \le m$ **then**
         $\{$ let g_i is the parent of p_i;
                 $S \leftarrow S \cup \{v_i\}$;
                 if $g_i \in D$ **then** $D \leftarrow D \cup \{v_i, p_i\}$ **else** $D \leftarrow D \cup \{p_i, g_i\}$;
                 $\ell \leftarrow m+1$;
         $\}$
$\}$
if $v_{n-1} \notin D$ **then** $S \leftarrow S \cup \{v_n\}$ and $D \leftarrow D \cup \{v_{n-1}, v_n\}$;

Theorem 3.41. *Algorithm PairedDomTreePD finds a maximum 3-stable set and a minimum paired-dominating set of a tree T in linear time. Furthermore, $2\alpha_3(T) = \gamma_{\mathrm{pr}}(T)$.*

Proof. By the testing "$N[v_i] \cap D = \emptyset$" and the last line, the final D is a paired-dominating set of T. As whenever two vertices are added into D one vertex is added into S, it is the case that $2|S| = |D|$.

Next to show that the final S is a 3-stable set. Suppose to the contrary that $(v_i, v_j) \le 3$ for two distinct $v_i, v_j \in S$.

Case 1. $i < j < n$.

By the algorithm, $p_i, g_i \in D$ after iteration i, no matter $g_i \in D$ or not at the beginning of iteration i. As v_j is chosen into S, it is the case that $N[v_j] \cap D = \emptyset$ before iteration j, in particular after iteration i. So, $p_i, g_i \in D$ and $d(v_i, v_j) \leq 3$ imply that v_j is a grand-child of p_i, which lead to $j < i$ by property (STO2) of a strong tree ordering, a contradiction.

Case 2. $i < j = n$.

By the last line of the algorithm, $v_n \in S$ is because $v_{n-1} \notin D$ after iteration $n - 3$. This then implies that children and grand-children of v_{n-1} are not in S, as a vertex in S will force its parent and grand-parent in D. Hence, $d(v_i, v_n) > 3$ as desired.

Finally, $2|S| \leq 2\alpha_3(T) \leq \gamma_{\mathrm{pr}}(T) \leq |D| = 2|S|$, and so all inequalities are equalities. Consequently, S is a maximum 3-stable set, D is a minimum paired-dominating set and $2\alpha_3(T) = \gamma_{\mathrm{pr}}(T)$. □

3.7 Tree-related graphs

Besides the methods demonstrated in the previous sections, the "transformation method" sometimes is also used in the study of variations of domination. Roughly speaking, the method transforms a domination problem in certain graphs to another well-known problem, which is solvable. As this method depends on the variation of domination and the class of graphs studied, it will be mentioned only when it is used when such a problem is surveyed in the later chapters.

An important reason for the variations of the domination problem in trees can be solved easily is that trees have leaves. In other words, every induced subgraph of a tree is either a vertex or has an end block that is an edge. It is expected that graphs for which every induced subgraph is simple or has an end block that is simple have the same advantage. The simplest blocks may be complete graphs and cycles, see the descriptions below.

A *block graph* is a graph whose blocks are complete graphs. Block graphs include forests.

A *cactus* is a connected graph whose blocks are edges or cycles. Cacti include trees.

Generalizations of trees include k-*trees*. For a positive integer k, k-*trees* are defined recursively by the following two rules; and k-trees are all generated from this way.

(i) A complete graph of k vertices is a k-tree.

(ii) The graph obtained from a k-tree by adding a new vertex adjacent to a clique of $k - 1$ vertices is a k-tree.

Partial k-trees are subgraphs of k-trees. Note that 2-trees are precisely trees, and partial 2-trees are exactly forests.

For more papers on variations of domination in block graphs, cacti, k-trees and partial k-trees, see the references at the end of the book marked by (3.7).

3.8   Exercises

(3.1) For every integer $n \geq 1$, let $t(n)$ denote the number of nonisomorphic trees with n vertices.

    (a) Determine $t(n)$ for $1 \leq n \leq 8$.

    (b) Find a lower bound for $t(n)$ that is better than $O(n)$.

(3.2) Show that a graph is acyclic if and only if every induced subgraph has a vertex of degree at most one.

(3.3) Show that for an integer $n \geq 2$, a sequence of positive integers $d_1, d_2, \ldots, d_n$ is the degree sequence of a tree if and only if $\sum_{i=1}^{n} d_i = 2n - 2$.

(3.4) Show that Erdős-Sós Conjecture is true for every tree of diameter at most 3.

(3.5) (a) Show that a tree T has at least $\Delta(T)$ leaves.

    (b) Determine all trees T having exactly $\Delta(T)$ leaves.

(3.6) Prove that conditions (4) and (5) are equivalent to others in Theorem 3.10.

(3.7) Calculate the number of edges for a graph G of $n \geq 3$ vertices in which $G - v$ is a tree for every $v \in V(G)$. Determine all graphs G in which $G - v$ is a tree for every $v \in V(G)$.

(3.8) Show that for every three vertices u, v, w in a tree, the three path $P(u, v), P(v, w), P(w, u)$ intersect at exactly one vertex.

(3.9) In a BFS tree, a vertex is on level i if the distance from it to the root is i. Prove that the end vertices of a cross edge are on the same level.

(3.10) Prove statement (2) in Theorem 3.12.

(3.11) Prove statement (2) in Theorem 3.13.

(3.12) Prove Theorem 3.14.

(3.13) Design a labeling algorithm for the mixed total domination problem in trees.

(3.14) Design a labeling algorithm for the new version of the mixed total domination problem in trees.

(3.15) Prove Theorem 3.15.

(3.16) Prove Theorem 3.16.

(3.17) Design a labeling algorithm for the edge domination problem in trees.

(3.18) Show that for every tree T there is a minimum edge dominating set which is also an independent edge dominating set.

(3.19) A dominating set D of a graph G is *locating* if $D(u) \neq D(v)$ for every two distinct vertices $u, v \in V(G) - D$, where $D(u)$ denotes the set of all vertices in D adjacent to u. The *locating-domination number* $\gamma_{\mathrm{loc}}(G)$ of a graph G is the minimum size of a locating-dominating set of G.

    (a) Determine $\gamma_{\mathrm{loc}}(P_n)$ for every integer $n \geq 1$. Justify your answer.

    (b) Design a labeling algorithm for the locating-domination problem in trees.

(3.20) Prove Lemma 3.19.

(3.21) Prove Theorem 3.20.

(3.22) Design a dynamic programming algorithm for the independent domination problem.

(3.23) Prove Lemma 3.21.

(3.24) Prove Theorem 3.22.

(3.25) Design a dynamic programming algorithm for the perfect domination problem in trees.

(3.26) For every integer $n \geq 2$, determine $\alpha_2(P_2 \square P_n)$ and $\gamma(P_2 \square P_n)$. Justify your answers.

(3.27) For every integer $n \geq 3$, verify that $\alpha_{t2}(C_n) = \lfloor \frac{n}{4} \rfloor + \lfloor \frac{n+1}{4} \rfloor$ and $\gamma_t(C_n) = \lfloor \frac{n+2}{4} \rfloor + \lfloor \frac{n+3}{4} \rfloor$. Determine all n for which $\alpha_{t2}(C_n) = \gamma_t(C_n)$.

(3.28) For every integer $n \geq 2$, determine $\alpha_{t2}(P_2 \square P_n)$ and $\gamma_t(P_2 \square P_n)$. Justify your answers.

(3.29) Prove Proposition 3.25.

(3.30) For every integer $n \geq 3$, determine $\alpha(C_n)$, $\Gamma(C_n)$ and $\Gamma_{\mathrm{f}}(C_n)$. Justify your answers.

(3.31) For every integer $n \geq 2$, determine $\alpha(P_2 \square P_n)$, $\Gamma(P_2 \square P_n)$ and $\Gamma_f(P_2 \square P_n)$. Justify your answers.

(3.32) Determine $\gamma_{\text{pow}}(P_5 \square P_6)$. Justify your answer.

(3.33) In a general graph G, a *generalized spider* is the subgraph induced by the set of all vertices which are propagated from a fixed vertex v, where v is called the *root* of this generalized spider. The *generalized spider number* $\text{gsp}(G)$ of a graph G is the minimum s such that $V(G)$ is the disjoint union of the vertex sets of s generalized spiders.

   (a) Prove that $\text{gsp}(G) \leq \gamma_{\text{pow}}(G)$ for every graph G.

   (b) Is it true that $\text{gsp}(G) = \gamma_{\text{pow}}(G)$ for every graph G? Justify your answer. If this is not true, provide examples of graphs G for which $\text{gsp}(G) < \gamma_{\text{pow}}(G)$.

(3.34) For $n \geq 1$, determine $\alpha_3(P_2 \square P_n)$ and $\gamma_{\text{pr}}(P_2 \square P_n)$. Justify your answers.

(3.35) For $n \geq 1$, determine $\alpha_3(P_3 \square P_n)$ and $\gamma_{\text{pr}}(P_3 \square P_n)$. Justify your answers.

(3.36) Determine the number of edges of a connected block graph of n vertices each of whose blocks has exactly k vertices. Justify your answer.

(3.37) Prove that a cactus of n vertices has at most $\lfloor 3(n-1)/2 \rfloor$ edges.

(3.38) Determine the number of edges of a k-tree of n vertices. Justify your answer.

Chapter 4

Chordal Graphs

4.1 Chordal graphs as classical perfect graphs

Trees/forests are graphs with the simplest structures. The next simplest graphs may be block graphs, obtained by blowing edges to cliques from forests. Cacti are also simple in the sense that they are obtained from trees by replacing some edges by cycles. They are simple since all their blocks are simple and easy to handle.

Chordal graphs are slightly more complex; these are graphs in which every cycle of length at least 4 has a chord. In a chordal graph, a chord of a cycle of length 5 divides the cycle into a triangle and a cycle of length 4; and then one more chord is needed for the cycle of length 4; thus in total there are at least 2 chords. In general, a cycle of length $\ell \geq 4$ has at least $\ell - 3$ chords.

In a chordal graph, chords make the vertices of cycles "closer" and so the graph has no large induced cycles. This makes chordal graphs somewhat resemble trees.

In the literature, chordal graphs have also been called *triangulated, rigid-circuit, monotone transitive* and *perfect elimination* graphs. They were studied by many researchers including Hajnal and Surányi (1958), Berge (1960), Dirac (1961), Lekkerkerker and Boland (1962) and Fulkerson and Gross (1965) for the theory of perfect graphs. For more details, see Chapter 4 of Golumbic's book (1980).

113

In 1958, Hajnal and Surányi proved that the complement of every chordal graph G is perfect, that is, $\alpha(H) = \theta(H)$ for every induced subgraph H of G. In 1960, Berge proved that every chordal graph G is perfect, that is, $\omega(H) = \chi(H)$ for every induced subgraph H of G. This is one indication of the Weak Perfect Graph Conjecture.

In a 1961 paper, Dirac used the concepts of minimal $a–b$ separating sets and simplicial vertices, under different names, to reprove these two theorems. Here is Dirac's first theorem.

Theorem 4.1 (Dirac, 1961). *A graph G is chordal if and only if every minimal $a–b$ separating set is a clique for every two nonadjacent vertices a and b.*

Proof.   Suppose G is chordal. Choose a minimal $a–b$ separating set S with $G[A]$ and $G[B]$ being the components of $G - S$ containing a and b, respectively. Since S is minimal, each vertex $x \in S$ is adjacent to some vertex in A and some vertex in B. Therefore, for every pair of vertices $x, y \in S$, there exist paths $(x, a_1, a_2, \ldots, a_r, y)$ and $(y, b_1, b_2, \ldots, b_s, x)$, where all $a_i \in A$ and all $b_j \in B$. We may assume that these paths are chosen so that $r + s$ is as small as possible. It follows that $(x, a_1, a_2, \ldots, a_r, y, b_1, b_2, \ldots, b_s, x)$ is a cycle of length at least 4, implying that it must have a chord. But $a_i a_j \notin E(G)$ and $b_i b_j \notin E(G)$ by the minimality of $r + s$, and $a_i b_j \notin E(G)$ by the definition of an $a–b$ separating set. Thus, the only possible chord is xy. This gives that the minimal a-b separating set S is a clique.

On the other hand, suppose every minimal $a–b$ separating set is a clique for every two nonadjacent vertices a and b. For every cycle $C = (v_0, v_1, \ldots, v_{n-1}, v_n)$ of length $n \geq 4$, where $v_n = v_0$, either $v_0 v_2$ is a chord of the cycle C or else $v_0 v_2 \notin E(G)$. In the later case, every v_0-v_2 separating set S must contain v_1 and some v_i with $3 \leq i \leq n - 1$. By the assumption that S is a clique, $v_1 v_i \in E(G)$ is a chord of the cycle C. Therefore, G is chordal. $\square$

Theorem 4.1 then implies the following result by Berge.

Theorem 4.2 (Berge, 1960). *The complement of a chordal graph is perfect. In other words, $\omega(H) = \chi(H)$ for every induced subgraph H of G.*

Proof.   Since every induced subgraph of a chordal graph is chordal, it only needs to prove that $\omega(G) = \chi(G)$ for every chordal graph G. This is clear if G is a complete graph. Suppose G has two nonadjacent vertices a and b.

Choose a minimal a–b separating set S of G. Let $G[A_1], G[A_2], \ldots, G[A_r]$ be the components of $G - S$, where $r \geq 2$.

Clearly, $\omega(G) \geq \omega(G[S \cup A_i])$ for every i, so $\omega(G) \geq \max_i \omega(G[S \cup A_i])$. Let C be a clique of G with $|C| = \omega(G)$. It is only possible that C is a subset of some $S \cup A_j$, since every two vertices of C are adjacent but no vertex in one A_p is adjacent to a vertex in another A_q. Hence, $\max_i \omega(G[S \cup A_i]) \geq \omega(G[S \cup A_j]) \geq |C| = \omega(G)$. Therefore,

$$\omega(G) = \max_i \omega(G[S \cup A_i]).$$

Clearly, $\chi(G) \geq \chi(G[S \cup A_i])$ for every i, so $\chi(G) \geq \max_i \chi(G[S \cup A_i])$. In fact, G can be colored using exactly $\max_i \chi(G[S \cup A_i])$ colors. First color S, then independently extend the coloring to every $G[S \cup A_i]$. This gives a proper coloring of G and so

$$\chi(G) = \max_i \chi(G[S \cup A_i]).$$

By the induction hypothesis, $\omega(G[S \cup A_i]) = \chi(G[S \cup A_i])$ for every i. Hence, $\omega(G) = \chi(G)$ as desired. $\qquad\square$

The proof of Theorem 4.2 may be turned into an algorithm for computing $\omega(G)$ and $\chi(G)$ of a chordal graph. To make it easier one may assume that $r = 2$ by considering $G[S \cup A_2 \cup A_3 \cup \ldots A_r]$ as one graph. However, this is still not efficient enough, as after splitting one graph into two graphs, it needs to split them into four. Continue this process, it needs to solve for exponential numbers of subproblems. To resolve the problem, one possible way is to find such a decomposition that the subproblems can be easily solved for $G[S \cup A_i]$, for instance making A_1 containing just one vertex x. In this case, $N(x)$, or equivalently $N[x]$ is a clique. This kind of vertex is called a *simplicial vertex*.

Here is Dirac's second theorem. The proof also shows that to prove a statement it is easier to prove a stronger one.

Theorem 4.3 (Dirac, 1961). *Every chordal graph has a simplicial vertex. Moreover, if it is not a complete graph, then it has two nonadjacent simplicial vertices.*

Proof. The theorem is clear if G is a complete graph. Assume G has two nonadjacent vertices a and b. Let S be a minimal a–b separating set for which $G[A]$ and $G[B]$ are components of $G - S$ containing a and b, respectively. By the induction hypothesis, either $G[S \cup A]$ has two nonadjacent simplicial vertices and one of which is in A (since S is a clique), or else $S \cup A$ is a clique and every vertex of A is a simplicial vertex of $G[S \cup A]$. Furthermore, since $N[A] \subseteq S \cup A$, a simplicial vertex of $G[S \cup A]$ is also a simplicial vertex of G. Similarly, B contains a simplicial vertex of G. The theorem then follows.   □

Theorem 4.3 then implies the following result by Hajnal and Surányi.

Theorem 4.4 (Hajnal and Surányi, 1958). *Every chordal graph is perfect. In other words, $\alpha(H) = \theta(H)$ for every induced subgraph H of G.*

Proof.   Since every induced subgraph of a chordal graph is chordal, it only needs to prove that $\alpha(G) = \theta(G)$ for every chordal graph G. The theorem is clear if G is a complete graph.

Suppose G is not a complete graph. By Theorem 4.3, G has a simplicial vertex x with $N[x] \neq V(G)$. Consider $G' = G - N[x]$, which is a chordal graph smaller than G. By the induction hypothesis, $\alpha(G') = \theta(G') = k$. Choose a stable set I' of G' of size k, and a clique cover $\mathcal{Q}'$ of G' of size k. Then $I := I' \cup \{x\}$ is a stable set of G of size $k+1$, and $\mathcal{Q} := \mathcal{Q}' \cup \{N[x]\}$ is a clique cover of G of size $k+1$. Consequently,

$$|I| \leq \alpha(G) \leq \theta(G) \leq |\mathcal{Q}| = k+1 = |I|,$$

and so all inequalities are equalities. The theorem then follows.   □

Unfortunately, there is no similar primal-dual equality between 2-stability and domination for chordal graphs, as shown by the 3-sun S_3 in Figure 1.11 that $\alpha_2(S_3) = 1 < 2 = \gamma(S_3)$. Also, $\alpha_2(C_n) = \lfloor n/3 \rfloor$ and $\gamma(C_n) = \lceil n/3 \rceil$ for all $n \geq 3$.

Note that leaves are simplicial vertices. Catching the spirit in the proof of Theorems 4.1 and 4.3, it is possible to get a theorem slightly more general than Theorem 4.3, which generalizes Propositions 3.1 for trees.

Theorem 4.5 (Voloshin, 1982; Farber-Jamison, 1986). *For every vertex x in a chordal graph G, there is a simplicial vertex y at maximum distance from x.*

Proof.   Without loss of generality, assume that G is connected. The theorem is proved by induction on the number n of graph G.

For $n = 1$, the only vertex of G is simplicial.

Suppose $n \geq 2$ and the theorem is true for $n' < n$. If x is adjacent to all other vertices, then apply the induction hypothesis to the chordal graph $G - x$. Each simplicial vertex y of $G - x$ is also a simplicial vertex of G, since x is adjacent to all vertices in $N_{G-x}[y]$.

Otherwise, let F be the set of vertices in G with maximum distance from x, and let H be a component of $G[F]$. Let S be the set of vertices in $G - F$ having neighbors in $V(H)$, and let H' be the component of $G - S$ containing x, see Figure 4.1.

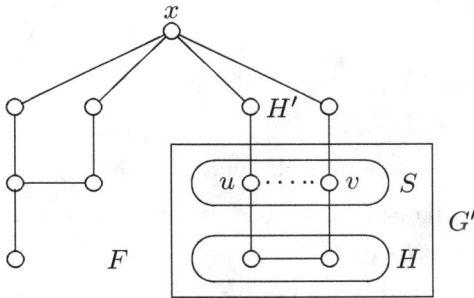

Fig. 4.1. A vertex at maximum distance from a fixed vertex x is simplicial.

First, claim that S is a clique. Note that every vertex in S has a neighbor in H and a neighbor in H'. Then for every two distinct vertices u and v in S, the union of shortest u–v paths through H and through H' is a cycle of length at least 4. Since there are no edges from $V(H)$ to $V(H')$, this cycle has no chord other than uv. Hence, u is adjacent to v, and so S is a clique.

Consider $G' = G[S \cup V(H)]$, which excludes x and therefore is smaller than G. Apply the induction hypothesis to G' and a vertex $x' \in S$ with minimum degree $\deg_{G'}(x')$ to get a simplicial vertex y' at maximum distance from x' in G'. If $y' \in V(H)$, then it is also a simplicial vertex of G, since $N_{G'}(y') = N_G(y')$. If $y' \in S$, then x' is adjacent to every vertex $v \in V(H)$, since $d_{G'}(x', v) \leq d_{G'}(x', y') = 1$. Then, y' is also adjacent to all vertices in $V(H)$ as $\deg_{G'}(x') \leq \deg_{G'}(y')$. Thus G' is a complete graph, since y' is a simplicial vertex in G'. In this case, every vertex in $V(H)$ is a simplicial vertex of G as desired. $\square$

4.2   More characterizations of chordal graphs

Using the fact that a chordal graph have a simplicial vertex and the hereditary property, Fulkerson and Gross (1965) suggested a recognition algorithm for chordal graphs: *Repeatedly locate a simplicial vertex and eliminate it from the graph, until either no vertices remains and the graph is chordal, or else at some stage no simplicial vertex exists and the graph is not chordal.* Note that this is not a linear-time algorithm. Next section will discuss two linear-time algorithms for recognizing chordal graphs.

The former case of Fulkerson and Gross's algorithm states that the vertices of the graph can be ordered into $[v_1, v_2, \ldots, v_n]$ such that v_i is a simplicial vertex of $G[\{v_i, v_{i+1}, \ldots, v_n\}]$ for $1 \le i \le n$. In other words,

$$i < j < k \text{ and } v_i v_j, v_i v_k \in E(G) \text{ imply } v_j v_k \in E(G). \qquad \text{(PEO)}$$

Such an ordering of vertices is called a *perfect elimination ordering*. Note that a tree ordering is a perfect elimination ordering.

Theorem 4.6 (Fulkerson and Gross, 1965). *A graph G is chordal if and only if it has a perfect elimination ordering.*

Proof.   Suppose G is chordal. First, by Theorem 4.3, there is a simplicial vertex v_1 of $G_1 := G$. Next, choose a simplicial vertex v_2 of $G_2 := G_1 - v_1$. Continue this process, once v_i is chosen but $i < n$, choose a simplicial vertex v_{i+1} of $G_{i+1} := G_i - v_i$. Continue this process until $i = n$ to obtain a perfect elimination ordering of G.

Suppose G has a perfect elimination ordering $[v_1, v_2, \ldots, v_n]$. For any cycle of length at least 4, choose the vertex of the cycle with the smallest index. By (PEO), the two neighbors of this vertex in the cycle are adjacent, which gives a chord of the cycle. This proves that G is chordal.   □

Next, we provide an alternative proof for the necessity of Theorem 4.6 without using Theorem 4.3. In fact, this proof implies Theorem 4.3.

Alternative proof for ($\Rightarrow$) of Theorem 4.6. For any vertex ordering $\sigma = [v_1, v_2, \ldots, v_n]$ of the vertices of G, define the vector

$$d(\sigma) = (d_1, d_2, \ldots, d_n),$$

where each $d_i = |\{j : j > i \text{ and } v_j \text{ is adjacent to } v_i\}|$ is called the *forward-degree* of v_i in the ordering σ. A vector $(d_1, d_2, \ldots, d_n)$ is said to be greater in the *back-lexicographic order* than another vector $(d_1', d_2', \ldots, d_n')$ if there exists some $r \leq n$ such that $d_i = d_i'$ for $i > r$, but $d_r > d_r'$. Choose a vertex ordering σ such that $d(\sigma)$ is back-lexicographically largest.

Suppose $i < j < k$ and $v_k \in N(v_i) \backslash N(v_j)$. Consider the ordering σ' obtained from σ by interchanging v_i and v_j. Then $d_r' = d_r$ for all $r > j$. Since $d(\sigma')$ is back-lexicographically smaller than or equal to $d(\sigma)$,

$$|\{s : s > j \text{ and } v_s \in N(v_i)\}| = d_j' \leq d_j = |\{t : t > j \text{ and } v_t \in N(v_j)\}|.$$

However, k is an index such that $k > j$ and $v_k \in N(v_i) \backslash N(v_j)$. Then, there exists some $\ell > j$ such that $v_\ell \in N(v_j) \backslash N(v_i)$. This gives

$$i < j < k \text{ and } v_k \in N(v_i) \backslash N(v_j)$$
$$\text{imply } v_\ell \in N(v_j) \backslash N(v_i) \text{ for some } \ell > j . \qquad \text{(MCSO)}$$

Note that, as will be seen in the next section, (MCSO) is a vertex ordering after a *maximum cardinality search*. For the case when G is chordal, condition (MCSO) implies the following claim.

Claim. There does not exist any chordless path $P = (v_{i_1}, v_{i_2}, \ldots, v_{i_r})$ with $r \geq 3$ and $i_s < i_r < i_1$ for $1 < s < r$.

Proof of the claim. Suppose to the contrary that such a path P exists. Choose one with the largest i_r. Since $i_2 < i_r < i_1$ and $v_{i_1} \in N(v_{i_2}) \backslash N(v_{i_r})$, by (MCSO), there exists some $i_{r+1} > i_r$ such that $v_{i_{r+1}} \in N(v_{i_r}) \backslash N(v_{i_2})$. Let r' be the minimum index such that $r' \geq 3$ and $v_{i_{r+1}} v_{i_{r'}} \in E(G)$. Note that r' exists.

For the case when $v_{i_1} v_{i_{r+1}} \notin E(G)$, $P' = (v_{i_1}, v_{i_2}, \ldots, v_{i_{r'-1}}, v_{i_{r'}}, v_{i_{r+1}})$ (or its inverse) is a chordless path of length at least three with $i_s < i_{r+1} < i_1$ (or $i_s < i_1 < i_{r+1}$) for $1 < s \leq r'$. Therefore, $i_r < i_{r+1}$ (or $i_1 < i_{r+1}$) is a contradiction to the choice of P.

For the case when $v_{i_1} v_{i_{r+1}} \in E(G)$, P' together with the edge $v_{i_1} v_{i_{r+1}}$ forms a chordless cycle of length at least 4, a contradiction to the fact that G is chordal. $\qquad \square$

To prove (PEO), if $i < j < k$ and $v_i v_j, v_i v_k \in E(G)$ but $v_j v_k \notin E(G)$, then (v_k, v_i, v_j) is a chordless path with $i < j < k$, a contradiction to the above claim. $\qquad \square$

Note that this proof is also useful for proving the correctness of the maximum cardinality search for recognizing chordal graphs in the next section.

Given a perfect elimination ordering $[v_1, v_2, \ldots, v_n]$ of a chordal graph G, it is clear that

$$C_i := \{v_j \in N[v_i] : j \geq i\}$$

is a clique for $1 \leq i \leq n$. Conversely, for every clique C of G, choose a minimum indexed vertex v_i in C. By (PEO), C is a subset of C_i. In the case when C is a maximal clique, $C = C_i$. Consequently, the following proposition holds.

Proposition 4.7 (Fulkerson and Gross, 1965). *A chordal graph of n vertices has at most n maximal cliques, with equality if and only if the graph is edgeless.*

There exist graphs of n vertices which are not chordal but also have n maximal cliques, such as C_n with $n \geq 4$, or more generally cacti with exactly one block being some C_i with $i \geq 4$.

There exist also graphs of n vertices with an exponential number of maximal cliques, such as the complete r-partite graph $K_{2,2,\ldots,2}$ with $n = 2r$ has $2^r = 2^{n/2}$ maximal cliques, since $r = n/2$.

Another well-known characterization of chordal graphs is in terms of intersection families.

Theorem 4.8 (Walter, 1972, 1978; Gavril, 1974; Buneman, 1974). *The following statements are equivalent for a graph G.*

(1) *G is chordal.*
(2) *G is the intersection graph of a family of subtrees of a tree.*
(3) *There is a tree T, called a clique tree of G, whose vertex set $V(T)$ is the set of all maximal cliques of G and $T[\mathcal{C}_v]$ is a subtree of T for $v \in V(G)$, where $\mathcal{C}_v = \{C \in V(T) : v \in C\}$.*

Proof. (3) $\Rightarrow$ (2). Suppose u and v are two distinct vertices in G. If $\mathcal{C}_u \cap \mathcal{C}_v \neq \emptyset$, then there is some maximal clique $C \in \mathcal{C}_u \cap \mathcal{C}_v$. By the definition, $u, v \in C$ and so $uv \in E(G)$. On the other hand, if $uv \in E(G)$, then u and v are both in some maximal clique C of G. Then the trees $\mathcal{C}_u$ and $\mathcal{C}_v$ intersect at C. Hence, G is the intersection graph of the family $\{T[\mathcal{C}_v] : v \in V(G)\}$ of subtrees of the tree T.

(2) $\Rightarrow$ (1). Suppose graph G is the intersection graph of a family $\{T_v : v \in V(G)\}$ of subtrees of a tree T. Choose a vertex r of T as a root. Let r_v be

the vertex of T_v nearest to r in T. Suppose $(u_0, u_1, \ldots, u_{m-1}, u_m)$ is a cycle of length $m \geq 4$ in G, where $u_m = u_0$. Assume that $d_T(r_{u_1}, r) \geq d_T(r_{u_i}, r)$ for $0 \leq i \leq m$. Since T_{u_1} and T_{u_0} have a common vertex x which is a descendent of both r_{u_1} and r_{u_0}. Then either r_{u_1} is an ancestor of r_{u_0} or r_{u_0} is an ancestor of r_{u_1}. Since $d_T(r_{u_1}, r) \geq d_T(r_{u_0}, r)$, it is the case that r_{u_0} is an ancestor of r_{u_1}, which implies that r_{u_1} is in T_{u_0}. Similarly, $d_T(r_{u_1}, r) \geq d_T(r_{u_2}, r)$ implies that r_{u_1} is in T_{u_2}. Hence, T_{u_0} and T_{u_2} have a common vertex r_{u_1}, which implies that $u_0 u_2$ is a chord of the cycle $(u_0, v_2, \ldots, u_m)$. This proves that G is chordal.

(1) $\Rightarrow$ (3). The implication holds for the case of G is a complete graph, since T and all subtrees can be chosen as K_1.

Suppose G is not a complete graph. Choose a simplicial vertex s of G and consider $G' := G - s$. By the induction hypothesis, there exists a tree T' whose vertex set $V(T')$ is the set of all maximal cliques of G' and $T'[\mathcal{C}_v]$ is a subtree of T' for $v \in V(G')$.

Note that $N_G[s]$ is a maximal clique in G. Consider the following two cases.

Case 1. $N_G(s)$ is a maximal clique in G'.

In this case, consider the tree T obtained from T' by renaming the vertex $N_G(s)$ with $N_G[s]$. Then the subtree $T'[\mathcal{C}_v]$ in T' is replaced by the subtree $T[\mathcal{C}_v]$ in T for $v \in V(G')$; and $\mathcal{C}_s = \{N_G[s]\}$ induces a tree in T.

Case 2. $N_G(s)$ is not a maximal clique in G'.

In this case, choose a maximal clique $C \supseteq N_G(s)$ in G'. Consider the tree T obtained from T' by adding a new leaf $N_G[s]$ adjacent to C. Then $T[\mathcal{C}_v] = T'[\mathcal{C}_v]$ for $v \in V(T') \backslash N_G(s)$, $T[\mathcal{C}_v]$ is obtained from the tree $T'[\mathcal{C}_v]$ by adding the leaf $N_G[s]$ adjacent to C for $v \in N_G(s)$, and $\mathcal{C}_s = \{N_G[s]\}$ induces a tree in T. $\qquad\square$

The subtrees $T_1, T_2, \ldots, T_n$ and the host tree T for the intersection graph representation of a chordal graph G are not unique. Sub-classes of chordal graphs with specific choices T or the subtrees $T_1, T_2, \ldots, T_n$ are discussed as follows.

An *interval graph* is the intersection graph of a finite nonempty family of closed intervals in the real line. Suppose G is the intersection graph of $\{[a_1, b_1], [a_2, b_2], \ldots, [a_n, b_n]\}$, where each $[a_i, b_i]$ is a closed interval with left endpoint a_i and right endpoint b_i in the real line. (By some small changes, one may assume that all endpoints of the intervals are distinct. The definition remains equivalent if open intervals are used instead of closed intervals.) Let T be the tree with

> vertex set $V(T) = \{a_1, b_1, a_2, b_2, \ldots, a_n, b_n\}$ and
>
> edge set $E(T) = \{uv : u \text{ and } v \text{ are two consecutive values in } V(T)\}$.

If $|V(T)| = m$, then T is in fact the m-path P_m. Let T_i be the subgraph of T induced by $\{v \in V(T) : a_i \le v \le b_i\}$. Then it is the $|V(T_i)|$-path. It is easy to see that G is the intersection graph of the family $\{T_1, T_2, \ldots, T_n\}$ of sub-paths of the path T.

Chordal graphs that are not interval graphs include directed path graphs and undirected path graphs. An *undirected path graph* is the intersection graph of a family of paths in a tree. A *directed path graph* is the intersection graph of a family of directed paths of a rooted tree, whose edges are directed from vertices to their children. The relations between these graphs are

$$\{\text{interval graphs}\} \subsetneq \{\text{directed path graphs}\}$$
$$\subsetneq \{\text{undirected path graphs}\}$$
$$\subsetneq \{\text{chordal graphs}\}.$$

Recall that S_3 is the 3-sun as in Figure 1.11. The following properties hold.

(1) $\overline{S}_3$ is not an interval graph but is a directed path graph.
(2) S_3 is not a directed path graph but is an undirected path graph.
(3) $S_3 + K_1$ is not an undirected path graph but is a chordal graph.

A graph is *split* if its vertex set is the disjoint union of a clique C and a stable set S. Note that a split graph is chordal as the ordering with the vertices in S first and the vertices in C next gives a perfect elimination ordering. The examples above are all split graphs.

A general property is as follows, whose proof is left to the readers.

Proposition 4.9. *If $G + K_1$ is an undirected path graph, then G is an interval graph.*

4.3 Algorithms for perfect graph theory in chordal graphs

Using perfect elimination orderings, Theorem 4.4 can be transformed into an algorithm for finding a maximum stable set and a minimum clique cover of a chordal graph, and also proving the strong duality equality $\alpha(G) = \theta(G)$.

Algorithm alpha-thetaChordal: Find a maximum stable set S and a minimum clique cover $\mathcal{Q}$ of a chordal graph G, providing a perfect elimination ordering $[v_1, v_2, \ldots, v_n]$ is given.

$S \leftarrow \emptyset; \quad \mathcal{Q} \leftarrow \emptyset;$
initially all vertices are unmarked;
for $i = 1$ **to** n **do**
    if v_i is unmarked **then**
    $\{ \quad S \leftarrow S \cup \{v_i\};$
        add $C_i := \{v_j : j \geq i, v_j \in N[v_i]\}$ into $\mathcal{Q}$;
        mark all vertices in C_i;
    $\}$

It is easy to see that the algorithm outputs a stable set S and a clique cover $\mathcal{Q}$ of the same size. Then

$$|S| \leq \alpha(G) \leq \theta(G) \leq |\mathcal{Q}| = |S|,$$

and so all inequalities are equalities. These imply that S is a maximum stable set, $\mathcal{Q}$ is a minimum clique cover of G and $\alpha(G) = \theta(G)$.

Theorem 4.10. *Algorithm alpha-thetaChordal gives a maximum stable set and a minimum clique cover of a chordal graph in linear time, and also proves the strong duality equality $\alpha(G) = \theta(G)$, providing a perfect elimination ordering is given. Hence complements of chordal graphs are perfect.*

The following algorithm finds a maximum clique and a minimum proper coloring for a chordal graph G, and also proves the strong duality equality $\omega(G) = \chi(G)$.

Algorithm omega-chiChordal: Find a maximum clique C and a minimum proper coloring f of a chordal graph G, providing a perfect elimination ordering $[v_1, v_2, \ldots, v_n]$ is given.

$m \leftarrow 0;$ //* The maximum color used so far. *//
$p \leftarrow 0;$ //* The position where color m is used, i.e., $f(v_p) = m$. *//
for $i = n$ **down to** 1 **do**
$\{$ $f(v_i) \leftarrow$ the smallest color not used by any
              neighbor v_j of v_i with $j > i$;
    if $f(v_i) > m$ **then** $\{$ $m \leftarrow f(v_i);$ $p \leftarrow i;$ $\}$
$\}$
$C \leftarrow \{v_j : j \geq p, v_j \in N[v_p]\};$

More detail implementation of finding $f(v_i)$ is as follows. Consider an array used$[1..n]$, where the value used$[c]$ is to indicate if the color c was used before. Initially, all used$[c]$ are set to 0, indicating that no colors are used. To find the least positive integer not equal to any $f(v_j)$ with $v_j \in N(v_i)$ and $j > i$, reset used$[f(v_j)] \leftarrow i$ for all such v_j. And then color v_i by the least t such that used$[t] \neq i$. Note that for the next iteration, we don't have to reset those used$[f(v_j)]$ from i to 0. More precisely, the implementation is as follows.

> **for** all $v_j \in N(v_i)$ with $j > i$ **do** used$[f(v_j)] \leftarrow i;$
> $t \leftarrow 1;$
> **while** used$[t] = i$ **do** $t \leftarrow t + 1;$
> $f(v_i) \leftarrow t;$

Theorem 4.11. *Algorithm omega-chiChordal gives a maximum clique and a minimum coloring of a chordal graph G, and also proves the strong duality equality $\omega(G) = \chi(G)$, providing a perfect elimination ordering is given. Hence chordal graphs are perfect.*

Proof. At the end of Algorithm omega-chiChordal, totally m colors are used, and the color of v_p is m. The reason for v_p to be colored by m is that for any $1 \leq i < m$ there is some vertex v_{j_i} with $j_i > p$ is colored by i. Hence, $|C| \geq m$ and so

$$|C| \leq \omega(G) \leq \chi(G) \leq m \leq |C|,$$

which leads to that all inequalities are in fact equalities. Therefore, C is a maximum clique in G, f is a minimum coloring of G and the strong duality equality $\omega(G) = \chi(G)$ holds. $\qquad\square$

Theorems 4.10 and 4.11 prove that the weak perfect graph conjecture is true for chordal graphs. This is one indication that the conjecture may possibly be true.

Sometimes it is useful not only to find a maximum clique of a chordal graph, but also to find all maximal cliques. Recall that $C_i = \{v_j \in N[v_i] : j \geq i\}$ is a clique for $1 \leq i \leq n$ and every maximal clique is some C_i. By the meaning of a maximal clique,

C_i is a maximal clique

$\Longleftrightarrow$ $C_i \nsubseteq C_k$ for every $k < i$ with v_k adjacent to v_i

$\Longleftrightarrow$ $C_i \nsubseteq C_k$ for every $k < i$ with $i = \min\{j : v_j \in N(v_k), j > k\}$

$\Longleftrightarrow$ $|C_i| > |C_k| - 1$ for every $k < i$ with $i = \min\{j : v_j \in N(v_k), j > k\}$

$\Longleftrightarrow$ $|C_i| > t_i$, where

$$t_i = \max\{|C_k| - 1 : k < i \text{ with } i = \min\{j : v_j \in N(v_k), j > k\}\}.$$

This leads to the following linear-time algorithm for finding all maximal cliques of a chordal graph.

Algorithm MaximalCliquesChordal: Find all maximal cliques of a chordal graph G, providing a perfect elimination ordering $[v_1, v_2, \ldots, v_n]$ is given.

$\mathcal{M} \leftarrow \emptyset$;
for $k = 1$ **to** n **do** **if** $|C_k| > 1$ **then** $s_k \leftarrow \min\{j : v_j \in N(v_k), j > k\}$;
for $i = 1$ **to** n **do** $t_i \leftarrow 0$;
for $i = 1$ **to** n **do**
$\{$ **if** $|C_i| > t_i$ **then** $\mathcal{M} \leftarrow \mathcal{M} \cup \{C_i\}$;
     if $|C_i| > 1$ **then** $t_{s_i} \leftarrow \max\{t_{s_i}, |C_i| - 1\}$;
$\}$

Theorem 4.12. *Algorithm MaximalCliquesChordal finds all maximal cliques of a chordal graph G in linear time.*

For an alternative method for finding the chromatic number and all maximal cliques of a chordal graph, see the book by Golumbic (1980, pp. 98–99).

At the end of this section, a linear-time algorithm is presented for testing if a graph is chordal. In case of a positive answer, a perfect elimination ordering is also given.

A well-known method is the lexicographic breadth-first search (LexBFS) method by Rose *et al.* (1976), which is a reverse BFS checking some extra conditions when visits the next vertex. During the process of the algorithm each vertex has a set as its label, which initially is empty and keeps updated. At iteration i, which is from n down to 1, a vertex with the lexicographical largest label is chosen as v_i to visit; and then i is added to the labels of its unvisited neighbors.

In order to compare two sets, the following definition is needed. A set $A = \{a_1 > a_2 > \cdots > a_r\}$ is *lexicographically smaller than* another set $B = \{b_1 > b_2 > \cdots > b_s\}$, denoted by $A \prec B$, if there is some $t \leq s$ such that $a_i = b_i$ for $1 \leq i < t$ but either $t = r + 1$ or $t \leq r$ with $a_t < a_t$. Also, $A \preceq B$ means $A \prec B$ or $A = B$. For instance, $\{8, 6\} \prec \{8, 6, 4, 2\}$ and $\{8, 6, 3, 2, 1\} \prec \{8, 6, 4, 2\}$.

Note that a naive method of comparing two sets costs time. A linear-time implementation, using doubly linked lists, of this algorithm is slightly complicated, see the book by Golumbic (1980, pp. 84–91).

Algorithm LexBFS: Find a lexicographic breadth-first search ordering $[v_1, v_2, \ldots, v_n]$ of a graph G.

assign the label $\emptyset$ to each vertex;
for $i = n$ **down to** 1 **do**
{   choose a vertex x with a largest label;
       $v_i \leftarrow x$;
       for each unnumbered neighbor y of x **do** add i to label(y);
}

In the above algorithm, we can in fact choose any vertex x to be numbered by n initially. For some purpose, we may need to do so and call the procedure as LexBFS(G, x).

Tarjan (1976) gave an alternative method using the *maximum cardinality search* (MCS). The only difference of MCS from LexBFS is that a vertex with the largest label size is chosen. As comparing the sizes of two sets is simpler than comparing the two sets, the implementation is easier. Here is the algorithm.

Algorithm MCS. Find a maximum cardinality search ordering $[v_1, v_2, \ldots, v_n]$ of a graph G.

$d(x) \leftarrow 0$ for each vertex x;   //* Number of visited neighbors of x. */
for $i = n$ **down to** 1 **do**;
{   choose a vertex x with largest $d(x)$ and set $d(x) \leftarrow -1$;
       $v_i \leftarrow x$;
       for each neighbor y of x with $d(y) \geq 0$ **do** $d(y) \leftarrow d(y) + 1$;
}

More detail implementations for finding a vertex x with a largest label $d(x)$ is as follows.

```
d(x) ← 0 for each vertex x;   //* Number of visited neighbors of x. *//
for i = 1 to Δ(G) do
      { set(i) ← ∅;   size(i) ← 0; }
set(0) ← V(G);
size(0) ← n;
ℓ ← 0;   //* The largest index for which set(ℓ) ≠ ∅; *//
for i = n down to 1 do;
{      choose a vertex x from set(ℓ);
       vᵢ ← x;
       d(x) ← −1;
       delete x from set(ℓ);
       size(ℓ) ← size(ℓ) − 1;
       for each neighbor y of x with d(y) ≥ 0 do
       {   delete y from set(d(y));
           size(d(y)) ← size(d(y)) − 1;
           d(y) ← d(y) + 1;
           add y to set(d(y));
           size(d(y)) ← size(d(y)) + 1;
       }
       ℓ ← ℓ + 1;
       while size(ℓ) = 0 do ℓ ← ℓ − 1;
}
```

In the algorithm, set(i) in fact only needs for $0 \leq i \leq \Delta(G)$.

The following shows how the MCS algorithm finds a maximum cardinality search ordering c, b, d, e, f, a of the graph in Figure 4.2.

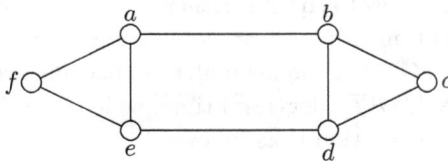

Fig. 4.2. A graph of six vertices and eight edges.

The following table lists $\text{set}(i)$, ℓ and v_i for every i, where $\text{set}(i)$ are only shown for $0 \leq i \leq 3 = \Delta(G)$ and those $\text{set}(i) = \emptyset$ are indicated by blank.

i	6	5	4	3	2	1
$\text{set}(0)$	$\{a,b,c,d,e,f\}$	$\{c,d\}$	$\{c,d\}$	$\{c\}$		
$\text{set}(1)$		$\{f,e,b\}$	$\{b\}$	$\{d,b\}$	$\{c\}$	
$\text{set}(2)$			$\{e\}$		$\{b\}$	$\{c\}$
$\text{set}(3)$						
ℓ	0	1	2	1	2	2
v_i	a	f	e	d	b	c

Theorem 4.13. *A graph G is chordal if and only if Algorithm MCS produces a perfect elimination ordering of G.*

Proof.　If Algorithm MCS produces a perfect elimination ordering, then G is chordal by Theorem 4.6.

On the other hand, suppose G is chordal. Consider the Algorithm MCS now. For every index i and every vertex x, let $d_i(x)$ denote the number of visited neighbors of x at the beginning of iteration i in Algorithm MCS. For example, $d_n(x) = 0$ for all x, and $d_{n-1}(x) = 1$ for $x \in N(v_n)$ while $d_{n-1}(x) = 0$ for $x \notin N[v_n]$ and $d_{n-1}(v_n) = -1$.

For $i < j < k$, by the choice of v_j in the algorithm, $d_j(v_j) \geq d_j(v_i)$ since v_j has most visited neighbors among all unvisited vertices. Then $v_k \in N(v_i) \backslash N(v_j)$ implies that there is some $\ell > j$ with $v_\ell \in N(v_j) \backslash N(v_i)$. In other words, (MCSO) holds and so (PEO) hold, as shown in the alternative proof for ($\Rightarrow$) of Theorem 4.6.　　□

A final problem is how to check whether a vertex ordering $[v_1, v_2, \ldots, v_n]$ is a perfect elimination ordering or not. A naive method is to check if $\{v_j : j > i, v_i v_j \in E(G)\}$ is a clique for all i by checking every pair of vertices in this set. This needs $O(\sum_i \deg(v_i)^2)$ time, which is $O(\Delta(G)|E(G)|)$ time. A linear-time implementation is as follows.

Algorithm TestChordality: Check if a vertex ordering $[v_1, v_2, \ldots, v_n]$ is a perfect elimination ordering.

for all vertices v **do** $\text{TEST}(v) \leftarrow 0$;
for all vertices v **do** $A(v) \leftarrow \emptyset$;
for $i = 1$ **to** $n - 1$ **do**
$\{$ $X \leftarrow \{v_j \in N(v_i) : j > i\}$;
     if $X \neq \emptyset$ **then**
     $\{$ let v_j be the minimum indexed vertex in X;
         concatenate $X \backslash \{v_j\}$ to $A(v_j)$;
     $\}$
     for $w \in N(v_i)$ **do** $\text{TEST}(w) \leftarrow i$;
     for $w \in A(v_i)$ **do**
         if $\text{TEST}(w) < i$
         then return the message "G is not chordal" and **stop**;
$\}$
return the message "G is chordal";

4.4 NP-completeness for domination

It was seen in Chapter 3 that there are linear-time algorithms for the domination problem and its variations in trees. People then try to generalize the results for general graphs. However, the NP-completeness theory raised by Cook suggests that this in general is quite impossible. In fact, David Johnson at a graph theory conference held in Qualicum Beach, Vancouver Island, British Columbia sometime during 1975–1976 showed that the domination problem is NP-complete via a reduction from 3-SAT.

This section first proves the NP-completeness of the domination problem via 3-SAT, which is closely related to SAT studied by Cook. In 3-SAT, each clause contains exactly 3 literals.

To see that 3-SAT is NP-complete, one can reduce SAT to 3-SAT by transforming a clause $a_{j,1} \vee a_{j,2} \vee \cdots \vee a_{j,n_j}$ with $n_j \geq 4$ to

$$(a_{j,1} \vee a_{j,2} \vee v_{j,1}) \wedge (a_{j,3} \vee \bar{v}_{j,1} \vee v_{j,2}) \wedge (a_{j,4} \vee \bar{v}_{j,2} \vee v_{j,3}) \wedge \cdots$$
$$(a_{j,n_j-2} \vee \bar{v}_{j,n_j-4} \vee v_{j,n_j-3}) \wedge (a_{j,n_j-1} \vee a_{j,n_j} \vee \bar{v}_{j,n_j-3})$$

for new logical variables $v_{j,1}, v_{j,2}, \ldots, v_{j,n_j-3}$.

The 3-Satisfiability Problem (3-SAT)
Parameters: A set $U = \{u_1, u_2, \ldots, u_n\}$ of logical variables and a family
$\{C_j = a_{j,1} \vee a_{j,2} \vee a_{j,3} : 1 \leq j \leq m\}$ of clauses using the variables in U and
their negations, each clause contains exactly three literals.
Problem: Is it possible to assign true/false values to U such that every
clause C_j is satisfiable?

Theorem 4.14. *The domination problem is NP-complete.*

Proof.   The proof of the theorem is by transforming the 3-satisfiability
problem to the domination problem.

Given a set $U = \{u_1, u_2, \ldots, u_n\}$ of logical variables and a family $\{C_j = a_{j,1} \vee a_{j,2} \vee a_{j,3} : j = 1, 2, \ldots, m\}$ of clauses using the variables in U, each
clause contains exactly three literals, construct a graph G consisting of a
triangle A_i for each variable u_i, a 4-cycle B_j for each clause C_j and an edge
between each literal $a_{j,i}$ and its corresponding u_k or $\overline{u}_k$. More precisely,
G is the graph with

$$\text{vertex set } V(G) = \{u_i, \overline{u}_i, v_i : 1 \leq i \leq n\} \cup \qquad \text{(for triangles } A_i\text{)}$$

$$\{a_{j,1}, a_{j,2}, a_{j,3}, a_{j,0} : 1 \leq j \leq m\} \quad \text{(for 4-cycles } B_j\text{)},$$

$$\text{edge set } E(G) = \{u_i\overline{u}_i, u_iv_i, \overline{u}_iv_i : 1 \leq i \leq n\} \cup$$

$$\{a_{j,1}a_{j,2}, a_{j,2}a_{j,3}, a_{j,3}a_{j,0}, a_{j,0}a_{j,1} : 1 \leq j \leq m\} \cup$$

$$\{a_{j,i}u_k : 1 \leq j \leq m, 1 \leq i \leq n, a_{j,i} = u_k\} \cup$$

$$\{a_{j,i}\overline{u}_k : 1 \leq j \leq m, 1 \leq i \leq n, a_{j,i} = \overline{u}_k\}.$$

Figure 4.3 shows an example of transformation.

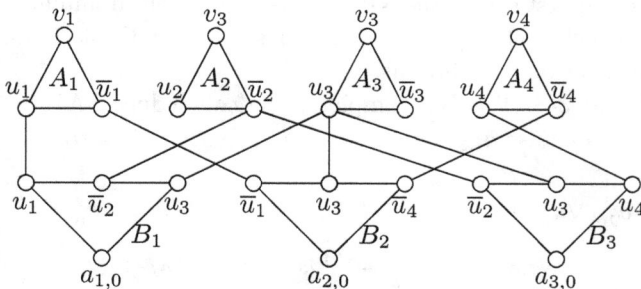

Fig. 4.3.   The graph corresponds to $(u_1 \vee \overline{u}_2 \vee u_3) \wedge (\overline{u}_1 \vee u_3 \vee \overline{u}_4) \wedge (\overline{u}_2 \vee u_3 \vee u_4)$.

Next claim that it is possible to assign true/false values to U such that every clause C_j is satisfiable if and only if the graph G has a dominating set of size $n + m$.

Suppose it is possible to assign true/false values to U such that every clause C_j is satisfiable. For $1 \leq j \leq m$, choose just one i such that $a_{j,i}$ is true. Let

$$D = \{u_i : \text{if } u_i \text{ is assigned true}\} \cup$$

$$\{\overline{u}_i : \text{if } u_i \text{ is assigned false}\} \cup$$

$$\{a_{j,i+2} : 1 \leq j \leq m, a_{j,i} \text{ is true}\},$$

where the index $i + 2$ is taken modulo 4. Then D is a dominating set of size $n + m$, since

(1) $u_i, \overline{u}_i, v_i$ are dominated by u_i or $\overline{u}_i$ depending on which one is true;

(2) for $1 \leq j \leq m$, the chosen $a_{j,i}$ is dominated by the true u_i or $\overline{u}_i$, and other $a_{j,i'}$'s are dominated by $a_{j,i+2}$.

On the other hand, suppose G has a dominating set of size $n + m$. Note that D must contain at least one vertex in A_i to dominate v_i for each i, and at least one vertex in B_j to dominate $a_{j,0}$ for each j. Hence, D contains exactly one vertex in A_i for each i, and exactly one vertex in B_j for each j. We may assume that A_i contains exactly one of u_i and $\overline{u}_i$ for each i. Assign u_i true if $u_i \in D$ and assign u_i false if $\overline{u}_i \in D$. For each j, as the vertex in $B_j \cap D$ dominates only three vertices in B_j, there is some $a_{j,i}$ that is dominated by some u_k or $\overline{u}_k$ in D, which is assigned true. Hence, the clause C_j is true.

Since the 3-satisfiability problem is NP-complete, the domination problem is also NP-complete. □

The above proof of the NP-completeness of domination is slightly complicated. This is due to the fact that the formula for 3-SAT is quite different from a graph. It takes efforts to transform a logical expression to a graph and the proof goes between them. The following uses a reduction from the vertex cover problem, which is much closer to the domination problem as they are both problems in graphs. In fact the proof gives a more general result that the domination problem is NP-complete for split graphs.

The Vertex Cover Problem (VC)
Parameter: A graph G and a positive integer k.
Problem: Is there a subset $C \subseteq V(G)$ of size at most k such that each edge xy of G has either $x \in C$ or $y \in C$?

Second proof of Theorem 4.14. The proof is given by transforming the vertex cover problem to the domination problem.

Given a graph G, construct the graph G' with

$$\text{vertex set } V(G') = V(G) \cup E(G),$$

$$\text{edge set } E(G') = \{uv : u, v \in V(G), u \neq v\} \cup$$

$$\{ve : v \in V(G), v \in e \in E(G)\}.$$

See Figure 4.4 for an example of transformation.

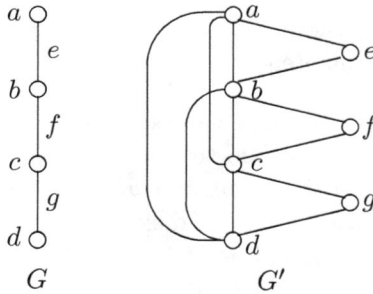

Fig. 4.4. A transformation from G to G'.

Next claim that G has a vertex cover of size at most k if and only if G' has a dominating set of size at most k.

If G has a vertex cover C of size at most k, then C is a dominating set of G' of size at most k, by the definition of G'.

On the other hand, suppose G' has a dominating set D of size at most k. If D contains any $e \in E(G)$, say $e = uv$, then replace e with u to get a new dominating set of size at most k. In this way, we may assume that D is a subset of $V(G)$. It is then clear that D is a vertex cover of G of size at most k.

Since the vertex cover problem is NP-complete, the domination problem is also NP-complete. □

Note that the dominating set D of G' in the proof above is in fact a clique, and so the total, the connected and the clique domination problems are also NP-complete. Note that for total domination, the construction of G' needs some modifications in order to avoid the case that $|D| = 1$. Now, G' needs two more vertices x^* and e^* such that x^* is adjacent to $V(G) \cup \{e^*\}$ in G'. In this case, the original D plus x^* is a total dominating set of the new G'.

Also, G' (respectively, the modified G' for total domination) is a split graph whose vertex set $V(G')$ is the disjoint union of the clique $V(G)$ (respectively, $V(G) \cup \{x^*\}$) and the stable set $E(G)$ (respectively, $E(G) \cup \{e^*\}$). Therefore, these four problems are NP-complete even when restricted to the special class of split graphs.

Corollary 4.15. *The usual, the total, the connected and the clique domination problems are NP-complete for split graphs.*

Although the graph G' in the above proof is not bipartite, a modification is possible to get the following result.

Theorem 4.16. *The domination problem is NP-complete for bipartite graphs.*

Proof. The vertex cover problem in general graphs is transformed to the domination problem in bipartite graphs as follows. Given a graph G, construct the graph G' with

$$\text{vertex set } V(G') = \{x, y\} \cup V(G) \cup E(G),$$

$$\text{edge set } E(G') = \{xy\} \cup \{yv : v \in V(G)\} \cup$$

$$\{ve : v \in V(G), v \in e \in E(G)\}.$$

Note that G' is a bipartite graph whose vertex set $V(G')$ is the disjoint union of two stable sets $\{x\} \cup V(G)$ and $\{y\} \cup E(G)$. See Figure 4.5 for an example of transformation.

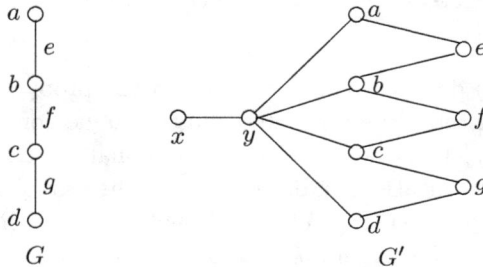

Fig. 4.5. A transformation from G to a bipartite graph G'.

Next claim that G has a vertex cover of size at most k if and only if G' has a dominating set of size at most $k + 1$.

If G has a vertex cover C of size at most k, then $\{y\} \cup C$ is a dominating set of G' of size at most $k + 1$.

On the other hand, suppose G' has a dominating set D of size at most $k + 1$. Since $N_{G'}[x] = \{x, y\}$, D must contain x or y. One may assume that D contains y but not x, as $(D \setminus \{x\}) \cup \{y\}$ is also a dominating set of size at most $k + 1$. Since $y \in D$, if D contains any $e \in E$, say $e = uv$, one can replace e with u to get a new dominating set of size at most $k + 1$. In this way, one may assume that $D \setminus \{y\}$ is a subset of $V(G)$. It is then clear that $D \setminus \{y\}$ is a vertex cover of G of size at most k.

Since the vertex cover problem is NP-complete, the domination problem is NP-complete for bipartite graphs.　　　　　　　　　　　　　□

Note that the dominating set of G' in the proof above is connected, and so the following results are also obtained.

Corollary 4.17. *The total and the connected domination problems are NP-complete for bipartite graphs.*

Booth and Johnson (1982) proved that the domination problem is NP-complete for undirected path graphs by the following NP-complete problem 3-DM.

The 3-Dimensional Matching Problem (3-DM)
Parameter: A subset P of the Cartesian product $W \times X \times Y$ of three disjoint sets W, X, Y each of size q.

Problem: Does P have a 3-dimensional matching $M \subseteq P$ of size q? The set M is a 3-dimensional matching means that for every two distinct triples $(w_1, x_1, y_1), (w_2, x_2, y_2) \in M$, we have $w_1 \neq w_2, x_1 \neq x_2, y_1 \neq y_2$.

Theorem 4.18. *The domination problem is NP-complete for undirected path graphs.*

Proof.   Consider an instance of the 3-dimensional matching problem in which there are three disjoint sets W, X, Y each of size q and a subset $P \subseteq W \times X \times Y$ of size p. The problem is to find a 3-dimensional matching $M \subseteq P$ of size q.

First construct a tree T having $6p + 3q + 1$ vertices from which an undirected path graph G is obtained. Every vertex of the tree is represented by a set as follows.

For each triple $Z_j \in P$, where $1 \leq j \leq p$, there are six vertices:

$$\{a_j, b_j, c_j, d_j\}, \quad \{a_j, b_j, d_j, f_j\}, \quad \{c_j, d_j, g_j\},$$

$$\{a_j, b_j, e_j\}, \quad \{a_j, e_j, h_j\}, \quad \{b_j, e_j, i_j\}.$$

Next, there is a vertex

$$\{w\} \cup \{a_j : w \in Z_j\} \quad \text{for each } w \in W,$$

$$\{x\} \cup \{b_j : x \in Z_j\} \quad \text{for each } x \in X,$$

$$\{y\} \cup \{c_j : y \in Z_j\} \quad \text{for each } y \in Y.$$

Finally, the vertex $r = \{a_j, b_j, c_j : 1 \leq j \leq p\}$. The tree T is shown in Figure 4.6, which is considered rooted at r. This results in an undirected path graph G with vertex set

$$V(G) = \{a_j, b_j, c_j, d_j, e_j, f_j, g_j, h_j, i_j : 1 \leq j \leq p\} \cup W \cup X \cup Y$$

of size $9p + 3q$, where the undirected path in T corresponding to a vertex $v \in V(G)$ consists of those vertices (sets) containing v in the tree T.

The theorem follows from the claim that G has a dominating set of size $2p + q$ if and only if the 3-dimensional matching problem has a solution.

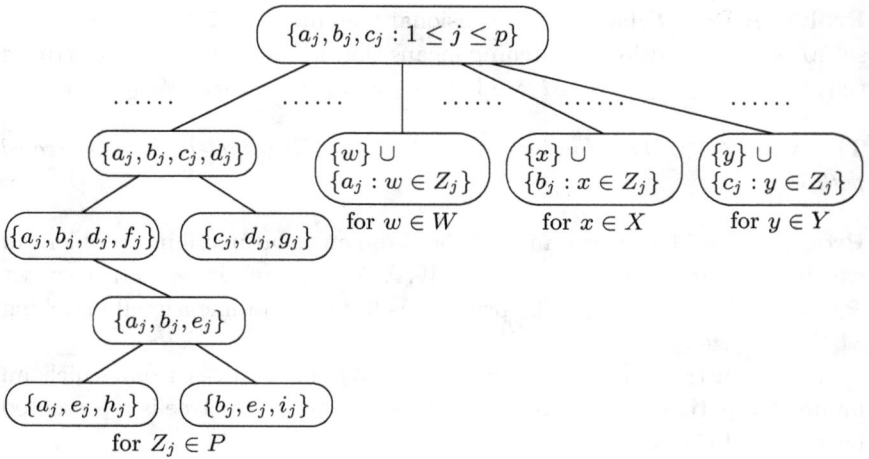

Fig. 4.6. A transformation of P to a tree T.

Suppose G has a dominating set D of size $2p + q$. First, the only way to dominate the subtree corresponding $Z_j \in P$ with two vertices is to choose d_j and e_j. And any larger dominating set might just as well consist of a_j, b_j and c_j, since none of the other possible vertices dominate any vertex outside of the subtree. Also, for $w \in W$ (similarly for $x \in X$ and $y \in Y$), a vertex a_j with $w \in Z_j$ dominates more vertices than w. Hence, we can assume without loss of generality that for each $Z_j \in P$ the dominating set D either contains all three of a_j, b_j and c_j or else both of d_j and e_j. Suppose t tuples use a_j, b_j and c_j; and $p - t$ tuples use d_j and e_j. Then $2p + q = |D| = 3t + 2(p - t)$ and so $t = q$. Picking the q triples Z_j for which a_j, b_j and c_j were chosen yields a solution to the 3-dimensional matching problem.

Conversely, suppose there exists a 3-dimensional matching $M \subseteq P$. Then it is easy to check that $D = \{a_j, b_j, c_j : Z_j \in M\} \cup \{d_j, e_j : Z_j \in P \setminus M\}$ is a dominating set of G of size $3q + 2(p - q) = 2p + q$.   □

Laskar *et al.* (1984) used a similar argument to prove that the total domination problem is NP-complete for undirected path graphs.

For more NP-complete results for variations of domination, see the references marked by (4.4).

4.5 Independent domination in chordal graphs

While most variations of the domination problem are NP-complete in chordal graphs, Farber (1982) showed the surprising result that the independent domination problem is polynomially solvable using a linear programming method.

On the other hand, Chang (2004) showed that the weighted independent domination problem is NP-complete for chordal graphs. The proof has the same spirit as the second proof for Theorem 4.14.

Theorem 4.19. *The weighted independent domination problem is NP-complete for chordal graphs.*

Proof. The vertex cover problem in general graphs is transformed to the weighted independent domination problem in chordal graphs as follows.

Given a graph G, construct the following graph G' (Figure 4.7) with

$$\text{vertex set } V(G') = \{v'', v', v : v \in V(G)\} \cup E(G),$$

$$\text{edge set } E(G') = \{v''v', v'v : v \in V(G)\} \cup$$

$$\{ve : v \in V(G), v \in e \in E(G)\} \cup$$

$$\{e_1 e_2 : e_1 \neq e_2 \text{ in } E(G)\}.$$

Note that G' is a chordal graph with a simplicial elimination ordering by listing all v'' first, then all v', then all v and finally all $e \in E(G)$. The weight of every $e \in E(G)$ is $3|V(G)|$ to discourage their selection in an optimal solution and the weight of every vertex in $V(G')\backslash E(G)$ is 1.

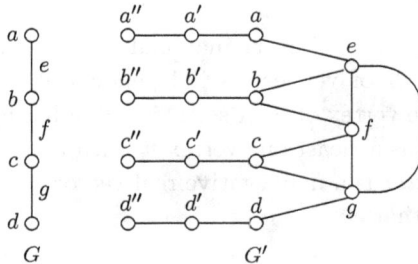

Fig. 4.7. A transformation from G to a chordal graph G'.

The theorem follows from the claim that G has a vertex cover of size at most k if and only if G' has an independent dominating set of weight at most $|V(G)| + k$.

If G has a vertex cover C of size at most k, then

$$\{v'', v : v \in C\} \cup \{v' : v \in V(G) \backslash C\}$$

is an independent dominating set of G' with weight at most $|V(G)| + k$.

On the other hand, suppose G' has an independent dominating set D of weight at most $|V(G)| + k$. As $k \le |V(G)|$, D contains no elements in $E(G)$. Let $C = D \cap V(G)$. It is clear that C is a vertex cover of G. Also, for each $v \in V(G)$ the set D contains exactly one vertex in $\{v'', v'\}$. Thus, C is of size at most k.

Since the vertex cover problem is NP-complete, the weighted independent domination problem is NP-complete for chordal graphs.   □

Farber's algorithm for the independent domination problem in chordal graphs is by means of a linear programming method. The following uses a method without linear programming.

Suppose G is a chordal graph with a perfect elimination ordering $[v_1, v_2, \ldots, v_n]$ and vertex weights $[w_1, w_2, \ldots, w_n]$ of real numbers. For convenience, the following notations are used:

$$i \sim j \quad \text{for } i = j \text{ or } v_i \text{ is adjacent to } v_j,$$

$$i \lesssim j \quad \text{for } i \le j \text{ and } i \sim j,$$

$$i \gtrsim j \quad \text{for } i \ge j \text{ and } i \sim j.$$

It follows from the definition of a perfect elimination ordering that every

$$C_j = \{v_i : i \gtrsim j\}$$

is a clique for $1 \le j \le n$; and every maximal clique is of some C_j.

Note that a set S of vertices is independent if and only if each C_j contains at most one vertex of S. Also, D is a dominating set if and only if for each j, D contains at least one vertex v_i with $i \sim j$.

A *w-2-stable vector* is a nonnegative real vector $(y, z) = (y_1, y_2, \ldots, y_n, z_1, z_2, \ldots, z_n)$ such that

$$f(i) = w_i - \sum_{j \sim i} y_j + \sum_{j \lesssim i} z_j \ge 0 \quad \text{for } 1 \le i \le n.$$

The *value* of (y, z) is $\mathrm{val}(y, z) = \sum_{j=1}^{n}(y_j - z_j)$. The *w-weight* of a set D is $w(D) = \sum_{v_i \in D} w_i$.

First, we have the following lemma for the weak duality inequality.

Lemma 4.20. *In a w-weighted chordal graph G with a perfect elimination ordering $[v_1, v_2, \ldots, v_n]$, if (y, z) is a w-2-stable vector and D is an independent dominating set, then $\mathrm{val}(y, z) \leq w(D)$. Moreover, $\mathrm{val}(y, z) = w(D)$ if and only if the following* complementary slackness conditions *hold.*

(CS1) *If $v_i \in D$, then $f(i) = 0$.*
(CS2) *For $1 \leq j \leq n$, if $y_j > 0$, then there is exactly one $v_i \in D$ with $i \sim j$.*
(CS3) *For $1 \leq j \leq n$, if $z_j > 0$, then there is exactly one $v_i \in D$ with $i \gtrsim j$.*

Proof. By the definition of independent domination, $\sum_{v_i \in D, i \sim j} 1 \geq 1$ and $\sum_{v_i \in D, i \gtrsim j} \leq 1$ for all j. Then

$$w(D) - \mathrm{val}(y, z)$$

$$= \sum_{v_i \in D} w_i - \sum_{j=1}^{n} y_j + \sum_{j=1}^{n} z_j$$

$$\geq \sum_{v_i \in D} w_i - \sum_{j=1}^{n}\left(y_j \sum_{v_i \in D, i \sim j} 1 \right) + \sum_{j=1}^{n}\left(z_j \sum_{v_i \in D, i \gtrsim j} 1 \right)$$

$$= \sum_{v_i \in D} w_i - \sum_{v_i \in D}\sum_{j \sim i} y_j + \sum_{v_i \in D}\sum_{j \lesssim i} z_j$$

$$= \sum_{v_i \in D}\left(w_i - \sum_{j \sim i} y_j + \sum_{j \lesssim i} z_j \right)$$

$$= \sum_{v_i \in D} f(i) \geq 0.$$

This gives $\mathrm{val}(y, z) \leq w(D)$. Moreover, $\mathrm{val}(y, z) = w(D)$ if and only if all inequalities in the above are equalities.

First of all, $\sum_{v_i \in D} f(i) = 0$ if and only if $v_i \in D$ implies $f(i) = 0$, since $f(i) \geq 0$ for all i. This gives (CS1).

Secondly, for all j, $y_j = y_j \sum_{v_i \in D, i \sim j} 1$ if and only if $y_j > 0$ implies $1 = \sum_{v_i \in D, i \sim j} 1$. This gives (CS2).

Finally, for all j, $z_j = z_j \sum_{v_i \in D, i \gtrsim j} 1$ if and only if $z_j > 0$ implies $1 = \sum_{v_i \in D, i \gtrsim j} 1$. This gives (CS3). □

The algorithm for the weighted independent domination on chordal graphs has two stages.

Initially, $y_j = z_j = 0$ for all j and so $f(i) = w_i$ for all i. Note that (y, z) may not be a w-2-stable vector, since some $f(i)$ may be negative. In stage one, at iteration j for j from 1 to n, assigns exactly one of y_j and z_j to a nonnegative value such that $f(i) \geq 0$ for all $i \leq j$. At the end of stage one, (y, z) is a w-2-stable vector.

The variables $x_1, x_2, \ldots, x_n$ are used to represent the dominating set D. Initially, $x_i = 2$ for all i indicating that no vertices have been scanned in stage two. In stage two, at iteration i, for i from n down to 1, if $x_i = 2$ and $f(i) = 0$ and $y_k = 0$ for all $v_k \in N(v_i)$ with $k > i$, then re-assigns $x_i = 1$ indicating $v_i \in D$ and $x_j = 0$ indicating $v_j \notin D$ for all $v_j \in N(v_i)$. At the end of stage two, let $D = \{v_i : x_i = 1\}$.

More precisely, the algorithm is as follows.

Algorithm wIndDomChordal: Find a maximum valued w-2-stable vector (y, z) and a minimum weighted independent dominating set D of a chordal graph G with a perfect elimination ordering $[v_1, v_2, \ldots, v_n]$.

each $y_j \leftarrow 0$;   each $z_j \leftarrow 0$;   each $f(i) \leftarrow w_i$;   each $x_i \leftarrow 2$;

Stage One: **for** $j = 1$ **to** n **do**
$\quad$ {   **if** $f(j) < 0$ **then** $z_j \leftarrow -f(j)$
$\quad\quad$ **else** $y_j \leftarrow \min\{f(i) : i \lesssim j\}$;
$\quad\quad$ $f(i) \leftarrow f(i) - y_j$ for $i \sim j$;
$\quad$ }

Stage Two: **for** $i = n$ **down to** 1 **do**
$\quad\quad$ **if** $x_i = 2$, $f(i) = 0$, $y_k = 0$ for all $k \sim i$ with $k > i$
$\quad\quad$ **then** { $x_i \leftarrow 1$; $x_j \leftarrow 0$ for all $v_j \in N(v_i)$; }

The algorithm does not solve the general case of weighted independent domination problem in chordal graphs, but does handle the case when all weights w_i are in $\{1, 0, -1, -2, \ldots\}$. In particular, it solves the independent domination problem in chordal graphs.

Theorem 4.21. *Suppose G is a chordal graph with a perfect elimination ordering $[v_1, v_2, \ldots, v_n]$ and vertex weights $[w_1, w_2, \ldots, w_n]$ in*

$\{1, 0, -1, -2, \ldots\}$. *Algorithm wIndDomChordal finds a maximum valued w-2-stable vector (y, z) and a minimum weighted independent dominating set D of G in linear time. Moreover, the strong duality equality val$(y, z) = w(D)$ holds.*

Proof. **(i) Feasibility of the dual solution:** Initially, all y_j and z_j are 0. For each j from 1 to n, by the instructions in stage one, z_j, y_j and $f(j)$ are nonnegative at the end of iteration j; then y_j and z_j keep the same after; and $f(j)$ keeps the same or decreases to a nonnegative value in later iterations. Hence, at the end of stage one, (y, z) is a w-2-stable vector.

(ii) Feasibility of the primal solution: For $1 \le j \le n$, by the choice of z_j and y_j, there is some minimum $i \lesssim j$ such that $f(i) = 0$ at the end of iteration j in stage one, and keeps the same after that time. Hence,

$$y_k = 0 \quad \text{for all } k \sim i \text{ with } k > j. \tag{4.1}$$

If $x_i = 2$ and $y_k = 0$ for all $k \sim i$ with $k > i$ at the beginning of iteration i in stage two, then x_i is assigned to 1 and so v_i is put into the dominating set D. In other words, v_j is dominated by $v_i \in D$. Otherwise, $x_i = 0$ or $y_{k'} > 0$ for some $k' \sim i$ with $k' > i$ at the beginning of iteration i.

For the first case when $x_i = 0$, by the last instruction in stage two, v_i is adjacent to some $v_\ell \in D$ with $\ell > i$. Then $j \gtrsim i$, $\ell > i$ and $\ell \sim i$ imply that $j \sim \ell$ as $[v_1, v_2, \ldots, v_n]$ is a perfect elimination ordering. In other words, v_j is dominated by $v_\ell \in D$.

For the second case when $y_{k'} > 0$ for some $k' \sim i$ with $k' > i$ at the beginning of iteration i, it is the case that $i < k' \le j$ by (4.1). Then $i \ne j$ and so $f(j) > 0$ at the end of iteration j in stage one, by the minimality of i, that is

$$0 < f(j) = w_j - \sum_{k \lesssim j} y_k + \sum_{k \lesssim j} z_k.$$

Since all vertex weights are integers, $y_{k'}$ is an integer and so $y_{k'} \ge 1$. This together with $w_j \le 1$ (it is important that w_j is an integer ≤ 1) imply that $z_{i'} > 0$ for some $i' \lesssim j$. Then $f(i') = 0$ and $y_k = 0$ for all $k \sim i'$ with $k > i'$. This i' plays a similar role as i that either $v_{i'} \in D$ or some $v_{\ell'} \in D$ with $\ell' > i'$ dominates v_j.

These prove that D is a dominating set of G. D is a stable set by the last instruction of the algorithm.

(iii) Complementary slackness: Finally, we need to check the three conditions (CS1), (CS2) and (CS3).

(CS1) If $v_i \in D$, then $x_i = 1$. By the instructions in stage two, $f(i) = 0$.

(CS2) Suppose $y_j > 0$ but there are two distinct $v_i, v_{i'} \in D$ with $i \sim j$ and $i' \sim j$. If $i \geq j$ and $i' \geq j$, then by (PEO) v_i is adjacent to v_i, which contradicts the stability of D. Hence, we may assume that $i < j$. Now, $x_i = 1$, by instructions in stage two $f(i) = 0$ and $y_k = 0$ for all $v_k \in N(v_i)$ with $k > i$. This contradicts to that $y_j > 0$, $v_j \in N(v_i)$ and $j > i$. Hence, $i \sim j$ for exactly one $v_i \in D$.

(CS3) If $z_j > 0$, then $i \gtrsim j$ for some $v_i \in D$ as shown in the primal feasibility argument. Such i is unique, since D is stable.   □

4.6   Perfect/efficient edge domination in chordal graphs

Recall that in a graph, an edge dominates itself and all edges adjacent to it. A *perfect* (respectively, *efficient*) *edge dominating set* of a graph G is a subset $D \subseteq E(G)$ such that every edge in $E(G) \backslash D$ (respectively, $E(G)$) is dominated by exactly one edge in D. The *perfect* (respectively, *efficient*) *edge domination number* $\gamma_{pe}(G)$ (respectively, $\gamma_{ee}(G)$) of G is the minimum size of a perfect (respectively, efficient) edge dominating set of G.

While every graph G always has a perfect edge dominating set, it may have no efficient edge dominating set, in which case denotes $\gamma_{ee}(G) = \infty$. Figure 4.8 shows a tree of six vertices with no efficient edge dominating sets. Other examples include C_n with n not a multiple of 3, K_n with $n \geq 4$ and $K_{r,s}$ with $r \geq s \geq 2$.

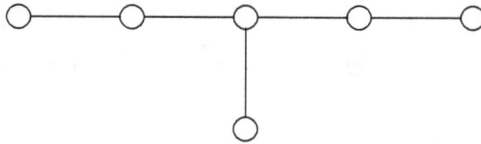

Fig. 4.8.   A tree with no efficient edge dominating sets.

Lu *et al.* (2002) gave linear-time algorithms for the weighted perfect/ efficient edge domination problems in chordal graphs. They first prove that the two problems are NP-complete for bipartite graphs and planar bipartite graphs. They then gave linear-time algorithms for the problems in generalized series-parallel graphs. Finally, linear-time algorithms for the problems in chordal graphs were obtained by reducing them to the problems in generalized series-parallel graphs.

A *generalized series-parallel* $G(s,t)$ is a graph G with a *source* s and a *sink* t; and is defined recursively as follows. Note that parallel edges are allowed in the definition.

(1) Complete graph K_2 with $V(K_2) = \{s,t\}$ is a generalized series-parallel graph $K_2(s,t)$ and is called the *basis graph* for the class of generalized series-parallel graphs.

(2) Given two generalized series-parallel graphs $G_1(s_1,t_1)$ and $G_2(s_2,t_2)$, the graph G obtained by applying one of the following three operations to G_1 and G_2 is also a generalized series-parallel graph.

   (a) *Series composition* $SC(G_1,G_2)$: Identify sink t_1 and source s_2 to obtain $G(s_1,t_2)$, see Figure 4.9(a).

   (b) *Parallel composition* $PC(G_1,G_2)$: Identify sources s_1 and s_2 and identify sinks t_1 and t_2 to obtain $G(s_1,t_1)$, see Figure 4.9(b).

   (c) *Dangling composition* $DC(G_1,G_2)$: Identify sink t_1 and source s_2 to obtain $G(s_1,t_1)$, see Figure 4.9(c).

(3) Only K_2 and the graphs constructed from generalized series-parallel graphs by a finite number of applications of series, parallel and dangling compositions are generalized series-parallel graphs.

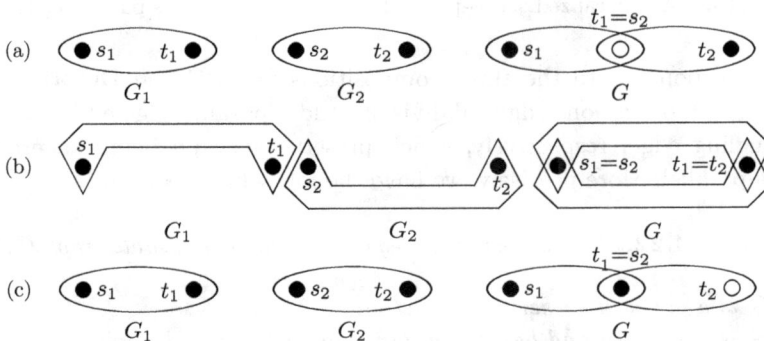

Fig. 4.9. (a) Series composition, (b) parallel composition, and (c) dangling composition.

In the definition, if the sequences of operations include only SC and PC compositions, then the obtained graphs are *series-parallel graphs*.

For every generalized series-parallel graph G, its structure can be represented by a parse tree. A *parse tree* $\mathrm{PT}(G)$ is defined as a binary tree in which each vertex of $\mathrm{PT}(G)$ represents a subgraph of G according to the following rules.

(1) Every leaf a of $\mathrm{PT}(G)$ has a label (s, t) and represents the generalized series-parallel graph $G_a(s_a, t_a) = K_2(s, t)$.
(2) Every internal vertex p of $\mathrm{PT}(G)$ has a label $C_p \in \{\mathrm{SC}, \mathrm{PC}, \mathrm{DC}\}$ and represents the generalized series-parallel graph $G_p(s_p, t_p) = C_p(G_a(s_a, t_a), G_b(s_b, t_b))$, where a is the left child and b is the right child of p. Note that $s_p = s_a$ and ($t_p = t_a$ or $t_p = t_b$).

Figure 4.10 shows a generalized series-parallel graph and its parse tree.

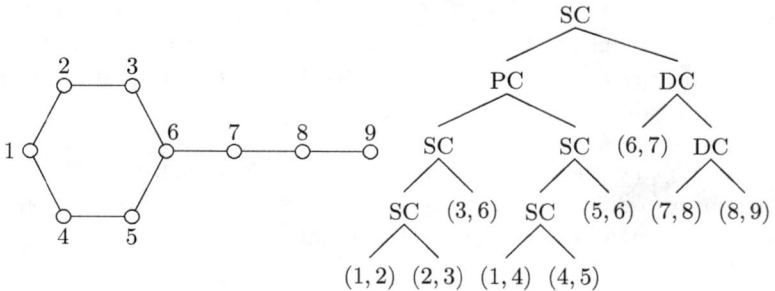

Fig. 4.10. A generalized series-parallel graph $G(1, 7)$ and its parse tree $\mathrm{PT}(G)$.

Corresponding to the three compositions SC, PC and DC, there are three graph operations edge subdivision, add a parallel edge and attaching a dangling edge, respectively, which preserve a graph being generalized series-parallel. More precisely, we have the following theorem.

Theorem 4.22. *The following are equivalent for every multigraph G.*

(1) *G is a generalized series-parallel graph.*
(2) *G is connected and has no subdivisions of K_4 as subgraphs.*
(3) *G is obtained from K_2 by a sequence of one of the following operations: subdivide an edge e, add an edge parallel to e, attach a dangling edge e.*

Proof. (1) $\Rightarrow$ (2). This will be proved by induction. Suppose $G(s,t)$ is a generalized series-parallel graph. The case of $G = K_2$ is obvious. Next consider the case of $G(s,t) = C(G_1(s_1,t_1), G_2(s_2,t_2))$, where $C \in \{SC, PC, DC\}$ and G_1 and G_2 are generalized series-parallel graphs. By the induction hypothesis, G_1 and G_2 are connected and have no subdivisions of K_4 as subgraphs. It is clear that G is connected. Suppose that G has a subgraph H which is a subdivision of K_4. Then H is not entirely in G_1 or G_2. And so the K_4 has a vertex x in G_1 but not in G_2, and a vertex y in G_2 but not in G_1. The existence of three internally vertex disjoint x–y paths in H gives a contradiction. This proves that G is generalized series-parallel.

(2) $\Rightarrow$ (3). This will be proved by induction. Suppose G is connected and has no subdivisions of K_4 as subgraphs. The case of $G = K_2$ is obvious. Now assume that G has at least 3 vertices. It is only necessary to prove that

$$G \text{ has a leaf, a vertex of degree 2 or two parallel edges.} \quad (4.2)$$

If G has a leaf, then (4.2) holds. Now assume that all vertices are of degree at least 2. Choose a 2-connected component H of G having at most one cut-vertex c. Choose a cycle C of largest length in H.

Claim 1. In the graph H, if $x \in V(C)$ is of degree at least 3, then there is an x–x' path P_x such that $V(P_x) \cap V(C) = \{x, x'\}$ and $E(P_x) \cap E(C) = \emptyset$, where x and x' are not consecutive in C.

Proof of Claim 1. First, H has an edge xz not in C. Choose an edge xy in C. Since G is 2-connected, there is a cycle C' containing xy and xz. Part of C' gives an x–x' path P_x with $V(P_x) \cap V(C) = \{x, x'\}$ and $E(P_x) \cap E(C) = \emptyset$. Note that x and x' are not consecutive in C, for otherwise either G has two parallel edges or P_x has at least two edges. For the former case, (4.2) holds. For the latter case, $(C \cup P_x) - xx'$ gives a cycle of length larger than the length of C, a contradiction to the choice of C. $\square$

We shall prove that C has a vertex of degree 2 other than c. Suppose to the contrary that all vertices of $V(C) \backslash \{c\}$ are of degree at least 3. Choose a vertex a of such one. By Claim 1, there exists an a–a' path P_a such that $V(P_a) \cap V(C) = \{a, a'\}$ and $E(P_a) \cap E(C) = \emptyset$, where a and a' are not consecutive in C. We may assume that one a–a' path P of C has length as small as possible and c is not an internal vertex of P. Since a and a' are not consecutive in C, P has a vertex b other than a and a'. Since b is of degree at least 3, then by Claim 1, there exists an b–b' path P_b such that $V(P_b) \cap V(C) = \{b, b'\}$ and $E(P_b) \cap V(C) = \emptyset$. By the minimality of P, vertex b' is not in $V(P)$.

For the case of $V(P_a) \cap V(P_b) = \emptyset$, the vertices a, a', b, b' have six paths connecting them to form a subdivision of K_4, a contradiction.

For the case of $V(P_a) \cap V(P_b) \neq \emptyset$, find subpath P' of P_b from b to b'' such that $V(P_a) \cap V(P') = \{b''\}$. Then a, a', b, b'' have six paths connecting them to form a subdivision of K_4, a contradiction.

Hence, b is of degree 2 other than c in H, and so is of degree 2 in G. Hence (4.2) holds as desired.

(3) $\Rightarrow$ (1). This will be proved by induction. Suppose G is a graph satisfying the condition in (3). We shall prove that G has a parse tree $\mathrm{PT}(G)$. If $G = K_2$, then the claim is obvious. Suppose G is obtained from G', which satisfies the conditions in (3), by subdividing an edge e, or adding an edge parallel to e, or adding a new vertex x adjacent to a vertex $y \in V(G')$. By the induction hypothesis, G' has a parse tree $\mathrm{PT}(G')$. Consider three cases.

(i) G is obtained from G' by subdividing $e = xz$ into xy and yz . Let e correspond to a node a of $\mathrm{PT}(G')$ labeled by (x, z). Relabel a by SC and add a left child of a labeled by (x, y) and a right child labeled by (y, z). This results in a parse tree $\mathrm{PT}(G)$ for G.

(ii) G is obtained from G' by adding an edge parallel to $e = xy$. Let e correspond to a node a of $\mathrm{PT}(G')$ labeled by (x, y). Relabel a by PC and add a left child of a labeled by (x, y) and a right child labeled by (x, y). This results a parse tree $\mathrm{PT}(G)$ for G.

(iii) G is obtained from G' by adding a new vertex x adjacent to a vertex $y \in V(G')$. For the case when $\mathrm{PT}(G')$ rooted at r' has a node a such that $G_a(s_a, t_a)$ has its sink $t_a = y$, add a new node a' labeled by DC with parent as the original parent of a, change a to be the left child of a', and add a new right child of a' labeled by (y, x). This results a parse tree $\mathrm{PT}(G)$ of G. Otherwise, $G'(s', t')$ has its source $s' = y$. In this case, add a new root r labeled by PC, add a new left child of r labeled by (x, y), and use r' as the right child of r. This results a parse tree $\mathrm{PT}(G)$ of G. $\qquad \square$

Condition (3) in the above theorem gives an efficient algorithm for recognizing generalized series-parallel graphs. It only needs to check if the connected graph has a vertex of degree at most 2. If a vertex of degree 1 is found, then delete it from the graph. If a vertex of degree 2 is found, then merge its two incident edges into one. If this creates a parallel edge, then delete it. Continue this process until two vertices left.

The same arguments without considering DC give the following result.

Theorem 4.23 (Duffin, 1965). *The following are equivalent for every 2-connected multigraph G.*

(1) G *is a series-parallel graph.*
(2) G *has no subdivisions of K_4 as subgraphs.*
(3) G *is obtained from K_2 by a sequence of one of the following operations: subdivide an edge e, add an edge parallel to e.*

The algorithm for the weighted perfect edge domination problem in generalized series-parallel graphs is by means of a dynamic programming method. Suppose $G(s, t)$ is a generalized series-parallel graph in which every edge e has a weight $w(e)$. To solve the problem some more generalizations of the problem are introduced.

For $a, b \in \{0, 1, 2, 3\}$, let $\gamma_{\mathrm{pe}}^{ab}(G, s, t; w)$ be the minimum weight $w(D)$ of a perfect edge dominating set D of the graph

$$
\begin{cases}
G, & \text{when } a \neq 3 \text{ and } b \neq 3; \\
G\backslash\{s\}, & \text{when } a = 3 \text{ and } b \neq 3; \\
G\backslash\{t\}, & \text{when } a \neq 3 \text{ and } b = 3; \\
G\backslash\{s, t\}, & \text{when } a = 3 \text{ and } b = 3
\end{cases}
$$

with the boundary conditions that

$$
\begin{cases}
D \text{ contains no edge incident to } s, & \text{when } a = 0; \\
D \text{ contains exactly one edge incident to } s, & \text{when } a = 1; \\
D \text{ contains all edges incident to } s, & \text{when } a = 2; \\
D \text{ contains no edge incident to } s \text{ or its neighbors}, & \text{when } a = 3
\end{cases}
$$

and

$$
\begin{cases}
D \text{ contains no edge incident to } t, & \text{when } b = 0; \\
D \text{ contains exactly one edge incident to } t, & \text{when } b = 1; \\
D \text{ contains all edges incident to } t, & \text{when } b = 2; \\
D \text{ contains no edge incident to } t \text{ or its neighbors}, & \text{when } b = 3.
\end{cases}
$$

Such a perfect edge dominating set is called an *ab-perfect edge dominating set* of $G(s,t)$.

It is clear that for any generalized series-parallel graph $G(s,t)$,

$$\gamma_{\mathrm{pe}}(G;w) = \min\{\gamma_{\mathrm{pe}}^{ab}(G,s,t;w) : a,b \in \{0,1,2\}\}.$$

For the basis graph $K_2(s,t)$,

$$\gamma_{\mathrm{pe}}^{ab}(K_2,s,t;w) = \begin{cases} w(s,t), & \text{if } ab \in \{11,12,21,22\}; \\ 0, & \text{if } ab \in \{03,30,33\}; \\ \infty, & \text{otherwise.} \end{cases} \qquad (4.3)$$

The algorithm then follows from the following three lemmas.

Lemma 4.24. *If $G(s_1,t_2)$ is the generalized series-parallel graph obtained from generalized series-parallel graphs $G_1(s_1,t_1)$ and $G_2(s_2,t_2)$ by applying a series composition, then the following holds for $a,b \in \{0,1,2,3\}$.*

$$\gamma_{\mathrm{pe}}^{ab}(G,s_1,t_2;w) = \min \left\{ \begin{array}{l} \gamma_{\mathrm{pe}}^{a0}(G_1,s_1,t_1;w) + \gamma_{\mathrm{pe}}^{0b}(G_2,s_2,t_2;w), \\ \gamma_{\mathrm{pe}}^{a1}(G_1,s_1,t_1;w) + \gamma_{\mathrm{pe}}^{3b}(G_2,s_2,t_2;w), \\ \gamma_{\mathrm{pe}}^{a3}(G_1,s_1,t_1;w) + \gamma_{\mathrm{pe}}^{1b}(G_2,s_2,t_2;w), \\ \gamma_{\mathrm{pe}}^{a2}(G_1,s_1,t_1;w) + \gamma_{\mathrm{pe}}^{2b}(G_2,s_2,t_2;w) \end{array} \right\}.$$

Let $\bar{0} = 0$, $\bar{1} = 3$, $\bar{3} = 1$ and $\bar{2} = 2$. Note that in the additive group $\mathbb{Z}_4$, the inverse of every i is $\bar{i}$. Then the above formula can be written as:

$$\gamma_{\mathrm{pe}}^{ab}(G,s_1,t_2;w) = \min_{0 \le i \le 3} \{\gamma_{\mathrm{pe}}^{ai}(G_1,s_1,t_1;w) + \gamma_{\mathrm{pe}}^{\bar{i}b}(G_2,s_2,t_2;w)\}.$$

This equivalent formula is convenient for proving the correctness of the lemma. The lemma follows from the fact that D is an *ab*-perfect edge dominating set of $G(s_1,t_2)$ if and only if it is the disjoint union of an *ai*-perfect edge dominating set D_1 of $G_1(s_1,t_1)$ and an $\bar{i}b$-perfect edge dominating set D_2 of $G_2(s_2,t_2)$ for some $i \in \{0,1,2,3\}$.

Lemma 4.25. *If $G(s_1,t_1)$ is the generalized series-parallel graph obtained from generalized series-parallel graphs $G_1(s_1,t_1)$ and $G_2(s_2,t_2)$ by applying a parallel composition, then the following hold for $a,b \in \{0,2,3\}$.*

(1) $\gamma_{\text{pe}}^{ab}(G, s_1, t_1; w) = \gamma_{\text{pe}}^{ab}(G_1, s_1, t_1; w) + \gamma_{\text{pe}}^{ab}(G_2, s_2, t_2; w)$.

(2) $\gamma_{\text{pe}}^{a1}(G, s_1, t_1; w) = \min \left\{ \begin{array}{l} \gamma_{\text{pe}}^{a1}(G_1, s_1, t_1; w) + \gamma_{\text{pe}}^{a3}(G_2, s_2, t_2; w), \\ \gamma_{\text{pe}}^{a3}(G_1, s_1, t_1; w) + \gamma_{\text{pe}}^{a1}(G_2, s_2, t_2; w) \end{array} \right\}$.

(3) $\gamma_{\text{pe}}^{1b}(G, s_1, t_1; w) = \min \left\{ \begin{array}{l} \gamma_{\text{pe}}^{1b}(G_1, s_1, t_1; w) + \gamma_{\text{pe}}^{3b}(G_2, s_2, t_2; w), \\ \gamma_{\text{pe}}^{3b}(G_1, s_1, t_1; w) + \gamma_{\text{pe}}^{1b}(G_2, s_2, t_2; w) \end{array} \right\}$.

(4) $\gamma_{\text{pe}}^{11}(G, s_1, t_2; w) = \min \left\{ \begin{array}{l} \gamma_{\text{pe}}^{11}(G_1, s_1, t_1; w) + \gamma_{\text{pe}}^{33}(G_2, s_2, t_2; w), \\ \gamma_{\text{pe}}^{13}(G_1, s_1, t_1; w) + \gamma_{\text{pe}}^{31}(G_2, s_2, t_2; w), \\ \gamma_{\text{pe}}^{31}(G_1, s_1, t_1; w) + \gamma_{\text{pe}}^{13}(G_2, s_2, t_2; w), \\ \gamma_{\text{pe}}^{33}(G_1, s_1, t_1; w) + \gamma_{\text{pe}}^{11}(G_2, s_2, t_2; w) \end{array} \right\}$.

Let $I_1 = \{1, 3\}$ and $I_a = \{a\}$ for $a = 0, 2, 3$. Then these four formulas can be combined into one:

$$\gamma_{\text{pe}}^{ab}(G, s_1, t_1; w) = \min_{\substack{\{a', a''\}=I_a \\ \{b', b''\}=I_b}} \{\gamma_{\text{pe}}^{a'b'}(G_1, s_1, t_1; w) + \gamma_{\text{pe}}^{a''b''}(G_2, s_2, t_2; w)\}.$$

Lemma 4.26. *If $G(s_1, t_1)$ is the generalized series-parallel graph obtained from generalized series-parallel graphs $G_1(s_1, t_1)$ and $G_2(s_2, t_2)$ by applying a dangling composition, then the following hold for $a \in \{0, 1, 2, 3\}$ and $b \in \{0, 2, 3\}$.*

(1) $\gamma_{\text{pe}}^{ab}(G, s_1, t_1; w) = \gamma_{\text{pe}}^{ab}(G_1, s_1, t_1; w) + \min_{0 \le j \le 2} \gamma_{\text{pe}}^{bj}(G_2, s_2, t_2; w)$.

(2) $\gamma_{\text{pe}}^{a1}(G, s_1, t_1; w) = \min \left\{ \begin{array}{l} \gamma_{\text{pe}}^{a1}(G_1, s_1, t_1; w) + \min_{0 \le j \le 2} \gamma_{\text{pe}}^{3j}(G_2, s_2, t_2; w), \\ \gamma_{\text{pe}}^{a3}(G_1, s_1, t_1; w) + \min_{0 \le j \le 2} \gamma_{\text{pe}}^{1j}(G_2, s_2, t_2; w) \end{array} \right\}$.

Similarly, these two formulas can be combined into one:

$$\gamma_{\text{pe}}^{ab}(G, s_1, t_1; w) = \min_{\substack{\{b', b''\}=I_b \\ 0 \le j \le 2}} \{\gamma_{\text{pe}}^{ab'}(G_1, s_1, t_1; w) + \gamma_{\text{pe}}^{b''j}(G_2, s_2, t_2; w)\}.$$

In conclusion, we have the following result.

Theorem 4.27. *There is a linear-time algorithm for the weighted perfect edge domination problem in generalized series-parallel graphs.*

Finally, we consider the weighted perfect edge domination problem in chordal graphs. The following lemma is useful for general graphs.

Lemma 4.28. *If D is a perfect edge dominating set of a graph G and C is a clique of G with $|C| > 3$, then all edges of $G[C]$ are in D.*

Proof.   Suppose there is an edge uv in $G[C]$ but not in D. By symmetry, uv is dominated by some ux in D. Since $|C| > 3$, the clique C contains some vertex y other than u, v, x. Since $uv \notin D$ is dominated by $ux \in D$, it is the case that $uy, vy \notin D$. Also, vy is dominated by some $e \in D$. Then either uv or uy, which is not in D, is dominated by two edges ux and e in D, a contradiction.   □

Corollary 4.29. *If M is the union of all cliques of size greater than 3 in G, then every perfect edge dominating set D of G contains every edge incident to some vertex in M.*

Now suppose G is a chordal graph with a perfect elimination ordering $[v_1, v_2, \ldots, v_n]$ and every edge e has a weight $w(e)$. Recall that

$$C_i = \{v_j \in N[v_i] : j \geq i\}$$

is a clique for $1 \leq i \leq n$. Every clique C is a subset of C_i, where i is the minimum index for which $v_i \in C$.

By Corollary 4.29, for the weighted perfect edge domination problem in chordal graphs it only needs to treat those whose cliques are of size at most three. Such a graph is the union of generalized series-parallel graphs.

Lemma 4.30. *A connected chordal graph G without cliques of size greater than three is generalized series-parallel.*

Proof.   Choose a perfect elimination ordering $[v_1, v_2, \ldots, v_n]$ of G. Suppose that G is not generalized series-parallel. By Theorem 4.22, there is a subdivision $S(K_4)$ of K_4, which has vertex set $V(K_4) = \{v_{i_1}, v_{i_2}, v_{i_3}, v_{i_4}\}$ with $i_1 < i_2 < i_3 < i_4$. For $j = 2, 3, 4$, let $P_j = (v_{i_1} = v_{j_0}, v_{j_1}, \ldots, v_{j_{j_r}} = v_{i_j})$ be the v_{i_1}-v_{i_j} path in $S(K_4)$. By shortening the paths using chords, we may assume that every P_j is chordless. Then for $j = 2, 3, 4$, $i_1 = j_0 < j_1$ for otherwise $j_1 < j_0$ implies that $j_{r+1} < j_r$ and $j_{r+1} < j_{r+2}$ for some r, leading to a chord $v_{j_r} v_{j_{r+1}}$, a contradiction. These give a 4-clique $\{v_{i_1}, v_{2_1}, v_{3_1}, v_{4_1}\}$, which is impossible. Hence, G is generalized series-parallel.   □

An alternative proof. The lemma is clear for $n = 2$. Suppose $n \geq 3$. Note that $G - v_1$ is also a connected chordal graph without cliques of size greater than three. By the induction hypothesis, $G - v_1$ is generalized series-parallel. Since $|C_1| \leq 3$, vertex v_1 is of degree 1 or 2. If v_1 is of degree 1, then G is obtained from $G - v_1$ by attaching a dangling edge. If v_1 is of degree 2, say adjacent to x and y, then G is obtained from $G - v_1$ by first adding an edge parallel to xy and secondly dividing this new edge at vertex v_1. Hence, by Theorems 4.22, G is generalized series-parallel.

Note that this proof also gives a parse tree of this graph. □

Let M be the union of cliques of size greater than 3. Then a perfect edge dominating set includes the set E_M of all edges incident to vertices in M. Also every clique of $G' := G - M$ has size at most 3, whose components $G_1, G_2, \ldots, G_m$ are generalized series-parallel. Then

$$\gamma_{\text{pe}}(G; w) = w(E_M) + \sum_{i=1}^{m} \gamma_{\text{pe}}^{M}(G_i; w),$$

where $\gamma_{\text{pe}}^{M}(G_i; w)$ is the minimum weight of a perfect edge dominating set of G_i subject to the condition that for every vertex v in $V(G_i)$ there are already n_v edges in E_M incident to v.

The calculation of $\gamma_{\text{pe}}^{M}(G_i; w)$ is the same as that for $\gamma_{\text{pe}}(G_i; w)$ by calculating $\gamma_{\text{pe}}^{abM}(G_i, s, t; w)$ for $a, b \in \{0, 1, 2, 3\}$. The only difference is to use revised values for the basis graph $K_2(s, t)$:

$$\gamma_{\text{pe}}^{abM}(K_2, s, t; w) = \begin{cases} w(s,t), & \text{if } n_s + n_t \geq 2 \text{ or } ab \in \{11, 12, 21, 22\}; \\ 0, & \text{if } ab = 00 \text{ with } n_s + n_t = 1 \\ & \text{or } ab \in \{03, 30, 33\} \text{ with } n_s = n_t = 0; \\ \infty, & \text{otherwise.} \end{cases}$$

(4.4)

For the weighted efficient edge domination problem in generalized series-parallel graphs, the supporting lemmas are nearly the same except that it is only necessary to consider $\gamma_{\text{ee}}^{rs}(G, u, v; w)$ for $r, s \in \{0, 1, 3\}$. This is due to the fact that the every vertex of the graph is incident to at most one edge in an efficient edge dominating set. The weighted efficient edge domination problem in chordal graphs is then solvable. In general, a chordal graph may have no efficient edge dominating set.

4.7   Exercises

(4.1) Determine all $\overline{C}_n$ which are chordal.

(4.2) Determine all complements of trees which are chordal.

(4.3) Show that in a chordal graph, every edge in every cycle C forms a triangle with a third vertex of C. Show the converse that if every edge in every cycle C of a graph G forms a triangle with a third vertex of C, then G is chordal.

(4.4) Show that if C is a cycle of length at least four in a chordal graph G, then G has a cycle whose vertex set consists all but one vertex of C. Show the converse that if for every cycle C of length at least four in a graph G there is a cycle whose vertex set consists of all but one vertex of C, then G is chordal (Hendry, 1990).

(Comment: Hendry asked for a chordal graph with a spanning cycle and a cycle C which is not spanning, is it possible to find a cycle whose vertex set consists of $V(G)$ plus a vertex?)

(4.5) The *Szekeres-Wilf number* of a graph G is $1 + \max_{H \subseteq G} \delta(H)$. Show that a graph G is chordal if and only if for every induced subgraph of G the Szekeres-Wilf number equals the clique number (Voloshin, 1982).

(4.6) Let $k_r(G)$ denote the number of r-cliques in a connected chordal graph G. Show that $\sum_{r \geq 1}(-1)^{r-1}k_r(G) = 1$. (Hint: By the binomial formula, $\sum_{i=0}^{m}(-1)^i\binom{m}{i} = 1$ for every positive integer m.)

(4.7) Let Q be a maximal clique in a chordal graph G. Show that if $G - Q$ is connected, then Q contains a simplicial vertex (Voloshin-Gorgos, 1982).

(4.8) Show that trees are directed path graphs but there exist some trees that are not interval graphs.

(4.9) Show that S_3 is not a directed path graph but is an undirected path graph.

(4.10) Show that S_3 adding a new vertex adjacent to all vertices is not an undirected path graph but is a chordal graph.

(4.11) Prove Proposition 4.9.

(4.12) Determine all split graphs which are also interval graphs.

(4.13) Determine all split graphs which are also directed path graphs.

(4.14) Determine all split graphs which are also undirected path graphs.

(4.15) Show that Algorithm LexBFS gives an ordering $[v_1, v_2, \ldots, v_n]$ of the vertices of a graph G satisfying then following condition (LexBFSO), which is stronger than the condition (MCSO).

$$i < j < k \text{ and } v_k \in N(v_i) \backslash N(v_j)$$
$$\text{imply } v_\ell \in N(v_j) \backslash N(v_i) \text{ for some } \ell > k. \quad \text{(LexBFSO)}$$

(4.16) Show that the independent domination problem is NP-complete for bipartite graphs.

(4.17) Show that the Roman domination problem is NP-complete for split graphs. Also show that the Roman domination problem is NP-complete for bipartite graphs.

(4.18) Show that the perfect domination problem is NP-complete for bipartite graphs. (Hint: Consider the following NP-complete problem.
The Exact Cover Problem (EC) Given a family of sets $\mathcal{F} = \{S_1, S_2, \ldots, S_n\}$ with $X = \cup_{1 \leq i \leq n} S_i$, does there exist a subfamily of pairwise disjoint sets whose union is equal to X? (Yen and Lee, 1996).

(4.19) Show that the perfect domination problem is NP-complete for chordal graphs (Yen and Lee, 1996).

(4.20) Design a linear-time algorithm for the weighted perfect domination problem in split graphs.

(4.21) Show that the edge domination problem is NP-complete for bipartite graphs.

(4.22) Design a linear-time algorithm for the weighted independent domination problem in split graphs.

(4.23) Provide generalized series-parallel graphs that are not series-parallel.

(4.24) Determine all trees that are series-parallel.

(4.25) Determine all split graphs that are series-parallel.

(4.26) Determine all split graphs that are generalized series-parallel.

(4.27) Design an efficient algorithm for recognizing generalized series-parallel graphs.

(4.28) Prove Lemma 4.25.

(4.29) Prove Lemma 4.26.

Chapter 5

Interval Graphs

5.1 Connection to chordal graphs and comparability graphs

Recall that an interval graph is the intersection graph of a family of closed intervals in the real line. More precisely, a graph G with $V(G) = \{v_1, v_2, \ldots, v_n\}$ is an interval graph if it has an interval representation, which is a family $\mathcal{F} = \{I_i : 1 \leq i \leq n\}$ of closed intervals in the real line such that $v_i v_j \in E(G)$ if and only if $i \neq j$ and $I_i \cap I_j \neq \emptyset$.

The Fourth Austrian Congress of Mathematicians was held in Vienna from September 17 to 22, 1956. According to the News of the Austrian Mathematical Society in 1957, there were a total of 207 lectures presented, each lasting 20 minutes. Among them, Hajós (1957) gave a lecture titled "About a type of graph." He asked the question of determining whether a given graph is interval or not. For the translation of the abstract of his lecture, see the book by Golumbic (1980, p. 171). This seems to be the earliest source of interval graphs.

The molecular biologist Benzer (1959) asked a related question independently during his investigations of the fine structure of the gene. The problem is whether the sub-elements of genes are linked together in a linear order. He could deal with the problem successfully for a certain microorganism. Of these microorganisms, there are a standard form and mutants, the latter arises if a certain connected portion of the genetic structure is blemished. By recombination tests, it is possible to decide whether the blemished parts of two given mutants overlap or not. Thus, for a large number of portions of the genetic structure true, the experiments lead to data as to whether any two of these portions overlap or not. The problem is to decide whether the data are compatible with a linear structure of the gene. Consider the graph with mutants as vertices and two vertices are adjacent if their corresponding mutants overlap. The problem is then to determine whether the graph is an interval graph.

It is surprising that interval graphs connect the two classical classes of perfect graphs: chordal graphs and comparability graphs. First, interval graphs are chordal.

Proposition 5.1 (Lekkerkerker and Boland, 1962). *Every interval graph G is chordal.*

Proof.   Choose an interval representation $\mathcal{F} = \{I_i = [a_i, b_i] : 1 \leq i \leq n\}$ of G. Suppose $C = (v_{i_0}, v_{i_1}, \ldots, v_{i_r})$ is a cycle in G with $r \geq 4$. By symmetry, assume $b_{i_1} = \min_{0 \leq j \leq r} b_{i_j}$. Since $[a_{i_0}, b_{i_0}]$ intersects $[a_{i_1}, b_{i_1}]$ and $b_{i_0} \geq b_{i_1}$, we have $b_{i_1} \in [a_{i_0}, b_{i_0}]$. Similarly, $b_{i_1} \in [a_{i_2}, b_{i_2}]$. Hence, $I_0 \cap I_2 \neq \emptyset$. This gives that $v_0 v_2$ is a chord of C and so G is chordal. $\square$

The insight of the above proof is the ordering of the right endpoints b_i of the intervals $I_i = [a_i, b_i]$. This in fact gives a vertex ordering which satisfies a condition stronger than a perfect elimination ordering. When designing algorithms for interval graphs, using this ordering is simpler than using the intervals for the vertices. Examples will be seen in the next section.

Theorem 5.2 (Ramalingam and Pandu Rangan, 1988b). *A graph G is an interval graph if and only if its vertices can be ordered into $[v_1, v_2, \ldots, v_n]$, called an* interval ordering, *satisfying*

$$i < j < k \text{ and } v_i v_k \in E(G) \text{ imply } v_j v_k \in E(G). \tag{IO}$$

Proof.   ($\Rightarrow$) Suppose G is the intersection graph of

$$\mathcal{F} = \{I_i = [a_i, b_i] : 1 \leq i \leq n\},$$

where interval I_i corresponds to vertex v_i for $1 \leq i \leq n$. Without loss of generality, assume that $b_1 \leq b_2 \leq \ldots \leq b_n$. Suppose $i < j < k$ and $v_i v_k \in E(G)$. First $b_i \leq b_j \leq b_k$, since $i < j < k$. Next, $v_i v_k \in E(G)$ implies $I_i \cap I_k \neq \emptyset$ and then $a_k \leq b_i$. Therefore, $a_k \leq b_j \leq b_k$ and so $I_j \cap I_k \neq \emptyset$, i.e., $v_j v_k \in E(G)$.

($\Leftarrow$) On the other hand, suppose (IO) holds. For every $v_i \in V(G)$, let i' be the minimum index such that $v_{i'} \in N[v_i]$ and let interval $I_i = [i', i]$ correspond to vertex v_i. If $v_i v_k \in E(G)$ with $i < k$, then $k' \leq i < k$ and so $i \in I_i \cap I_k \neq \emptyset$. On the other hand, if $I_i \cap I_k \neq \emptyset$ with $i < k$, say $j \in I_i \cap I_k$, then $i' \leq j \leq i$ and $k' \leq j \leq k$ imply $k' \leq i < k$. Since $v_{k'} v_k \in E(G)$, by (IO), $v_i v_k \in E(G)$. Therefore, G is an interval graph with

$$\mathcal{F}' = \{I_i = [i', i] : 1 \leq i \leq n\}$$

as an interval representation. $\qquad \square$

Note that (IO) implies, and in fact is stronger than, (PEO). This also shows that interval graphs are chordal.

On the way to characterize interval graphs in terms of chordless 4-cycles and comparability graphs, Gilmore and Hoffman (1964) implicitly gave the following theorem, also see Golumbic (1980, pp. 172–174) and Section 7.5. Here is a proof using interval orderings rather than going through comparability graphs.

Theorem 5.3 (Gilmore and Hoffman, 1964). *A graph G is an interval graph if and only if it has a clique tree that is a path, in other words, the maximal cliques of G can be linearly ordered such that, for every vertex v of G, the maximal cliques containing v occur consecutively.*

Proof. ($\Rightarrow$) Suppose G is an interval graph. Choose an interval ordering $[v_1, v_2, \ldots, v_n]$ of G, which is also a perfect elimination ordering. Note that a maximal clique of G is of the form

$$C_i := \{v_j \in N[v_i] : j \geq i\}.$$

Consider all maximal cliques $C_{i_1}, C_{i_2}, \ldots, C_{i_m}$ of G with $i_1 < i_2 < \cdots < i_m$.

For every vertex v_j of G, suppose $v_j \in C_{i_p} \cap C_{i_r}$ and $p < q < r$. Then $i_p < i_q < i_r \leq j$ and $v_{i_p} v_j \in E(G)$. By (IO), $v_{i_q} v_j \in E(G)$ and so $v_j \in C_{i_q}$, as desired.

($\Leftarrow$) Suppose all maximal cliques of G can be ordered into $C_{i_1}, C_{i_2}, \ldots, C_{i_m}$ such that, for every vertex v_j of G, the maximal cliques containing v_j occur consecutively. For every vertex v_j of G, let interval $I_j = [p_j, q_j]$, where p_j is the minimum index and q_j the maximum index for which $v_j \in C_{i_{p_j}} \cap C_{i_{q_j}}$.

If $v_j v_k \in E(G)$ with $j < k$, then $v_j, v_k \in C_{i_{p_j}} \cap C_{i_{q_j}}$. By the definition, $p_k \leq p_j \leq q_j \leq q_k$. Hence, $I_j \cap I_k \neq \emptyset$.

Conversely, suppose $I_j \cap I_k \neq \emptyset$, where $j < k$. Choose $r \in I_j \cap I_k$. By the assumptions, $v_j, v_k \in C_{i_r}$ and so $v_j v_k \in E(G)$.

Therefore, G is an interval graph.   □

This theorem has an interesting matrix formulation. A zero-one matrix has the *consecutive 1's property for columns* if its rows can be permuted into such a way that the 1's in each column occur consecutively. The matrix M_1 below has the consecutive 1's property since its rows can be permuted to obtain M_2. But matrix M_3 does not possess the property.

$$
\begin{array}{c}
①\\②\\③\\④\\⑤\\⑥
\end{array}
\begin{pmatrix}
1 & 0 & 0 & 1 & 0\\
1 & 1 & 1 & 0 & 0\\
0 & 1 & 0 & 0 & 1\\
1 & 0 & 1 & 1 & 0\\
1 & 1 & 0 & 0 & 1\\
0 & 1 & 0 & 0 & 1
\end{pmatrix}
\quad \rightarrow \quad
\begin{array}{c}
①\\④\\②\\⑤\\③\\⑥
\end{array}
\begin{pmatrix}
1 & 0 & 0 & 1 & 0\\
1 & 0 & 1 & 1 & 0\\
1 & 1 & 1 & 0 & 0\\
1 & 1 & 0 & 0 & 1\\
0 & 1 & 0 & 0 & 1\\
0 & 1 & 0 & 0 & 1
\end{pmatrix}
\quad
\begin{pmatrix}
1 & 1 & 1 & 1 & 1\\
1 & 0 & 0 & 0 & 0\\
0 & 1 & 0 & 0 & 0\\
0 & 0 & 1 & 0 & 0\\
0 & 0 & 0 & 1 & 0\\
0 & 0 & 0 & 0 & 1
\end{pmatrix}
$$
$$
\qquad M_1 \qquad\qquad\qquad\qquad M_2 \qquad\qquad\qquad\qquad M_3
$$

The *clique matrix* of a graph G is the zero-one matrix M, whose rows correspond to maximal cliques and columns correspond to vertices of G such that an entry $m_{ij} = 1$ if and only if the ith clique contains the jth vertex. Theorem 5.3 then can be re-stated in term of the clique matrix.

Theorem 5.4 (Fulkerson and Gross, 1965). *A graph is an interval graph if and only if its clique matrix M has the consecutive 1's property for columns.*

The consecutive 1's property for columns of the maximal cliques gives a base for recognizing interval graphs, see the papers by Booth and Lueker (1976), Korte and Mohring (1989), Hsu (1993, 2002), Hsu and Ma (1999) and Hsu and McConnell (2003) among many others.

There are also some interval graph recognition algorithms without using maximal cliques, such as the 6-sweep LBFS algorithm by Corneil, Olariu and Stewart (1998, 2009) and the 4-sweep algorithm by Li and Wu (2014), who believed that a 3-sweep LBFS algorithm is possible.

There are some special interval graphs that are interested by people. A *proper interval graph* is an interval graph that has an interval representation in which no interval properly contains any other interval, call such a representation a *proper-interval representation*. A *unit interval graph* is an interval graph that has an interval representation in which each interval has unit length, call such a representation a *unit-interval representation*. These two classes of graphs look different from the definitions. However, Robert (1969) proved that they are precisely the same by means of semi-orders.

Roberts in fact used the term of indifference graphs. A graph G is an *indifference graph* if it has an *indifference function*, which is a real-valued function $f : V(G) \to \mathbb{R}$ such that $uv \in E(G)$ if and only if $|f(u) - f(v)| < 1$. This is an equivalent definition of unit interval graphs in a slightly different formulation.

The following uses a different approach to prove Roberts' results avoiding the arguments through semi-orders. The fourth statement is added for the similarity to (IO).

Theorem 5.5 (Roberts, 1969). *The following statements are equivalent for a graph G.*

(1) *The graph G is a proper interval graph.*
(2) *The graph G is a unit interval graph.*
(3) *The graph G is an indifferent graph.*
(4) *The vertex set $V(G)$ has an ordering $[v_1, v_2, \ldots, v_n]$ satisfies*

$$i < j < k \text{ and } v_i v_k \in E(G) \text{ imply } v_i v_j, v_j v_k \in E(G). \qquad \text{(TIO)}$$

Proof. (1) $\Rightarrow$ (2). Suppose G is a proper interval graph with $V(G) = \{v_1, v_2, \ldots, v_n\}$. Choose a proper-interval representation $\{I_i = [a_i, b_i] : 1 \le i \le n\}$ of G, where interval I_i represents vertex v_i. Without loss of generality, assume that $a_1 \le a_2 \le \cdots \le a_n$ and $b_1 \le b_2 \le \cdots \le b_n$.

Add intervals $[a_0, b_0] = [a_1 - 2, a_1 - 1]$ and $[a_{n+1}, b_{n+1}] = [b_n + 1, b_n + 2]$. If there is some $a_i = b_i$, say $b_{r-1} < a_r = b_r = a_{r+1} = b_{r+1} = \cdots = a_s = b_s < a_{s+1}$. Replacing each $[a_i, b_i]$ by $[a_s, (b_s + a_{s+1})/2]$ for $r \le i \le s$ results a new proper-interval representation of G. Hence, we may assume that $a_i < b_i$ for all i. Do the following revisions on $[a_i, b_i]$ for i from 1 to n to obtain a unit-interval representation for G.

First, $[a_0, b_0]$ is of unit length. Suppose $i \ge 1$ and $[a_j, b_j]$ is of unit length for $0 \le j < i$.

If $a_i = a_{i-1}$, then $b_i = b_{i-1}$ by the proper-interval condition. Hence, $[a_i, b_i]$ is of unit length. Advance i by one in this case. Now suppose $a_{i-1} < a_i$. Then $b_{i-1} < b_i$ by the proper-interval condition.

For the case of $a_{i-1} < b_{i-1} \leq a_i < b_i$, scale the part $[b_{i-1}, \infty)$ of the real line by a factor $1/(b_i - a_i)$. In this way, the new intervals also form a proper-interval representation of G. And, the new $[a_j, b_j]$ is of unit length for $1 \leq j \leq i$.

For the case of $a_{i-1} < a_i < b_{i-1} < b_i$, scale the part $[b_{i-1}, \infty)$ of the real line by a factor $(a_i + 1 - b_{i-1})/(b_i - b_{i-1})$. In other words, the point $x \in [b_{i-1}, \infty)$ is mapped to $b_{i-1} + (x - b_{i-1})(a_i + 1 - b_{i-1})/(b_i - b_{i-1})$ for $x \in [b_{i-1}, \infty)$. In this way, the new intervals also form a proper-interval representation of G. Also, $[a_j, b_j]$ remains unchanged for $0 \leq j \leq i-1$. And the left part $[a_i, b_{i-1}]$ of $[a_i, b_i]$ remains unchanged, while the right part $[b_{i-1}, b_i]$ of $[a_i, b_i]$ is changed to $[b_{i-1}, a_i + 1]$. That is, the interval $[a_i, b_i]$ is changed to $[a_i, a_i + 1]$.

At end of iteration n, a unit-interval representation of G is obtained.

(2) $\Rightarrow$ (3). Suppose G is a unit-interval graph. Choose a unit-interval representation $\mathcal{F} = \{I_i = [a_i, a_i + 1] : 1 \leq i \leq n\}$ of G, where I_i represents vertex v_i. Define $f : V(G) \to \mathbb{R}$ by $f(v_i) = a_i$ for $v_i \in V(G)$. Then

$$v_i v_j \in E(G) \iff |f(v_i) - f(v_j)| \leq 1.$$

Let $\varepsilon = \min\{2, |f(v_i) - f(v_j)| : v_i v_j \in E(G), |f(v_i) - f(v_j)| > 1\}$. Then $\varepsilon > 1$. And

$$v_i v_j \in E(G) \iff |f(v_i) - f(v_j)| < \varepsilon \iff \left|\frac{1}{\varepsilon}f(v_i) - \frac{1}{\varepsilon}f(v_j)\right| < 1.$$

Hence, $\frac{1}{\varepsilon}f$ is an indifference function and so G is an indifference graph.

(3) $\Rightarrow$ (4). Suppose G is an indifference graph. Choose an indifference function f of G. Assume the vertices of G are sorted by

$$f(v_1) \leq f(v_2) \leq \cdots \leq f(v_n).$$

If $i < j < k$ and $v_i v_k \in E(G)$, then $f(v_i) \leq f(v_j) \leq f(v_k)$ and $f(v_k) - f(v_i) < 1$. Hence, $f(v_j) - f(v_i) \leq f(v_k) - f(v_i) < 1$ and so $v_i v_j \in E(G)$, and $f(v_k) - f(v_j) \leq f(v_k) - f(v_i) < 1$ and so $v_j v_k \in E(G)$, as desired.

(4) $\Rightarrow$ (1). Suppose the vertices of G has an ordering $[v_1, v_2, \ldots, v_n]$ satisfying (TIO). A proof similar to that for Theorem 5.2 is as follows. For every $v_i \in V(G)$, let i' be the minimum index such that $v_{i'} \in N[v_i]$ and i'' the maximum index such that $v_{i''} \in N[v_i]$; and let interval $I_i = [i', i'']$ correspond to vertex v_i. Note that $i \in I_i$.

If $v_i v_k \in E$ with $i < k$, then $k' \leq i < k \leq k''$ and so $i \in I_i \cap I_k \neq \emptyset$.

Conversely, if $I_i \cap I_k \neq \emptyset$ with $i < k$, say $j \in I_i \cap I_k$, then $i' \leq j \leq i''$ and $k' \leq j \leq k''$ implying $k' \leq i < k \leq k''$. Since $v_{k'}v_k \in E(G)$, by (TIO), $v_i v_k'' \in E(G)$ and in turn $v_i v_k \in E(G)$.

Therefore, G is a proper interval graph with $\mathcal{F}' = \{I_i = [i', i''] : 1 \leq i \leq n\}$ as a proper-interval representation. $\square$

5.2 Total domination and domination in interval graphs

It is not clear why people worked on total domination before working on domination in interval graphs.

Bertossi (1986) gave an $O(n^2)$-time algorithm for total domination in interval graphs. He used an interval representation to design the algorithm. Suppose $\mathcal{F} = \{I_i = [a_i, b_i] : 1 \leq i \leq n\}$ is an interval representation of the interval graph G, where interval I_i represents vertex v_i. By small changes of the $2n$ endpoints of the intervals, one may assume that they are distinct. So an interval including another interval is the same as the interval properly including another. Also assume that the left endpoints of the intervals are sorted in increasing order: $a_1 < a_2 < \cdots < a_n$.

To study total domination, we further assume the following conditions. If necessary, one may change a_1 to $\min_{1 \leq i \leq n} a_i - 1$ and b_n to $\max_{1 \leq i \leq n} b_i + 1$.

- *Graph G is connected and has at least two vertices.*
- *In $\mathcal{F}$, there is no interval includes all other intervals.*

A (total) dominating set is *proper* if it contains no two distinct vertices v_i and v_j such that $I_i \subseteq I_j$.

Lemma 5.6. *Graph G has a proper minimum total dominating set.*

Proof. Choose a minimum total dominating set D such that $s = \sum_{v_i \in D}(b_i - a_i)$ is as large as possible. Also assume that the number p of pairs of distinct vertices v_i and v_j in D with $I_i \subseteq I_j$ is minimum.

If $p = 0$, then the lemma holds. Now suppose $p > 1$. Choose distinct vertices v_i and v_j in D with $I_i \subseteq I_j$.

By the assumption, there is some v_h other than v_i and v_j such that $I_h \cap I_j \neq \emptyset$ and $I_h \not\subseteq I_j$. Then $v_h \notin D$, for otherwise $v_h \in D \setminus \{v_i, v_j\}$ would imply that $D \setminus \{v_i\}$ is a total dominating set of size smaller than $|D|$, a contradiction.

First, there is no $v_k \in D$ such that $I_k \subseteq I_h$ and so $I_k \subsetneqq I_h$, for otherwise $D' = (D \backslash \{v_k\}) \cup \{v_h\}$ is a minimum total dominating set with $s' > s$, contradicting that s is maximal. Also, there is no $v_k \in D \backslash \{v_i, v_j\}$ such that $I_h \subseteq I_k$, for otherwise $I_k \cap I_j \neq \emptyset$, and then $D \backslash \{v_i\}$ is a total dominating set of size smaller than $|D|$, a contradiction.

Consequently, $D'' = (D \backslash \{v_i\}) \cup \{v_h\}$ is a minimum total dominating set with $p'' < p$, contradicting the minimality of p.

The lemma then follows. $\square$

Let G' be the interval graph obtained from G by adding two isolated vertices v_0 and v_{n+1} with $I_0 = [m-2, m-1]$ and $I_{n+1} = [M+1, M+2]$, where $m = \min_{1 \le i \le n} a_i$ and $M = \max_{1 \le i \le n} b_i$.

Consider the directed graph H with vertex set $V(H) = V(G')$ and edge set $E(H) = A \cup B$, where

$$A = \{v_i v_j : i < j, a_i < a_j < b_i < b_j\} \quad \text{and}$$

$$B = \{v_i v_j : i < j, \text{ there is no } h \text{ with } a_i < b_i < a_h < b_h < a_j < b_j\}.$$

Note that H is acyclic with v_0 as a source and v_{n+1} as a sink.

Lemma 5.7. *A proper total dominating set D of G corresponds to a v_0-v_{n+1} path P in H with the condition that P does not contain two consecutive edges of B, where $V(P) = D \cup \{v_0, v_{n+1}\}$.*

Proof.   Suppose D is a proper total dominating set of G. Then $D' = D \cup \{v_0, v_{n+1}\}$ is a proper dominating set of G'. Order the vertices of D' in increasing indices. Consider two consecutive vertices v_i and v_j of D' with $i < j$. If $v_i v_j \in E(G')$, then $a_i < a_j < b_i < b_j$ since D' is a proper dominating set of G'; and so $v_i v_j \in A \subseteq E(H)$. Now suppose $v_i v_j \notin E(G')$, and so $a_i < b_i < a_j < b_j$. Then $v_i v_j \in B \subseteq E(H)$. Otherwise there is some h with $a_i < b_i < a_h < b_h < a_j < b_j$, and so $v_h \in V(G) \backslash D$. Suppose v_h is dominated by $v_k \in D$. If $k \le i$, then $a_k \le a_i$ and so $b_k < b_i < a_h$, since D' is a proper dominating set of G', contradicting $v_k v_h \in E(G')$. If $j \le k$, then $b_h < a_j \le a_k$, contradicting $v_k v_h \in E(G')$. Therefore, vertices in D' form a v_0-v_{n+1} path P in H.

Also, the path P has no two consecutive edges in B. Suppose to the contrary that P has two consecutive edges $v_i v_h$ and $v_h v_j$ in B, where $i < h < j$ and so $b_i < a_h < b_h < a_j$. The same arguments as in the second half of the last paragraph give a contradiction. Hence, P has no two consecutive edges in B as desired.

On the other hand, suppose P is a v_0-v_{n+1} path in H with the condition that P does not contain two consecutive edges in B. As for every edge $v_i v_j$ in P with $i < j$, v_i and v_j dominate all vertices $v_{i+1}, v_{i+2}, \ldots, v_{j-1}$ with indices between i and j, $D = V(P) \backslash \{v_0, v_{n+1}\}$ is a dominating set of G. To see that D is a total dominating set, suppose to the contrary that there is a $v_j \in D$ not adjacent to any vertex in D. Choose the two consecutive edges $v_i v_j$ and $v_j v_k$ in P, where $i < j < k$. By the assumption, either $v_i v_j$ or $v_j v_k$ is in A and so v_j is adjacent to v_i or v_k in D, a contradiction. In summary, D is a total dominating set of G. Finally, D is proper by the definitions of sets A and B. □

In order to implement Lemma 5.7, it needs to reduce the shortest path problem with constraints on certain edges to an ordinary shortest path problem. For this purpose, construct a directed graph H' from H as follows. First, splitting every vertex v_i into two vertices v_i^{in} and v_i^{out} joined by an edge $v_i^{\text{in}} v_i^{\text{out}}$. For every edge $v_i v_j$ in A, introduce an edge $v_i^{\text{out}} v_j^{\text{in}}$; and for every edge $v_i v_j$ in B, an edge $v_i^{\text{in}} v_j^{\text{out}}$. One can check that a v_0-v_{n+1} path P in H corresponds to a v_0^{in}-v_{n+1}^{out} path P' in H'. Note that two consecutive edges $v_i v_j \in A$ and $v_j v_k \in A$ in H correspond to the portion $v_i^{\text{out}}, v_j^{\text{in}}, v_j^{\text{out}}, v_k^{\text{in}}$ in H'; while two consecutive edges $v_i v_j \in A$ and $v_j v_k \in B$ in H correspond to the portion $v_i^{\text{out}}, v_j^{\text{in}}, v_k^{\text{out}}$ in H'; and two consecutive edges $v_i v_j \in B$ and $v_j v_k \in A$ in H correspond to the portion $v_i^{\text{in}}, v_j^{\text{out}}, v_k^{\text{in}}$ in H'. Unlike Bertossi's approach, no edge weights are introduced here. Instead, it only needs to see that for every v_i in P, either v_i^{in} or v_i^{out} or both is/are in P'.

As for the time complexity of the algorithm, first H' can be constructed from $\mathcal{F}$ in $O(n^2)$ time from $\mathcal{F}$ (in fact $O(n + m)$ time), where $m = |E(G)|$. Secondly, a shortest path in H' can be found in $O(n^2)$ time (in fact $O(n+m)$ time for the cardinality case), since H and H' are acyclic. So the overall time complexity is $O(n^2)$ (in fact $O(n + m)$).

Note that Keil (1986) also gave a linear-time algorithm for the total domination problem in interval graphs. Ramalingam and Pandu Rangan (1988a) pointed out a flaw in the paper by Keil, and then presented a linear-time algorithm for the same problem.

The domination problem in interval graphs can also be solved by similar, in fact simpler, arguments as above. First, the two conditions before Lemma 5.6 are not necessary, since an interval graph always has a proper minimum dominating set without having these conditions. And a lemma similar to Lemma 5.7 for domination now is as follows.

Lemma 5.8. *A proper dominating set of the interval graph corresponds to a v_0-v_{n+1} path P in H.*

Since P has no extra condition on B, it is not necessary to construct H'. A shortest v_0-v_{n+1} path in H then gives a minimum dominating set of G.

The following gives a primal-dual algorithm for the domination problem in interval graphs by using interval orderings. Note that the total domination problem in interval graphs has no similar primal-dual relation.

Algorithm DomIntervalPD: Find a minimum dominating set D and a maximum 2-stable set S of an interval graph G, providing an interval ordering $[v_1, v_2, \ldots, v_n]$ is given.

$D \leftarrow \emptyset$; $S \leftarrow \emptyset$;

for $i = 1$ **to** n **do**

if $N[v_i] \cap D = \phi$ **then**

{ choose the maximum $j \gtrsim i$;

  $D \leftarrow D \cup \{v_j\}$;

  $S \leftarrow S \cup \{v_i\}$;

}

$D \leftarrow \emptyset$ $S \leftarrow \emptyset$;

for $i = 1$ **to** n **do** label v_i unscanned;

for $i = 1$ **to** n **do**

if v_i is unscanned **then**

{ choose the maximum $j \gtrsim i$;

  $D \leftarrow D \cup \{v_j\}$;

  $S \leftarrow S \cup \{v_i\}$;

  label v_j and all its neighbors scanned;

}

Theorem 5.9. *Algorithm DomIntervalPD gives a minimum dominating set D and a maximum 2-stable set S of an interval graph G in linear time providing an interval ordering is given. Moreover, the strong duality equality $\alpha_2(G) = \gamma(G)$ holds.*

Proof.   To verify the correctness of the algorithm, it is sufficient to prove that at the end of the algorithm D is a dominating set, S is a 2-stable set and $|D| \leq |S|$. This is because

$$|S| \leq \alpha_2(G) \leq \gamma(G) \leq |D| \leq |S|$$

imply that all inequalities are equalities.

D is clearly a dominating set as the if-then statement does.

Suppose S is not a 2-stable set, i.e., there exist v_i and $v_{i'}$ in S such that $i < i'$ but $d_G(v_i, v_{i'}) \leq 2$. Let j be the maximum index with $j \gtrsim i$. Since $d(v_i, v_{i'}) \leq 2$, either $v_i v_{i'} \in E(G)$ or else there is a path $v_i v_k v_{i'}$.

(1) If $v_i v_{i'} \in E(G)$, then $i < i' \leq j$ and so $v_j \in N[v_{i'}]$ by (IO).
(2) If $(v_i, v_k, v_{i'})$ is a path and $k \leq i$, then $k \leq i < i'$ and $v_k v_{i'} \in E(G)$. By (IO), $v_i v_{i'} \in E(G)$. This reduces to case (1) to get $v_j \in N[v_{i'}]$.
(3) If $(v_i, v_k, v_{i'})$ is a path and $i < k < i'$, then $i < k \leq j$ and $v_i v_j \in E(G)$. By (IO), $v_j \in N[v_k]$. For the case of $j \leq i'$, by (IO), $k \leq j \leq i'$ and $v_k v_{i'}$ imply $v_j \in N[v_{i'}]$. For the case of $i' < j$, by (IO), $k < i' < j$ and $v_k v_j \in E(G)$ imply $v_j \in N[v_{i'}]$.
(4) If of $(v_i, v_k, v_{i'})$ is a path and $i < i' \leq k$, then $i < i' \leq k \leq j$ and $v_i v_j \in E(G)$. By (IO), $v_j \in N[v_{i'}]$.

In any case, $v_j \in N[v_{i'}]$. By the algorithm, $v_j \in D$ at the end of iteration i. When the algorithm processes $v_{i'}$ later, $N[v_{i'}] \cap D \neq \emptyset$ which causes that S does not contain $v_{i'}$, a contradiction. This proves that S is a 2-stable set.

The inequality $|D| \leq |S|$ follows from the fact that when v_j, which may already be in D, is added into D a new vertex v_i is added into S. □

The following is an interesting question, see Exercise 5.14.

Question 2. *Whether replacing statement* "**if** $N[v_i] \cap D = \phi$ **then**" *in Algorithm DomIintervalPD with* "**if** $N(v_i) \cap D = \phi$ **then**" *results an algorithm solving the total domination problem in interval graphs?*

5.3 Weighted domination and variations in interval graphs

Ramalingam and Pandu Rangan (1988) provided a unified approach to the weighted independent domination, the weighted domination, the weighted total domination and the weighted connected domination problems in interval graphs. This is the first paper the concept of interval ordering appeared. By using interval orderings, dynamic programming algorithms are designed for the problems mentioned above in interval graphs.

In this chapter, a vertex v_i is often referred to as i for simplicity. This and some of the following sections use this simplification.

Suppose G is an interval graph with an interval ordering $[1, 2, \ldots, n]$, in which every vertex i is associated with a real number w_i as its weight. According to Lemma 1.3, assume that the weights w_i are nonnegative except for independent domination.

Besides $V_0 = \emptyset$ and G_0, the null graph, consider the following notation for $1 \leq i \leq n$.

(1) Let $V_i = \{1, 2, \ldots, i\}$.

(2) Let G_i be the subgraph $G[V_i]$ induced by V_i.

(3) Let $low(i)$ be the minimum vertex in $N[i]$.

(4) Let $maxlow(i) = \max_{low(i) \leq j \leq i} low(j) = low(m_i)$, where $low(m_i) \leq m_i \leq i$.

(5) Let $L_i = \{maxlow(i), maxlow(i) + 1, \ldots, i\}$.

(6) Let $M_i = \{j : j > i$ and j is adjacent to $i\}$.

(7) For any family X of sets of vertices, $\min X$ denotes a minimum-weighted set in X. If $X = \emptyset$, then $\min X$ denotes a set of infinite weight.

Note that all $low(i)$ and $maxlow(i)$ can be calculated in linear time.

Lemma 5.10. *The following hold for* $1 \leq i \leq n$.

(1) $N[i] \cap V_i = \{low(i), low(i) + 1, \ldots, i\}$.

(2) $low(i) \leq maxlow(i) = low(m_i) \leq m_i \leq i$.

(3) L_i *is a maximal clique in* G_i.

(4) $N[j] \cap V_i = \{low(j), low(j) + 1, \ldots, i\}$ *for* $j \in L_i$.

(5) $N[m_i] \cap V_i = L_i$.

Proof.   (1) This follows from the definition of $low(i)$ and property (IO).

(2) This follows from definitions of $low(i)$, $maxlow(i)$ and m_i.

(3) This follows from the definition of $maxlow(i)$ and property (IO).

(4) This follows from (1) and (3).

(5) This is a special case of (4).   □

Having all of these in mind, it is ready to present the algorithms for the weighted independent domination, the weighted domination, the weighted total domination and the weighted connected domination problems in interval graphs given by Ramalingam and Pandu Rangan.

5.3.1   *Weighted independent domination in interval graphs*

For weighted independent domination, the weights are real numbers which may be negative.

Let ID_i denote an independent dominating set of the graph G_i and let MID_i denote the minimum weighted ID_i.

For $i \geq 1$, vertex $m_i \in V_i$ is dominated by some $j \in ID_i$ in G_i. By Lemma 5.10 (5), $N[m_i] \cap V_i = L_i$ and so $j \in L_i$. As ID_i is stable and L_i is a clique,

$ID_i \cap L_i = \{j\}$ and L_i is dominated by j. By Lemma 5.10 (4), $N[j] \cap V_i = \{\text{low}(j), \text{low}(j) + 1, \ldots, i\}$. It is then necessary and sufficient that $ID_i \backslash \{j\}$ dominates $V_{\text{low}(j)-1}$. Therefore, $ID_i \backslash \{j\}$ is an independent dominating set of $G_{\text{low}(j)-1}$. In other words, ID_i is of the form $ID_{\text{low}(j)-1} \cup \{j\}$ for some j in L_i.

These give the following lemma. Note that MID_n is a minimum weighted independent dominating set of G.

Lemma 5.11. (a) $MID_0 = \emptyset$. (b) *For* $1 \leq i \leq n$,

$$MID_i = \min\{MID_{\text{low}(j)-1} \cup \{j\} : j \in L_i\}.$$

A linear-time algorithm for the weighted independent domination problem in interval graphs then follows from the above lemma. The detailed description of the algorithm is omitted.

5.3.2 *Weighted domination in interval graphs*

For weighted domination, according to Lemma 1.3 one may assume that the weights are nonnegative.

Let D_i denote a subset of $V(G)$ that dominates V_i. Unlike independent domination, it is not necessary to restrict D_i as a subset of V_i. Let MD_i denote a minimum weighted D_i.

For $i \geq 1$, vertex $m_i \in V_i$ is dominated by some $j \in D_i$. If $j \leq i$, then $j \in L_i$ by Lemma 5.10 (5). If $j > i$, then $m_i \leq i < j$ and $m_i j \in E(G)$ imply that $ij \in E(G)$ by property (IO); and so $j \in M_i$. Hence, $j \in L_i \cup M_i$ and so $\text{low}(j) \leq i$. Since j dominates all vertices in $V_i \backslash V_{\text{low}(j)-1}$ and no vertex in $V_{\text{low}(j)-1}$, it is necessary and sufficient that the remaining set $D_i \backslash \{j\}$ dominates $V_{\text{low}(j)-1}$. In other words, D_i is of the form $D_{\text{low}(j)-1} \cup \{j\}$ for some j in $L_i \cup M_i$.

These give the following lemma. Note that MD_n is a minimum weighted dominating set of G.

Lemma 5.12. (a) $MD_0 = \emptyset$. (b) *For* $1 \leq i \leq n$,

$$MD_i = \min\{MD_{\text{low}(j)-1} \cup \{j\} : j \in L_i \cup M_i\}.$$

A linear-time algorithm for the weighted domination problem in interval graphs then follows from the above lemma. The detailed description of the algorithm is omitted.

5.3.3 *Weighted total domination in interval graphs*

For weighted total domination, according to Lemma 1.3 one may assume that the weights are nonnegative.

Let TD_i denote a subset of $V(G)$ that total dominates V_i and let MTD_i be a minimum weighted TD_i. Again, it is not necessary to restrict D_i as a subset of V_i.

In this variation, in order to solve the total domination problem one needs also to solve a variant total domination problem. More precisely, let PD_i denote a subset of $V(G)$ that total dominates $V_{\mathrm{low}(i)-1} \cup \{i\}$ and let MPD_i be a minimum weighted PD_i.

Similar to the arguments for domination, for $i \geq 1$ the set TD_i includes some vertex j in $L_i \cup M_i$. If $j \in L_i$, then it is necessary and sufficient that the set $\mathrm{TD}_i \backslash \{j\}$ total dominates $V_{\mathrm{low}(j)-1} \cup \{j\}$. If $j \in M_i$, then it is necessary and sufficient that the set $\mathrm{TD}_i \backslash \{j\}$ total dominates $V_{\mathrm{low}(j)-1}$.

Similarly, for $i \geq 1$ the set PD_i includes some vertex j adjacent to i. By the definition, $j \geq \mathrm{low}(i)$. Hence, it is necessary and sufficient that the set $\mathrm{PD}_i \backslash \{j\}$ total dominates $V_{\min\{\mathrm{low}(i)-1,\mathrm{low}(j)-1\}}$.

These give the following lemma. Note that MTD_n is a minimum weighted total dominating set of G.

Lemma 5.13. (a) $\mathrm{MTD}_0 = \emptyset$. (b) *For* $1 \leq i \leq n$,

$$\mathrm{MTD}_i = \min(\{\mathrm{MPD}_j \cup \{j\} : j \in L_i\} \cup \{\mathrm{MTD}_{\mathrm{low}(j)-1} \cup \{j\} : j \in M_i\}),$$

$$\mathrm{MPD}_i = \min\{\mathrm{MTD}_{\min\{\mathrm{low}(j)-1,\mathrm{low}(i)-1\}} \cup \{j\} : j \in N(i)\}.$$

A linear-time algorithm for the weighted total domination problem in interval graphs then follows from the above lemma. The detailed description of the algorithm is omitted.

5.3.4 *Weighted connected domination in interval graphs*

For weighted connected domination, according to Lemma 1.3 one may assume that the weights are nonnegative.

Let CD_i denote a connected dominating set of G_i that contains the vertex i and let MCD_i denote a minimum weighted CD_i.

If $\mathrm{low}(i) = 1$, then MCD_i is $\{i\}$ since all vertices have nonnegative weights. If $\mathrm{low}(i) > 1$, then every CD_i contains vertices other than i, and hence some vertex adjacent to i in G_i. Let j be the maximum vertex in

$CD_i\backslash\{i\}$. Assume that $\text{low}(j) < \text{low}(i)$. Otherwise j is removed to get a CD_i with the same or smaller weight. If $\text{low}(j) < \text{low}(i)$, then any other vertex of G_j adjacent to i is also adjacent to j. Thus, it is necessary and sufficient that $CD_i\backslash\{i\}$ is a CD_j.

These give the following lemma.

Lemma 5.14. (a) *If* $\text{low}(i) = 1$, *then* $\text{MCD}_i = \{i\}$. (b) *For* $\text{low}(i) > 1$,

$$\text{MCD}_i = \min\{\text{MCD}_j \cup \{i\} : j \in N(i) \text{ and } j < i \text{ and } \text{low}(j) < \text{low}(i)\}.$$

(c) $\min\{\text{MCD}_i : i \in L_n\}$ *is a minimum weighted connected dominating set of the graph* G.

Part (c) of the above lemma follows from the fact that any dominating set of G contains at least one vertex in L_n. Thus, any set is a connected dominating set of G if and only if it is CD_i for some i in L_n.

A linear-time algorithm for the weighted connected domination problem in interval graphs then follows from the above lemma. The detailed description of the algorithm is omitted.

5.4 Paired-domination in interval graphs

Cheng *et al.* (2007) gave a linear-time algorithm for the paired-domination problem in interval graphs without isolated vertices. In this section, two methods are presented. The first method is similar to the dynamic programming approach in the preceding section for variations of domination. This is a modification of the presentation by Chang *et al.* (2007), who use interval representations with sorted right endpoints of the intervals, while the following approach uses interval orderings. The second method is similar to the primal-dual approach for the paired-domination in trees, see Section 3.6.

For the first method, suppose G is an interval graph without isolated vertices and $[1, 2, \ldots, n]$ is an interval ordering of G. Besides the notations in the previous section, consider the following notations for $1 \leq i \leq n$.

(1) Let $\text{minlow}(i) = \min\{\text{low}(k) : k \in N[i]\}$.
(2) Let ℓ_i be a $k' \in N(i)$ such that $\text{low}(k') = \min\{\text{low}(k) : k \in N(i)\}$.
(3) Let $L_i' = \{\text{minlow}(i), \text{minlow}(i) - 1, \ldots, i\}$.

Lemma 5.15. *The following properties hold for $1 \le i \le n$.*

(1) $\mathrm{minlow}(i) = \min\{\mathrm{low}(i), \mathrm{low}(\ell_i)\}$.
(2) $N(\{i, \ell_i\}) \cap V_i = L'_i$.

Proof.   (1) This follows from the definition of ℓ_i.
(2) This follows from (1) and property (IO).   □

Let PRD_i denote a subset of $V(G)$ that paired-dominates V_i. Similar to domination, it is not necessary to restrict PRD_i as a subset of V_i. Let MPRD_i denote a minimum sized PRD_i.

Similar to the arguments in the previous section, all neighbors of $m_i \in L_i \subseteq V_i$ are all in $L_i \cup M_i$, and so the set PRD_i contains some vertex j in $L_i \cup M_i$. Suppose PRD_i includes $\{j, j'\}$ as a pair. We claim that we may assume that $j' = \ell_j$. Suppose to the contrary that $j' \ne \ell_j$.

First, $\mathrm{low}(j') \ge \mathrm{low}(\ell_j)$ by the definition of ℓ_j, and so $\{j, \ell_j\}$ dominates more vertices of V_j than $\{j, j'\}$ does. In fact ℓ_j is the best possible one to dominate V_j among all j'. It is necessary and sufficient that the remaining set $\mathrm{PRD}_i \backslash \{j, \ell_j\}$ dominates $V_{\mathrm{minlow}(j)-1}$, since $\{j, \ell_j\}$ dominates all vertices in $V_i \backslash V_{\mathrm{minlow}(j)-1}$ but no vertex in $V_{\mathrm{minlow}(j)-1}$. In other words, a set is a PRD_i if and only if it is of the form $\mathrm{PRD}_{\mathrm{minlow}(j)-1} \cup \{j, \ell_j\}$ for some j in $L_i \cup M_i$.

These give the following lemma. Note that MPRD_n is a minimum paired-dominating set of G.

Lemma 5.16. (a) $\mathrm{MPRD}_0 = \emptyset$. (b) *For $1 \le i \le n$,*

$$\mathrm{MPRD}_i = \min\{\mathrm{MPRD}_{\mathrm{minlow}(j)-1} \cup \{j, \ell_j\} : j \in L_i \cup M_i\}.$$

A linear-time algorithm for the paired-domination problem in interval graphs then follows from the above lemma. The detailed description of the algorithm is omitted.

Next, a primal-dual algorithm for the paired-domination problem in interval graphs is as follows. Recall that a 3-stable set of a graph G is a subset $S \subseteq V(G)$ such that $d(x, y) > 3$ for every two distinct $x, y \in S$. The 3-stability number $\alpha_3(G)$ of G is the maximum size of a 3-stable set of G. The weak duality for paired-domination says that $2\alpha_3(G) \le \gamma_{\mathrm{pr}}(G)$.

Algorithm PairedDomIntervalPD: For an interval graph G without
isolated vertices, find a maximum 3-stable set S and a minimum paired-
dominating set D, providing an interval ordering $[v_1, v_2, \ldots, v_n]$ is given.

$S \leftarrow D \leftarrow \emptyset$;
for $i = 1$ **to** n **do**
$\{$ **if** $N[v_i] \cap D = \emptyset$ **then**
     $\{$ let j be the largest index such that $v_j \in N[v_i]$;
      let k be the largest index such that $v_k \in N(v_j)$;
      $S \leftarrow S \cup \{v_i\}$;
      $D \leftarrow D \cup \{v_j, v_k\}$;
     $\}$
$\}$

For the correctness of the algorithm, the following theorem is useful.

Theorem 5.17. *Suppose* $[v_1, v_2, \ldots, v_n]$ *is an interval ordering of an interval graph* G. *If* $d(v_{i_0}, v_{i_r}) = r$ *with* $i_0 < i_r$, *then there exists a shortest*
v_{i_0}-v_{i_r} *path* $(v_{i_0}, v_{i_1}, \ldots, v_{i_r})$ *such that* v_{i_s} *is the largest indexed neighbor of*
$v_{i_{s-1}}$ *for* $1 \le s \le r-1$, *in particular* $i_0 < i_1 < \ldots < i_{r-1}$.

Proof. The case of $r = 1$ is trivial. Suppose $r \ge 2$.

Choose a shortest v_{i_0}-v_{i_r} path $P = (v_{i_0}, v_{i_1}, \ldots, v_{i_r})$ such that the sum
of the indices of the vertices of P is as large as possible. Suppose the theorem
is not true, say $s \le r-1$ is the smallest index such that v_{i_s} is not the largest
indexed neighbor of $v_{i_{s-1}}$. In particular, $i_0 < i_1 < \ldots < i_{s-1}$.

First, $i_{s-1} < i_{s+1}$. Otherwise, $i_{s+1} < i_{s-1}$. Let t be the index with
$1 \le t \le s-1$ and $i_{t-1} < i_{s+1} < i_t$. By (IO), $v_{i_{s+1}} v_{i_t} \in E(G)$, contradicts
that P is a shortest path.

Secondly, $i_{s-1} < i_s$. Otherwise, $i_s < i_{s-1} < i_{s+1}$. By (IO), $v_{i_{s-1}} v_{i_{s+1}} \in
E(G)$, contradicts that P is a shortest path.

Let v_j be the largest indexed neighbor of v_{s-1}. Then $i_{s-1} < i_s < j$ and
so by (IO) $v_{i_s} v_j \in E(G)$. There are two cases. If $j < i_{s+1}$, then $i_s < j < i_{s+1}$
and so by (IO) $v_j v_{i_{s+1}} \in E(G)$. If $i_{s+1} < j$, then $i_s < i_{s+1} < j$ and so by
(IO) $v_{i_{s+1}} v_j \in E(G)$. In any case, replacing v_{i_s} by v_j in P leads to a new
v_{i_0}-$v_{i_{s+1}}$ path P' in which the sum of indices of the vertices is larger than
that of P, a contradiction. $\square$

Theorem 5.18. *Algorithm PairedDomIntervalPD finds a maximum
3-stable set and a minimum paired-dominating set of an interval graph in
linear time. Moreover, the strong duality equality* $2\alpha_3(G) = \gamma_{\mathrm{pr}}(G)$ *holds.*

Proof.   First to show that the final S is a 3-stable set. Suppose to the contrary that $d(v_i, v_i') = r \leq 3$ for two distinct $v_i, v_{i'} \in S$ with $i < i'$. By Theorem 5.17, there is a shortest v_i-$v_{i'}$ path $(v_{i_0}, v_{i_1}, \ldots, v_{i_r})$, where $v_{i_0} = v_i$, $v_{i_r} = v_{i'}$, $1 \leq r \leq 3$ and v_{i_s} is the largest indexed neighbor of $v_{i_{s-1}}$ for $1 \leq s \leq r - 1$.

Recall that in iteration i, j is the largest index such that $v_j \in N[v_i]$ and k is the largest index such that $v_k \in N(v_j)$.

If $r = 1$, then $v_i v_{i'} \in E(G)$. In this case, $i < i' \leq j$ and $v_i v_j \in E(G)$. By (IO), $v_{i'} \in N[v_j]$. Since at end of iteration i and so at the beginning of iteration i', $v_j \in D$. This leads to $N[v_{i'}] \cap D \neq \emptyset$ and so $v_{i'}$ cannot be put into S, a contradiction.

If $2 \leq r \leq 3$, then at end of iteration i and so at the beginning of iteration i', either $r = 2$ with $v_{i_1} \in D$ or $r = 3$ with $v_{i_1}, v_{i_2} \in D$. This leads to $N[v_{i'}] \cap D \neq \emptyset$ and so $v_{i'}$ cannot be put into S, a contradiction.

By the testing "$N[v_i] \cap D = \emptyset$," the final D is a dominating set of G.

By (IO), at end of iteration i, if $m = \max\{s : v_s \in D\}$, then $v_1, v_2, \ldots, v_m$ are all dominated by D. So in the next iteration i' with $N[v'] \cap D = \emptyset$, it is the case that $i' > m$, $j' \geq i'$ and $k' > m$. Hence, $v_{j'}$ and $v_{k'}$ are two adjacent vertices not in D. Therefore, D remains the set of vertices of a matching and then finally is a paired-dominating set of G.

In fact, it is then also the case that $|D| = 2|S|$ at any time during the algorithm. Hence,

$$2|S| \leq 2\alpha_3(G) \leq \gamma_{\text{pr}}(G) \leq |D| = 2|S|,$$

and so the inequalities are equalities. Consequently, S is a maximum 3-stable set, D is a minimum paired-dominating set and the strong duality equality $2\alpha_3(G) = \gamma_{\text{pr}}(G)$ holds. $\square$

Goddard *et al.* (2014) introduced a variation of domination called *disjunctive domination*. A *disjunctive dominating set* of a graph G is a set $D \subseteq V(G)$ such that every vertex not in D either has a neighbor in D or has two vertices in D at a distance two from it. The *disjunctive domination problem* is to find the *disjunctive domination number* $\gamma^{\text{d}}(G)$ of a graph G that is the minimum size of a disjunctive dominating set of G.

Henning *et al.* (2023) initiated the paired version of disjunctive domination. More precisely, a *paired disjunctive dominating set* of an isolate-free graph G is a disjunctive dominating set D such that $G[D]$ has a perfect matching. The *paired disjunctive domination problem* is to find the *paired disjunctive domination number* $\gamma_{pr}^d(G)$ of a graph G that is the minimum size of a paired disjunctive dominating set of G. Among many interesting results, they designed a linear-time algorithm for the paired disjunctive domination problem in interval graphs.

There is also a primal-dual formulation as follows. A 4-*stable set* in a graph G is a subset $S \subseteq V(G)$ such that $d(u, v) > 4$ for every two distinct vertices $u, v \in S$. The following inequality then holds.

Weak duality inequality (paired disjunctive domination).
$2\alpha_4(G) \leq \gamma_{pr}^d(G)$ for any isolate-free graph G.

To see the inequality, choose a minimum 4-stable set S and a maximum paired disjunctive dominating set D. The fact that $G[D]$ has a perfect matching implies that "D is a disjunctive dominating set" is the same as "every vertex in $V(G)$ has 2 vertices in D at a distance at most two from it in G." Consider the following mapping: every vertex $v \in S$ corresponds to two vertices $v', v'' \in D$ at distance at most two from v. Note that every vertex in D is mapped by at most two vertices in S, for otherwise some vertex of D is mapped by two vertices in S implying that these two vertices are at distance at most 4, a contradiction. The inequality then follows.

Note that the inequality may be strict as shown by C_6 that

$$2\alpha_4(C_6) = 2 < 4 = \gamma_{pr}^d(C_6).$$

For more examples, see Exercise 5.21.

It is an interesting question as follows.

Question 3. *Is there a primal-dual algorithm for the paired disjunctive domination problem in isolate-free interval graphs? In particular, whether the strong duality equality* $2\alpha_4(G) = \gamma_{pr}^d(G)$ *holds for every isolate-free interval graph G?*

5.5 Domatic numbers of interval graphs

The *domatic number problem* seeks to find the largest number of pairwise disjoint dominating sets $D_1, D_2, \ldots, D_r$ of a graph G, where r is the domatic number $d(G)$.

The domatic number problem in interval graphs was first studied by Bertossi (1988). He transformed the problem to a network flow problem. More precisely, an interval representation of an interval graph was used to construct a directed graph H. Then he used the network flow algorithm to find the maximum number of internally disjoint s–t directed paths in H, each of which corresponds to a dominating set of G. This solves the domatic number problem in $O(n^{2.5})$-time for interval graphs and $O(n \log n)$-time for proper interval graphs.

Bertossi modified the arguments for the total domination problem in interval graphs as shown in Section 5.2. He first constructs a directed graph H with source v_0 and sink v_{n+1} from an interval graph G. He then proved that "a proper dominating set of G corresponds to a v_0-v_{n+1} path in H," see Lemma 5.8. Hence finding the domatic number of G is the same as finding the maximum number of internally vertex-disjoint v_0-v_{n+1} paths of H. The domatic number problem is then solved in $O(n^{2.5})$-time by an algorithm for the network flow problem. For the problem in proper interval graphs, he further reduced the problem to the maximum cardinality matching problem on convex bipartite graphs. Then an $O(n \log n)$-time algorithm was obtained.

Later, several linear-time algorithms were given by Srinvasa Rao and Pandu Rangan (1989/90), Lu *et al.* (1990), Peng and Chang (1991), Chang *et al.* (1993, 1999) and Manacher and Mankus (1996).

Manacher and Mankus (1996) gave a linear-time algorithm for the domatic number problem in interval graphs without using the network flow algorithm. Suppose G is an interval graphs with an interval representation $\{I_i = [a_i, b_i] : 1 \leq i \leq n\}$, where interval I_i represents vertex v_i. Assume the endpoints of the intervals are distinct and sorted into

$$a_1 < a_2 < \cdots < a_n \quad \text{and} \quad b_{i_1} < b_{i_2} < \cdots < b_{i_n}.$$

The algorithm employs a primal-dual approach. First, consider the following weak duality inequality.

Proposition 5.19 (Weak duality inequality (domatic numbers)). *For any graph G, the domatic number $d(G) \leq \delta(G) + 1$.*

Proof.   Choose a vertex v of minimum degree in G, i.e., $\|N[v]\| = \delta(G)+1$. The proposition follows from that every dominating set contains at least one vertex in $N[v]$.   □

In the sequel, we will denote $\delta(G)$ simply by δ. The algorithm partitions the vertices into $\delta+1$ partition sets, each of which is a dominating set. Each vertex v_i is assigned a *partition designation number* PART$[v_i] = k$ if it is in the kth partition set, where $1 \leq k \leq \delta + 1$. In case when vertex v_i can be placed into any of the $\delta + 1$ partition sets, we denote this by $\text{PART}[v_i] = \infty$.

The partitioning process goes as follows. Vertex v_c is placed into a partition set one by one for c from 1 to n. The vertex last placed into a partition set is referred to as a *last*. We use $L[v_i] = 1$ to indicate that v_i is a *last*, and $L[v_i] = 0$ to indicate that v_i is not a *last*.

(1) Initially, let $\text{PART}[v_i] = i$ for $1 \leq i \leq \delta + 1$ and all other $\text{PART}[v_i] = 0$. Now there are $\delta + 1$ *lasts*, i.e., $L[v_i] = 1$ for $1 \leq i \leq \delta + 1$ and all other $L[v_i] = 0$.

(2) To determine which partition set the next v_c will be placed into, consider vertex v_{i_j} for j from 1 to n. Let j' be the smallest number such that $v_{i_{j'}}$ is a *last*, i.e., $L[v_{i_{j'}}] = 1$. Note that for each *last* v_{i_j}, $i_j < c$ and so $b_{i_j} < b_c$, since all v_i with $i \geq c$ are not yet placed. Also, $j' \leq j$ and so $b_{i_{j'}} \leq b_{i_j}$.

(3) If $b_{i_{j'}} \leq b_c$, then place v_c into the same partition as $v_{i_{j'}}$ in by setting $\text{PART}[v_c] \leftarrow \text{PART}[v_{i_{j'}}]$. Now v_c becomes the new *last* of that partition set; and so set $L[v_{i_{j'}}] = 0$ and $L[v_c] = 1$.

If $b_c < b_{i_{j'}}$, then v_c can be placed into any partition set; and so set $\text{PART}[v_c] = \infty$.

The formal description of the algorithm is as follows.

Algorithm DomaticInterval: Suppose G is an interval graph of minimum degree δ and has an interval representation $\{I_i = [a_i, b_i] : 1 \leq i \leq n\}$, where interval I_i represents vertex v_i, with distinct end points sorted into $a_1 < a_2 < \cdots < a_n$ and $b_{i_1} < b_{i_2} < \cdots < b_{i_n}$. Partition $V(G)$ into $\delta + 1$ dominating sets.

for $i = 1$ **to** $\delta + 1$ **do** { $\text{PART}[v_i] \leftarrow i$; $L[v_i] \leftarrow 1$; }
for $i = \delta + 2$ **to** n **do** $L[v_i] \leftarrow 0$;
$j' \leftarrow 1$;
for $c = \delta + 2$ **to** n **do**
{   **while** $L[v_{i_{j'}}] = 0$ **do** $j' \leftarrow j' + 1$;
      if $b_{i_{j'}} < b_c$
      then { $\text{PART}[v_c] \leftarrow \text{PART}[v_{i_{j'}}]$; $L[v_{i_{j'}}] \leftarrow 0$; $L[v_c] \leftarrow 1$; }
      else $\text{PART}[v_c] \leftarrow \infty$;
}

The following demonstrates the algorithm for the interval graph with $n = 10$, $\delta = 3$ and the interval representation in Figure 5.1. By Algorithm DomaticInterval, the partition array

$$\text{PART}[v_1, \ldots, v_{10}] = [1 \quad 2 \quad 3 \quad 4 \quad 2 \quad 2 \quad \infty \quad 4 \quad 1 \quad 1].$$

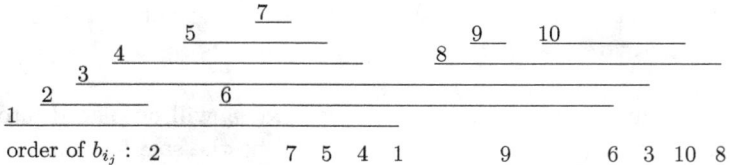

order of b_{i_j} : 2   7 5 4 1   9   6 3 10 8

Fig. 5.1.   An interval representation of an interval graph.

Theorem 5.20. *Algorithm DomaticInterval produces a minimum domatic partition for an interval graph in linear time. Moreover, the strong duality equality $d(G) = \delta(G) + 1$ holds for every interval graph G.*

Proof.   It suffices to show that every vertex v_i is dominated by all the $\delta + 1$ partition sets.

First, this is true for $v_1, v_2, \ldots, v_{\delta+1}$, since they form a clique. Otherwise, there exist $i < j \le \delta + 1$ such that $I_i \cap I_j = \emptyset$, which implies that $a_i < b_i < a_j < b_j$. Then I_i only intersects intervals $I_1, I_2, \ldots, I_{j-1}$. Thus $\deg(v_i) \le j - 2 \le \delta + 1 - 2 = \delta - 1$, a contradiction.

For $\delta + 2 \le i \le n$, consider the following cases.

Case 1. $\text{PART}[v_c] = \infty$ for some $v_c \in N[v_i]$.

At iteration c of the algorithm, $b_c < b_{i_{j'}}$ for the *last* $v_{i_{j'}}$ with minimum j'. Then $b_c < b_{i_j}$ and so I_c is a sub-interval of I_{i_j} for every *last* v_{i_j}, since $a_{i_j} < a_c$. As $I_i \cap I_c \ne \emptyset$, then $I_i \cap I_{i_j} \ne \emptyset$ for every last v_{i_j} and so v_i is dominated by all partition sets as desired.

Case 2. $\text{PART}[v_c] \ne \infty$ for every $v_c \in N[v_i]$.

By the algorithm, intervals for two distinct vertices in a same partition set are disjoint. If v_i is not dominated by all partition sets, then

$N[v_i]$ contains v_j and v_k, with $j < k$, such that $\text{PART}[v_j] = \text{PART}[v_k] \neq \text{PART}[v_i]$. In this case, $a_i < b_j < a_k < b_i$.

Suppose at iteration c, vertex v_c replaces v_j as the *last* of the partition set containing them. Then $j < c \le k$. Hence, for every last v_s other than v_j, i.e., $\text{PART}[v_s] \neq \text{PART}[v_j]$, $b_j < b_s$. Either $b_s < a_c$ implying $a_i < b_j < b_s < a_c < b_i$, or else $a_c < b_s$ implying $a_s < a_c < b_s$ and $a_i < a_c < b_i$. In any case $I_i \cap I_s \neq \emptyset$ and so v_i is dominated by all partition sets.

The running time is clearly linear. The strong duality follows from the weak duality inequality and the above property. $\qquad\square$

5.6 Domination in circular-arc graphs

In parallel to interval graphs, Klee (1969) raised the question "what are the intersection graphs of arcs in a circle?" In particular, he asked for an intrinsic, combinatorial characterization of graphs with special interest leading to an efficient algorithm for deciding whether a given graph is of the indicated form.

The intersection graphs of arcs in a circle are called *circular-arc graphs*. If no arc properly contains another, they are called *proper circular-arc graphs*.

Klee observed when the arcs do not cover the circle, the circular-arc graph is an interval graph. This is because that we can choose a point p of the circle not covered by any chosen arc, and then cut the circle at p and straighten it out to a line, the arcs becoming intervals.

A characterization of circular-arc graphs was given by Tucker (1970, 1971), originally formulated in terms of the augmented adjacency matrix of a graph. In his book, Golumbic (1980, p. 191) rewrote it in an equivalent formulation as follows. We revise the ordering reversely so that it is comparable to interval ordering of an interval graph.

This section also applies the convention of using its index i for the vertex v_i, where $1 \le i \le n$.

Sometimes we may have some vertex of the form $r \pm s$, which is out of the range from 1 to n. In this case, $r \pm s$ is taken modulo n. For instance, $-1, 0, n+1, n+2$ stand for $n-1, n, 1, 2$, respectively. Also, if $r > s$, then $r, r+1, \ldots, s-1$ stand for $r, r+1, \ldots, n, 1, 2, \ldots, s-1$.

Theorem 5.21 (Tucker, 1970, 1971). *A graph G is a circular-arc graph if and only if its vertices can be ordered into $[1, 2, \ldots, n]$ so that*

$$ij \in E(G) \text{ implies} \begin{cases} \text{either} & i, i+1, \ldots, j-1 \in N(j); \\ \text{or} & j, j+1, \ldots, i-1 \in N(i), \end{cases} \quad \text{(CaO)}$$

where the indices are taken modulo n.

Proof. ($\Rightarrow$) Suppose G has a circular-arc representation (closed arcs). We may assume, without loss of generality, that no pair of arcs share a common endpoint. Traversing clockwise from an arbitrary starting point once around the circle, index the vertices according to the order in which the second endpoints of their corresponding arcs occur. Let A_i denote the arc corresponding to vertex i. Clearly, vertex i is adjacent to vertex j if and only if the second endpoint of A_i lies within A_j or vice versa. In the former case, each of $A_i, A_{i+1}, \ldots, A_{j-1}$ intersects A_j, and in the latter case each of $A_j, A_{j+1}, \ldots, A_{i-1}$ intersects A_i. Hence, (CaO) holds.

($\Leftarrow$) Conversely, suppose the vertices are ordered as required in (CaO). Choose a circle and n points $p_1, p_2, \ldots, p_n$ of the circle in clockwise order. For each vertex i, let $m_i - 1$ be the vertex not adjacent to vertex i but vertices $m_i, m_i + 1, \ldots, i - 1$ are adjacent to vertex i. Let A_i be the arc clockwise from p_{m_i} to p_i. By construction, A_i intersects A_j ($i \neq j$) if and only if either $p_i \in A_j$ or $p_j \in A_i$. If $p_i \in A_j$, then vertex i is adjacent to vertex j; and if $p_j \in A_i$, then vertex j is adjacent to vertex i. On the other hand, if $ij \in E(G)$, then by (Cao) either $p_i \in A_j$ or $p_j \in A_i$. Hence, G is a circular-arc graph. □

The ordering $[1, 2, \ldots, n]$ satisfying (CaO) is called a *circular-arc ordering*. By Theorem 5.21, a graph is a circular-arc graph if and only if it has a circular-arc ordering. Note that interval graphs are circular-arc graphs, and interval orderings are circular-arc orderings.

A graph is a *unit circular-arc graph* if it is the intersection graph of a family of arcs each having length one in a circle. Parallel to proper interval graphs, the following is also true.

Theorem 5.22. *The following statements are equivalent for a graph G.*

(1) *The graph G is a proper circular-arc graph.*
(2) *The graph G is a unit circular-arc graph.*

(3) *The graph G has a vertex ordering $[1, 2, \ldots, n]$ such that for $1 \leq i \leq n$ there exist i' and i'' satisfying*

$$N[i] = \{i', i' + 1, \ldots, i - 1, i, i + 1, \ldots, i''\}.$$

Bonuccelli (1985) gave an $O(nm)$-time algorithm for the domination problem in circular-arc graphs with n vertices and m edges. His approach is as follows: for each vertex v, find a minimum dominating set containing v by using an $O(m)$-time algorithm for the domination problem in interval graphs. In fact for a fixed vertex u of minimum degree, it is only necessary to find a minimum dominating set containing v for each $v \in N[u]$. This gives the running time for the algorithm being (δm), where δ is the minimum degree of the graph.

Latter, Hsu and Tsai (1988, 1991), Yu *et al.* (1989) and Rhee (1990) presented linear-time algorithms for the domination problem in circular-arc graphs. The algorithm by Hsu and Tsai made it possible to run the algorithm for interval graphs in constance time, leading to a linear-time algorithm in circular-arc graphs. Their algorithm used the sorted endpoints of the arcs in a circular-arc representation of the graph. The following presentation uses circular-arc ordering of circular-arc graphs, but keeps the spirit of Hsu and Tsai's method.

If G has a vertex adjacent to all other vertices, then $\gamma(G) = 1$. In the following, assume that G has no vertex adjacent to all other vertices. Suppose G is a circular-arc graph with a circular-arc ordering $[1, 2, \ldots, n]$.

If $i, i + 1, \ldots, j - 1 \in N(j)$, then vertex j is called a *forward neighbor* of vertex i; and vertex i a *backward neighbor* of vertex j. Theorem 5.21 is the same as to say that a graph is circular-arc if and only if its vertices can be ordered such that every neighbor is either a forward neighbor or a backward neighbor. It is possible that a neighbor can be both a forward neighbor and a backward neighbor as shown by the 4-path P_4 with vertex set $V(P_4) = \{2, 3, 1, 4\}$, edge set $E(P_4) = \{23, 31, 14\}$ and circular-arc ordering $[1, 2, 3, 4]$, in which 3 is a forward neighbor and a backward neighbor of 1.

A *forward* (respectively, *backward*) *closed neighbor* of a vertex is the vertex itself or a forward (respectively, backward) neighbor.

For every vertex i, define the following notations.

(1) Let low(i) be the vertex $j \in N[i]$ such that vertices $j, j + 1, \ldots, i$ are the only backward closed neighbors of i.

(2) Let $u(i)$ be the vertex $j \notin N[i]$ such that $i, i+1, \ldots, j-1$ are forward closed neighbors of i. Note that $u(i)$ exists, since $\deg(i) < n-1$.

(3) Let high(i) be the forward closed neighbor j of i such that all forward closed neighbors of i are in $\{i, i+1, \ldots, j\}$, which may contain some nonforward neighbors of i.

(4) Let $S_i = \{i+1, i+2, \ldots, u(i), u(i)+1, \ldots, \text{high}(u(i))\}$.

First, consider the following useful lemma.

Lemma 5.23. *For every vertex i, if vertex j dominates $u(i)$ then $j \in S_i$.*

Proof. Since j dominates $u(i)$, either j is a forward neighbor of $u(i)$ or else j is a backward neighbor of $u(i)$. For the former case, $j \in \{u(i), u(i)+1, \ldots, \text{high}(u(i))\} \subseteq S_i$ by the definition of high($u(i)$). For the later case, $j \in \{\text{low}(u(i)), \text{low}(u(i))+1, \ldots, u(i)\} \subseteq S_i$, since $u(i)$ is not adjacent to i. $\square$

For every vertex i consider the following sequence of vertices $i_1, i'_2, i_2, i'_3, i_3, \ldots$ in circularly increasing order (it is possible that $i_s = i'_s$): $i_1 = i$, $i'_s = u(i_{s-1})$ and $i_s = \text{high}(i'_s)$ for $s \geq 2$.

Define GD(i) $= \{i = i_1, i_2, \ldots, i_r\}$ such that r is the smallest number satisfying that GD(i) is a dominating set of G. Note that $r \geq 2$, since $\gamma(G) \neq 1$.

Lemma 5.24. *For $2 \leq s \leq r$, vertex i_s is the only vertex in GD(i) $= \{i = i_1, i_2, \ldots, i_r\}$ that dominates $i'_s = u(i_{s-1})$.*

Proof. If $u(i_{s-1})$ is dominated by $i_p \in$ GD(i), then $i_p \in S_{i_{s-1}}$ by Lemma 5.23. Hence, $i_p = i_s$ as desired. $\square$

In the definition of GD(i), although r is chosen to be minimum, it does not imply that GD(i) is a minimal dominating set of G. However, Lemma 5.24 shows that GD(i) is a minimal dominating set subject to the condition of containing i, i.e., there is no proper subset of GD(i) that contains i and is a dominating set.

Hsu and Tsai had a misunderstanding at this point. An example is shown by the following circular-arc graph with $[1, 2, 3, 4, 5, 6]$ as a circular-arc ordering.

Note that the following table gives not only the neighborhood $N(i)$ but also an arc presentation arc(i) for every vertex i, where 12 points $p_1, p_2, \ldots, p_{12}$ of a circle are in clockwise order.

vertex i	1	2	3	4	5	6
$N(i)$	$\{6,3\}$	$\{3\}$	$\{4,1,2\}$	$\{5,3\}$	$\{6,4\}$	$\{5,1\}$
$\text{arc}(i)$	$[p_1,p_4]$	$[p_5,p_6]$	$[p_3,p_8]$	$[p_7,p_{10}]$	$[p_9,p_{12}]$	$[p_{11},p_2]$
$\text{low}(i)$	6	2	1	3	4	5
$u(i)$	2	4	5	6	1	2
$\text{high}(i)$	3	3	4	5	6	1
$\text{GD}(i)$	$\{1,3,6\}$	$\{2,5,3\}$	$\{3,6\}$	$\{4,1,3\}$	$\{5,3\}$	$\{6,3\}$

As shown in the table, $\text{GD}(1)$ and $\text{GD}(2)$ are not minimal dominating sets, while $\text{GD}(4)$ is a minimal but not minimum dominating set.

Lemma 5.25. *If D is a dominating set containing vertex i, then $|\text{GD}(i)| \leq |D|$.*

Proof.   Let $\text{GD}(i) = \{i = i_1, i_2, \ldots, i_r\}$. For every $2 \leq s \leq r$, the set D has a vertex j_s that dominates $u(i_{s-1})$. By Lemma 5.23, $j_s \in S_{i_{s-1}}$. Since $S_{i_1}, S_{i_2}, \ldots, S_{i_{r-1}}$ are pairwise disjoint and i is not in any of these $r-1$ sets, $|\text{GD}(i)| = r \leq |D|$. $\square$

Consequently, minimum dominating sets of the form $\text{GD}(i)$ do exist.

Corollary 5.26. *For any minimum dominating set D, if $i \in D$ then $\text{GD}(i)$ is also a minimum dominating set.*

Construct the directed graph H with $V(H) = V(G)$ in which $i \to j$ is a (directed) edge if and only if $j = \text{high}(u(i))$. The graph represented in the above table has the following edges in H:

$$2 \to 5, \quad 5 \to 3, \quad 3 \to 6, \quad 6 \to 3, \quad 4 \to 1, \quad 1 \to 3.$$

Using this notation, $\text{GD}(i) = \{i = i_1, i_2, \ldots, i_r\}$ corresponds to the $r-1$ edges $i_1 \to i_2, i_2 \to i_3, \ldots, i_{r-1} \to i_r$ in H.

Since every vertex of H has out-degree 1, H has at least one directed cycle. Also, two different directed cycles have no common vertex. Starting from any vertex i, there is some directed path ending at a vertex of some directed cycle. Hence, a minimum dominating set is of the form $\text{GD}(i)$ for some vertex i in some directed cycle.

In fact, for every cycle $(j_0, j_1, \ldots, j_t)$, by Lemma 5.25, all $\text{GD}(j_s)$ has the same size. Furthermore, we have the following result.

Lemma 5.27. *For every vertex v of a directed cycle C, $\text{GD}(v)$ is a minimum dominating set.*

Proof. Let C', C'' be two different directed cycles in H. Choose vertex $i \in V(C')$ and $j \in V(C'')$ such that $\{i, i+1, \ldots, j\}$ contains no vertex in $(V(C') \cup V(C''))\backslash\{i, j\}$. Let $\text{GD}(j) = \{j = j_1, j_2, \ldots, j_r\}$. For every $2 \leq s \leq r$, the set $\text{GD}(i)$ has a vertex i_s dominating $u(j_{s-1})$. By Lemma 5.24, $i_s \in S_{j_{s-1}}$. Since $S_{j_1}, S_{j_2}, \ldots, S_{j_r-1}$ are pairwise disjoint and i is not in any of these $r - 1$ sets, $|\text{GD}(i)| \geq r = |\text{GD}(j)|$. Similarly, $|\text{GD}(j)| \geq \text{GD}(i)|$ and so $|\text{GD}(i)| = |GD(j)|$. The lemma then follows. $\square$

The calculation of $\text{low}(i), u(i), \text{high}(i)$ and the constructions of $\text{GD}(i)$'s and H can be done in linear time. These give a linear-time algorithm for finding $\gamma(G)$ of a circular-arc graph G.

5.7   Exercises

(5.1) There are n classes, where the ith class starts from time a_i and ends at time b_i. Translate the following numbers into graph parameters.

  (a) The minimum number of classrooms needed for these classes.
  (b) The maximum number of classes that pairwise have no overlapping time.

(5.2) Show that the tree T with $V(T) = \{r, u_i, v_i : i = 1, 2, 3\}$ and $E(T) = \{ru_i, u_i v_i : i = 1, 2, 3\}$ is not an interval graph.

(5.3) Determine all trees which are also interval graphs.

(5.4) Show that $\overline{S}_3$ is not an interval graph.

(5.5) Determine all graphs with exactly one cycle which are interval graphs.

(5.6) An *asteroidal triple* is a set of three distinct vertices such that every two of them are connected by a path that avoids the neighbors of the remaining vertex. A graph is *AT-free* if it has no astroidal triples.

Show that an interval graph is AT-free.

(5.7) Show that matrix M_3 does not possess the consecutive 1's property for columns, where

$$M_3 = \begin{pmatrix} 1 & 1 & 1 & 1 & 1 \\ 1 & 0 & 0 & 0 & 0 \\ 0 & 1 & 0 & 0 & 0 \\ 0 & 0 & 1 & 0 & 0 \\ 0 & 0 & 0 & 1 & 0 \\ 0 & 0 & 0 & 0 & 1 \end{pmatrix}.$$

(5.8) Provide examples of $(0, 1)$-valued matrices having the consecutive 1's property for columns but are not clique matrices of interval graphs.

(5.9) Suppose A and B are two $(0, 1)$-valued $m \times n$ matrices. Show that if $A^T A = B^T B$, then either both A and B have the consecutive 1's property for columns or neither has it (Fulkerson and Gross, 1965).

(5.10) Suppose A and B are two $(0, 1)$-valued $m \times n$ matrices. Show that if $A^T A = B^T B$ and A does not contain either of the following submatrices, then $A = PB$ for some permutation matrix P (Ryser, 1969).

$$\begin{pmatrix} 1 & 1 & 1 \\ 1 & 0 & 0 \\ 0 & 1 & 0 \\ 0 & 0 & 1 \end{pmatrix}, \quad \begin{pmatrix} 0 & 0 & 0 \\ 0 & 1 & 1 \\ 1 & 0 & 1 \\ 1 & 1 & 0 \end{pmatrix}.$$

(5.11) Show that an interval graph is a proper interval graph if and only if it is $K_{1,3}$-free (Hint: For the sufficiency, choose an interval representation with a minimum number of interval pairs I_i and I_j, one properly includes the other.) (Roberts, 1969).

(5.12) Show that a chordal graph is a proper interval graph if and only if it is $\{K_{1,3}, S_3, \overline{S}_3\}$-free (Wegner, 1967).

(5.13) • An interval representation $\mathcal{F}$ of a graph is *over-unit* (respectively, *under-unit*) if $|I_i| \geq 1$ (respectively, $|I_i| \leq 1$) for all $I_i \in \mathcal{F}$.
  • The *total-length* of $\mathcal{F}$ is $\mathrm{Tlen}(G) = \sum\{|I_i| : I_i \in \mathcal{F}\}$.
  • The *infimum total-length* of an interval graph G over all *over-unit* interval representations $\mathcal{F}$ is $\mathrm{infTover}(G) = \inf\{\mathrm{Tlen}(\mathcal{F}) : \mathcal{F}$ is an over-unit interval representation of $G\}$.
  • The *supremum total-length* of an interval graph G over all *under-unit* interval representations $\mathcal{F}$ is $\mathrm{supTunder}(G) = \sup\{\mathrm{Tlen}(\mathcal{F}) : \mathcal{F}$ is an under-unit interval representation of $G\}$.
  • The *max-disjoint-union-length* of $\mathcal{F}$ is $\mathrm{mDul}(\mathcal{F}) = \max\{\mathrm{Tlen}(\mathcal{F}') : \mathcal{F}' \subseteq \mathcal{F}$ in which every pairs of intervals are disjoint$\}$.
  • The *infimum max-disjoint-union-length* of an interval graph G over all *over-unit* interval representations $\mathcal{F}$ is $\mathrm{infDover}(G) = \inf\{\mathrm{mDul}(\mathcal{F}) : \mathcal{F}$ is an over-unit interval representation of $G\}$.
  • The *supremum max-disjoint-union-length* of an interval graph G over all *under-unit* interval representations $\mathcal{F}$ is $\mathrm{supDunder}(G) = \sup\{\mathrm{mDul}(\mathcal{F}) : \mathcal{F}$ is an under-unit interval representation of $G\}$.

(a) Determine $\mathrm{infTover}(P_n)$, $\mathrm{supTunder}(P_n)$, $\mathrm{infDover}(P_n)$ and $\mathrm{supDunder}(P_n)$ for $n \geq 1$.
(b) Determine $\mathrm{infTover}(K_{1,n})$, $\mathrm{supTunder}(K_{1,n})$, $\mathrm{infDover}(K_{1,n})$ and $\mathrm{supDunder}(K_{1,n})$ for $n \geq 1$.
(c) For any interval graph G, determine $\mathrm{infTover}(G)$, $\mathrm{supTunder}(G)$, $\mathrm{infDover}(G)$ and $\mathrm{supDunder}(G)$.

(5.14) Discuss whether the algorithm obtained from Algorithm DomIntervalPD in Section 5.2 by replacing the statement "**if** $N[v_i] \cap D = \phi$ **then**" with "**if** $N(v_i) \cap D = \phi$ **then**", as shown as follows, is valid or not for the total domination problem in interval graphs.

Recall that a total 2-stable set of G is a subset $S \subseteq V(G)$ such that every two distinct vertices $u, v \in S$ are not adjacent to the same vertex of G, see Section 1.5. Note that the algorithm gives a minimum total dominating set for K_n with $n \geq 2$, but does not give a total 2-stable set for K_n with $n \geq 3$. The strong duality equality $\alpha_{t2}(G) = \gamma_t(G)$ is not always true for an interval graph G.

What can one say for a general interval graph?

Algorithm TotDomIntervalPD: Find a minimum total dominating set D and a maximum total 2-stable set S of an interval graph G with $\delta(G) \geq 1$, providing an interval ordering $[v_1, v_2, \ldots, v_n]$ is given.

$D \leftarrow \emptyset; \quad S \leftarrow \emptyset;$

for $i = 1$ **to** n **do**

if $N(v_i) \cap D = \phi$ **then**

$\{$ choose the maximum $j \gtrsim i;$

$\quad D \leftarrow D \cup \{v_j\};$

$\quad S \leftarrow S \cup \{v_i\};$

$\}$

$D \leftarrow \emptyset \quad S \leftarrow \emptyset;$

for $i = 1$ **to** n **do** label v_i unscanned;

for $i = 1$ **to** n **do**

if v_i is unscanned **then**

$\{$ choose the maximum $j \gtrsim i;$

$\quad D \leftarrow D \cup \{v_j\};$

$\quad S \leftarrow S \cup \{v_i\};$

$\quad$ label all neighbors of v_j scanned;

$\}$

(5.15) Suppose G is an interval graph with an interval ordering $[1, 2, \ldots, n]$. Design an algorithm to find $\text{low}(i)$, m_i and $\text{maxlow}(i)$ for $1 \leq i \leq n$.

(5.16) Prove or disprove that a connected interval graph has a minimum connected dominating set which is the vertex set of a path.

(5.17) Design an efficient algorithm for the clique domination problem in interval graphs.

(5.18) Design an efficient algorithm for the cycle domination problem in interval graphs.

(5.19) A *double dominating set* of a graph G is a subset $D \subseteq V(G)$ such that every vertex $v \in V(G) \backslash D$ is dominated by at least two vertices in D. Design an efficient algorithm for the double domination problem in interval graphs.

(5.20) Determine $\gamma_{\text{pr}}(P_2 \square P_n)$.

(5.21) Determine $\alpha_4(C_n)$ and $\gamma_{\text{pr}}^{\text{d}}(C_n)$ for $n \geq 3$.

(5.22) Determine $\alpha_4(P_n)$ and $\gamma_{\text{pr}}^{\text{d}}(P_n)$ for $n \geq 2$.

(5.23) Design an efficient algorithm for the paired disjunctive domination problem in trees.

(5.24) Determine the domatic number $d(P_2 \square P_n)$.

(5.25) Provide examples of graphs G whose domatic number $d(G) < \delta(G) + 1$.

(5.26) For a circular-arc representation of a graph using closed arcs, prove that we may assume that no pair of arcs share a common endpoint.

(5.27) Prove Theorem 5.22.

(5.28) Provide examples of circular-arc graphs which are not unit circular-arc graphs.

(5.29) Suppose G is a circular-arc graph with a circular-arc ordering $[1, 2, \ldots, n]$. Design an algorithm to find low(i), $u(i)$ and high(i) for $1 \leq i \leq n$.

Chapter 6

Strongly Chordal Graphs

6.1　Powers of chordal graphs

It is demonstrated in Section 1.5 that domination has close relation with clique cover. In particular, for any graph G,

$$\alpha_2(G) = \alpha(G^2) \leq \theta(G^2) \leq \gamma(G)$$

as shown in (1.2). Although C_4, C_5 and the 3-sun S_3 are examples of graphs G with $\theta(G^2) < \gamma(G)$, Theorem 1.4 shows that

$$\theta(G^2) = \gamma(G) \text{ for any } \{C_4, C_5, C_6, S_3\}\text{-free graph } G.$$

As chordal graphs are well studied in the perfect graph theory, it is natural to ask if the square of a chordal graph is also chordal. If this were the case, then for any S_3-free chordal graph G, first $\theta(G^2) = \gamma(G)$ can be computed by using Algorithm alpha-theta-Choral in Section 4.1, and secondly $\alpha_2(G) = \alpha(G^2) = \theta(G^2) = \gamma(G)$. However, as we will see below, squares of chordal graphs are not always chordal.

Powers of chordal graphs were investigated extensively by Chang (1982), Balakrishnan and Paulraja (1983), Farber (1983), Laskar and Shier (1983), Chang and Nemhauser (1984, 1985) and Duchet (1984), among others. In particular, Duchet proved that if G^k is chordal, then so is G^{k+2}.

This section derives some properties of powers of chordal graphs related to the domination problem. For the purpose of simplicity, new concepts of semi-edge, semi-walk and semi-chord etc are introduced.

A *semi-edge* of a graph G is a pair uv of vertices u and v such that $u = v$ or uv is an edge, i.e., $u \sim v$. For two vertices u and v in G, a u–v *semi-walk* of *length* r is a sequence $(u = v_0, v_1, \ldots, v_r = v)$ of vertices, where $v_{i-1}v_i$ is a semi-edge (that is, $v_{i-1} \sim v_i$) for $1 \leq i \leq r$. A semi-walk is *open* if $v_0 \neq v_r$; otherwise, it is *closed*. A closed semi-walk $(v_0, v_1, \ldots, v_r)$ is considered circularly, and so may also be written as $(v_i, v_{i+1}, \ldots, v_{r-2}, v_{r-1}, v_0, v_1, \ldots, v_{i-1}, v_i)$ for $1 \leq i \leq r-1$. In a semi-walk $(v_0, v_1, \ldots, v_r)$, a *semi-chord* is a semi-edge $v_i v_j$ with $|i - j| \geq 2$, except for the case of $v_0 = v_r$ the semi-edges $v_0 v_r, v_0 v_{r-1}, v_1 v_r$ are not semi-chords.

With these new terms in mind, we are now ready to begin the derivation. First, a useful property for chordal graphs in terms of semi-walks.

Lemma 6.1. *If* $W = (v_0, v_1, v_2, \ldots, v_r)$ *is a closed semi-walk of length* $r \geq 3$ *in a chordal graph* G, *then for any* i *there exists some index* $j \neq i, i+1$ *such that* $v_j \sim v_i$ *and* $v_j \sim v_{i+1}$, *where the indices are taken modulo* r.

For the case when W *is a cycle,* v_j *is adjacent to both* v_i *and* v_{i+1}.

Proof.   The proof is by induction on r. If $r = 3$, then $j = i+2$ is an index with $v_j \sim v_i$ and $v_j \sim v_{i+1}$ as desired.

Suppose $r \geq 4$ and the lemma is true for all $r' < r$. If W has a semi-chord $v_p v_q$, then it divides W into two shorter closed semi-walks each of length at least 3. One of the two closed semi-walks, say W', contains $v_i v_{i+1}$ as a semi-edge. By the induction hypothesis, in W' there is some index $j \neq i, i+1$ such that $v_j \sim v_i$ and $v_j \sim v_{i+1}$.

Now suppose that W has no semi-chords. If $v_i = v_{i'}$ for some $i \neq i'$, then $v_i v_{i'}$ is a semi-edge and so it is only possible that $i' = i \pm 1$, say $i' = i+1$ by symmetry. Then $j = i+2$ is an index with $v_j \sim v_i$ and $v_j \sim v_{i+1}$ as desired. Therefore, W is a cycle of length $r \geq 4$ without chords, contradiction to that G is chordal. The lemma then follows.   □

As a consequence, there is a new characterization of chordal graphs in terms of semi-walks.

Theorem 6.2 (An equivalent definition of a chordal graph). *A graph is chordal if and only if every closed semi-walk of length at least 4 has a semi-chord.*

Proof.   If the graph is chordal, then by Lemma 6.1 in any closed semi-walk $(v_0, v_1, v_2, \ldots, v_r)$ of length $r \geq 4$, there exits some $j \neq 1, 2$ such that $v_j \sim v_1$ and $v_j \sim v_2$. Either $v_1 v_j$ or $v_2 v_j$ is a semi-chord as desired.

The converse is also true, since a cycle is a closed semi-walk and a chord is a semi-chord.　　　□

Suppose G is a graph and $S_1, S_2, \ldots, S_m$ are m subsets of $V(G)$. Denote by $G(S_1, S_2, \ldots, S_m)$ the graph whose vertex set is $\{S_1, S_2, \ldots, S_m\}$ and edge set is

$$\{S_i S_j : i \neq j, x \sim y \text{ in } G \text{ for some } x \in S_i \text{ and some } y \in S_j\}.$$

Lemma 6.3. *If G is a chordal graph in which S_i is a subset of $V(G)$ with $G[S_i]$ connected for $1 \leq i \leq m$, then $G(S_1, S_2, \ldots, S_m)$ is chordal.*

Proof. Suppose $C = (S_{i_0}, S_{i_1}, \ldots, S_{i_r})$ is a cycle of length $r \geq 4$ in the graph $G(S_1, S_2, \ldots, S_m)$, where $i_0 = i_r$. Choose vertices $x_j, y_j \in S_{i_j}$ such that $x_j \sim y_{j+1}$ for $0 \leq j \leq r-1$, where indices are taken modulo r. Choose a shortest y_j-x_j path P_j in $G[S_{i_j}]$ for $0 \leq i \leq r-1$. Consider the closed semi-walk $W = (P_0, P_1, \ldots, P_{r-1}, y_0)$ in G of length at least 4. Assume that x_j's, y_j's and P_j's are chosen so that W is of minimum length.

By Theorem 6.2, W has a semi-chord uv. Assume that $u \in V(P_j)$ and $v \in V(P_k)$.

First, $j \neq k$ since P_j is a shortest path. And then $S_{i_j} S_{i_k}$ is an edge in the graph $G(S_1, S_2, \ldots, S_m)$.

Suppose $k = j \pm 1$. By symmetry, assume that $k = j + 1$. Then $u \in V(P_j)\backslash\{x_j\}$ or $v \in V(P_{j+1})\backslash\{y_{j+1}\}$. One may replace x_j by u and y_{j+1} by v to shorten W, contradicting the minimality of the length of W.

In conclusion, $k \notin \{j-1, j, j+1\}$ and so $S_{i_j} S_{i_k}$ is a chord of C. This verifies the lemma.　　　□

Theorem 6.4 (Duchet, 1984). *If G^k is chordal, then so is G^{k+2}.*

Proof. If $V(G) = \{v_1, v_2, \ldots, v_n\}$, then the theorem follows from Lemma 6.3 and the fact that $G^{k+2} = G^k(N_G[v_1], N_G[v_2], \ldots, N_G[v_n])$.　　　□

Corollary 6.5. *The odd powers of a chordal graph are chordal.*

Although the odd powers of a chordal graph are chordal, the square of a chordal graph may not be chordal. Figure 6.1 shows two chordal graphs whose squares are not chordal.

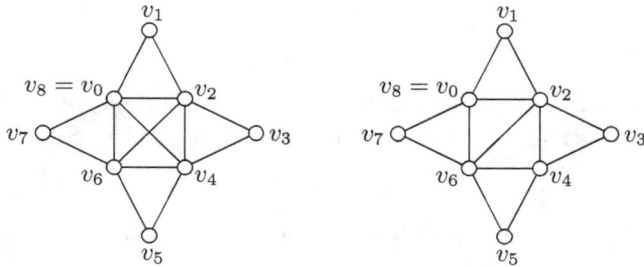

Fig. 6.1.   Two examples of chordal graphs whose squares are not chordal.

More general examples are as follows. For every integer $n \geq 3$, an *n-sun* is a chordal graph with a hamiltonian cycle $(v_0, v_1, v_2, v_3, \ldots, v_{2n-1}, v_{2n})$ such that $(v_0, v_2, \ldots, v_{2n-2}, v_{2n})$ is a cycle and $\deg(v_i) = 2$ for odd i with $1 \leq i \leq 2n-1$. The vertices v_i's with even i are called the *inner vertices* and the vertices v_i's with odd i are the *outer vertices* of the *n-sun*. A *complete n-sun* is an *n-sun* in which the set $\{v_0, v_2, \ldots, v_{2n-2}\}$ of inner vertices is a clique. The square of an *n-sun* contains a chordless *n-cycle*.

Suns were studied by Chang (1982) and Chang and Nemhauser (1984, 1985) for the domination problem. Suns are called sunflowers by Laskar and Shier (1983) and incomplete trampoline by Farber (1983).

Figure 6.1 shows two 4-suns. The left one is a complete 4-sun. The 4-sun at right is not complete. It contains two 3-suns.

In a cycle $(u_0, u_1, u_2, u_3, \ldots, u_{2n-1}, u_{2n})$ of even length, a chord $u_i u_j$ is called an *even-odd chord* (respectively, *odd-odd chord*) if one of i and j is even and the other is odd (respectively, both i and j are odd).

In the definition of an *n-sun*, the condition that $\deg(v_i) = 2$ for odd i with $1 \leq i \leq 2n-1$ implies that the cycle $(v_0, v_1, v_2, v_3, \ldots, v_{2n-1}, v_{2n})$ has no even-odd chords and no odd-odd chords. In fact, under the chordality of the graph, only the condition of without even-odd chords is essential, as shown below.

Theorem 6.6 (Chang (1982) Equivalent definition 1 of an n-sun).
For $n \geq 3$, a chordal graph G is an n-sun if and only if it has a hamiltonian cycle $W = (u_0, u_1, u_2, u_3, \ldots, u_{2n-1}, u_{2n})$ without even-odd chords.

Proof.   It only needs to prove the necessity.

Suppose G has a hamiltonian cycle $W = (u_0, u_1, u_2, u_3, \ldots, u_{2n-1}, u_{2n})$ without even-odd chords. By Lemma 6.1, there is some index $j \neq 1, 2$ such that u_j is adjacent to both u_1 and u_2. Since G has no even-odd chords, either $j = 0$ or $j = 3$. For the case of $j = 0$, let $v_i = u_i$ for all i; for the case of $j = 3$, let $v_i = u_{i+1}$ for all i. Then $(v_0, v_1, v_2, v_3, \ldots, v_{2n-1}, v_{2n})$ is a hamiltonian cycle of G without even-odd chords. It is also the case that $v_0 v_2$ is an edge.

Let r be the largest integer such that $(v_0, v_2, v_4, \ldots, v_{2r})$ is a walk. Then $r = n$ and so $(v_0, v_2, v_4, \ldots, v_{2n})$ is a cycle. Suppose to the contrary that $r < n$. Then in the cycle $(v_0, v_2, v_4, \ldots, v_{2r}, v_{2r+1}, v_{2r+2}, \ldots, v_{2n-1}, v_{2n})$ there exists some v_j adjacent to both v_{2r} and v_{2r+1} by Lemma 6.1. As W has no even-odd chords, $j = 2r + 2$. This gives that $(v_0, v_2, v_4, \ldots, v_{2r+2})$ is longer than $(v_0, v_2, v_4, \ldots, v_{2r})$, which contradicts the maximality of r.

Next prove that $\deg(v_i) = 2$ for all odd i with $1 \le i \le 2n - 1$. Suppose to the contrary that $\deg(v_i) \ge 3$ for some odd i. Then there exists a minimum index j such that $v_i v_j$ is a chord. As W has no even-odd chord, j is odd. By symmetry, assume that $i \ne j + 2$. Then the cycle $(v_i, v_{i+1}, v_{i+3}, v_{i+5}, \ldots, v_{j-3}, v_{j-1}, v_j, v_i)$ is a cycle of length at least 4. By Lemma 6.1, there is some even k with $i + 1 \le k \le j - 1$ such that $v_k v_i$ or $v_k v_j$ is an edge. One of them is an even-odd chord of W, a contradiction. $\square$

Moreover, in the definition of an n-sun, the hamiltonian cycle can be replaced by a hamiltonian semi-walk, which is a closed semi-walk passing through all vertices of the graph. In this definition, a vertex of the semi-walk may appear more than once. More precisely, for integer $n \ge 3$, an *n-semi-sun* is a graph having a hamiltonian semi-walk $(v_0, v_1, v_2, v_3, \ldots, v_{2n-1}, v_{2n})$ with no even-odd semi-chords.

Theorem 6.7 (Chang (1982) Equivalent definition 2 of an n-sun). *A chordal graph is an n-sun if and only if it is an n-semi-sun.*

Proof. An n-sun is clearly an n-semi-sun, since a hamiltonian cycle is a hamiltonian semi-walk and even-odd chords are even-odd semi-chords.

Conversely, suppose the chordal graph is an n-semi-sun. Then it has a hamiltonian semi-walk $W = (v_0, v_1, v_2, v_3, \ldots, v_{2n-1}, v_{2n})$ with no even-odd semi-chords.

We claim that W is in fact a hamiltonian cycle and so the graph is an n-sun. Suppose to the contrary that $v_i = v_j$ for some $i \ne j$, say $i < j$.

Case 1. i and j have the same parity.

First, $v_i = v_j$ and $v_j \sim v_{j-1}$ imply $v_i \sim v_{j-1}$, where $i < j - 1 < j$. Since W has no even-odd semi-chords, $i + 1 = j - 1$ and so $j = i + 2$.

Also, $v_i = v_j$ and $v_j \sim v_{j+1}$ imply $v_i \sim v_{j+1} = v_{i+3}$. This gives that $v_i v_{i+3}$ is an even-odd semi-chord, a contradiction.

Case 2. i and j have different parities.

Since W has no even-odd semi-chords, $j = i+1$ (indices in this case are taken modulo $2n$). Then $v_i = v_{i+1} \sim v_{i+2}$. Deleting v_{i+1} from W results a closed semi-walk W' in which v_i and v_{i+2} are consecutive. By Lemma 6.1, W' has a vertex $v_k \sim v_i$ and $v_k \sim v_{i+2}$, where $k \notin \{i, i+1, i+2\}$. If k has a different parity from i, then either $v_k v_i$ or $v_k v_{i+2}$ is an even-odd semi-chord of W, a contradiction. If k has the same parity as i, then $k \notin \{i-1, i\}$ leads to that $v_k v_{i+1}$ is an even-odd semi-chord of W, a contradiction.

In conclusion, W is a hamiltonian cycle and so the graph is an n-sun.

$\square$

The equivalent definitions of an n-sun are quite useful as can be seen in the following.

First, a result weaker than Theorem 1.4, which is a bridge connecting domination and the perfect graph theory.

Theorem 6.8. *If G is a 3-sun-free chordal graph, then every clique C in G^2 is a subset of $N_G[v]$ for some vertex v. Consequently, $\theta(G^2) = \gamma(G)$.*

Proof. Suppose to the contrary that G^2 has a clique $C \nsubseteq N_G[v]$ for every vertex v. Choose such a clique C of minimum cardinality. Then $|C| \geq 3$, since every two vertices of distance at most 2 is in a same closed neighborhood. Choose three vertices v_0, v_2, v_4 from C. For $i \in \{0, 2, 4\}$, by the minimality of C, there is some vertex v_{i+3} with $C \setminus \{v_i\} \subseteq N_G[v_{i+3}]$ but $v_i \nsim v_{i+3}$, where indices are taken modulo 6. This gives a closed semi-walk $(v_0, v_1, v_2, v_3, v_4, v_5, v_0)$ with no even-odd semi-chords, which gives a 3-semi-sun, and so is a 3-sun by Theorem 6.7, a contradiction. Hence, $C \subseteq N_G[v]$ for some vertex v.

It is known that $\theta(G^2) \leq \gamma(G)$. Also, $\theta(G^2) \geq \gamma(G)$ follows from that a clique C in G^2 is dominated by some vertex in G. Hence, $\theta(G^2) = \gamma(G)$.

$\square$

Theorem 6.9. *If G is a sun-free chordal graph, then G^2 is chordal.*

Proof. It suffices to prove that a cycle $C = (v_0, v_2, \ldots, v_{2n-2}, v_{2n})$ of length $n \geq 4$ in G^2 has a chord. For $0 \leq i \leq n-1$, since $d_G(v_{2i}, v_{2i+2}) \leq 2$, there is some v_{2i+1} such that $v_{2i} \sim v_{2i+1}$ and $v_{2i+1} \sim v_{2i+2}$ in G. Then $W =$

$(v_0, v_1, v_2, v_3, \ldots, v_{2n-1}, v_{2n})$, obtained from C by inserting odd indexed v_i, is a closed semi-walk in G. Since G has no n-suns, by Theorem 6.7, G has no n-semi-suns. Hence, W has an even-odd semi-chord $v_i v_j$. Without loss of generality assume that i is even and j is odd. Since $v_i \sim v_j$, $v_j \sim v_{j-1}$ and $v_j \sim v_{j+1}$ in G, both $v_i v_{j-1}$ and $v_i v_{j+1}$ are edges in G^2. One of them is a chord of C. Hence, G^2 is chordal. $\qquad\square$

Suppose G is a sun-free chordal graph. By Theorem 6.8, $\theta(G^2) = \gamma(G)$. By Theorem 6.9, G^2 is chordal and so $\alpha(G^2) = \theta(G^2)$. So, the domination problem in a sun-free chordal graph G is transformed to the clique cover problem in the chordal graph G^2, which can be computed in linear time as shown in Chapter 4.

The above derivations from domination to clique cover can also be extended to a distance variation of domination as shown by Chang (1982) and Chang and Nemhauser (1984, 1985).

Recall that for a positive integer k, a distance k-dominating set of a graph G is a subset D of $V(G)$ such that every vertex in $V(G)$ is within distance k from some vertex in D.[1] The distance k-domination number $\gamma_{\text{disk}}(G)$ is the minimum size of a distance k-dominating set D in G. A natural dual of distance k-domination is $2k$-stability. Namely, a $2k$-*stable set* of G is a subset S of $V(G)$ for which every two distinct vertices u and v in S has $d(u, v) > 2k$. It is easy to see the weak duality inequalities that for any graph G,

$$\alpha_{2k}(G) = \alpha(G^{2k}) \leq \theta(G^{2k}) \leq \gamma_k(G). \tag{6.1}$$

In the following, it is shown that if G is sun-free chordal then so is G^{2k} and the inequalities in (6.1) are equalities. Therefore, the distance k-domination problem in sun-free chordal graphs can be solved by transforming to the clique cover problem in the chordal graph G^{2k}.

The following gives an equivalent definition of a sun-free chordal graph, which makes many derivations easier. In fact it can also establish other related results.

---

[1]In the papers by Chang and Nemhauser, the term k-domination were used. As there are many other papers already used k-domination for a many-neighbor variation of domination, this book now uses the term distance k-domination for their work.

Lemma 6.10. *The n-cycle C_n with $n \geq 4$ is an m-semi-sun for some odd m.*

Proof.   Suppose the n-cycle C_n has vertex set $V(C_n) = \{v_0, v_1, \ldots, v_{n-1}\}$ and edge set $E(C_n) = \{v_0v_1, v_1v_2, \ldots, v_{n-1}v_n\}$, where $v_n = v_0$.

Four cases are considered as follows.

If $n = 4k$ with $k \geq 1$, then C_n is a $(2k+1)$-semi-sun, since the hamiltonian semi-walk $(v_0, v_0, v_1, v_2, v_2, v_3, v_4, \ldots, v_{n-1}, v_n)$ has no even-odd semi-chords.

If $n = 4k + 1$ with $k \geq 1$, then C_n is a $(2k+1)$-semi-sun, since the hamiltonian semi-walk $(v_0, v_0, v_1, v_2, \ldots, v_{n-1}, v_n)$ has no even-odd semi-chords.

If $n = 4k + 2$ with $k \geq 1$, then C_n is a $(2k+1)$-semi-sun, since the hamiltonian semi-walk $(v_0, v_1, \ldots, v_{n-1}, v_n)$ has no even-odd semi-chords.

If $n = 4k + 3$ with $k \geq 1$, then C_n is a $(2k+3)$-semi-sun, since the hamiltonian semi-walk $(v_0, v_0, v_1, v_2, v_2, v_3, v_4, v_4, v_5, v_6, \ldots, v_{n-1}, v_n)$ has no even-odd semi-chords.   □

Theorem 6.11. *A graph is sun-free chordal if and only if it is semi-sun-free.*

Proof.   This follows from Theorem 6.7 and Lemma 6.10.   □

Theorem 6.12. *A graph is odd-sun-free chordal if and only if it is odd-semi-sun-free.*

Proof.   This follows from Theorem 6.7 and Lemma 6.10.   □

Finally, the powers of sun-free chordal graphs are also sun-free chordal. The following result was proved by Lubiw (1982, 1987) in terms of totally balanced matrices and Γ-free matrices.

Theorem 6.13 (Lubiw, 1982, 1987). *If G is a sun-free chordal graph, then G^k is also sun-free chordal for every positive integer k.*

Proof.   First prove the case of $k = 2$. By Theorem 6.9, G^2 is chordal. Suppose G^2 has an n-sun $C = (v_0, v_1, v_2, v_3, \ldots, v_{2n-1}, v_{2n})$, where vertices of odd indices are outer vertices. For $0 \leq i \leq n-1$, the set $\{v_{2i}, v_{2i+1}, v_{2i+2}\}$

is a clique in G^2. By Theorem 6.8, $\{v_{2i}, v_{2i+1}, v_{2i+2}\} \subseteq N_G[v'_{2i+1}]$ for some vertex v'_{2i+1}. Then $W = (v_0, v'_1, v_2, v'_3, \ldots, v'_{2n-1}, v_{2n})$, obtained from C by replacing odd indexed v_i with v'_i, is a closed semi-walk in G. Since G has no n-suns, by Theorem 6.7, G has no n-semi-suns. Hence, W has an even-odd semi-chord $v'_i v_j$, where i is odd and j is even. Since $v_i \sim v'_i$ and $v'_i \sim v_j$ in G, then $v_i \sim v_j$ in G^2 and so $v_i v_j$ is a chord of C, a contradiction. Hence, G^2 has no n-suns, and so G^2 is sun-free chordal.

Suppose $k \geq 3$ and the theorem is true for $k' < k$.

Case 1. $k = 2k'$ is even. In this case $k' < k$.

By the induction hypothesis, $G^{k'}$ is sun-free chordal and then $(G^{k'})^2 = G^k$ is sun-free chordal.

Case 2. $k = 2k' - 1$ is odd. In this case, $k' < k$.

Construct the graph H with vertex set $V(H) = \{v, v' : v \in V(G)\}$ and edge set $E(H) = E(G) \cup \{vv' : v \in V(G)\}$. Note that every vertex v' with $v \in V(G)$ has degree 1, which can not be in any cycle of H. Since G is sun-free chordal, so is H. By Case 1, $H^{k+1} = H^{2k'}$ is sun-free chordal and then by Theorem 6.11 it is semi-sun-free.

For every two vertices v and u in $V(G)$, $d_H(v', u) = 1 + d_G(v, u)$. Then G^k has an n-semi-sun $(u_0, u_1, u_2, u_3, \ldots, u_{2n-1}, u_{2n})$ if and only if H^{k+1} has an n-semi-sun $(u_0, u'_1, u_2, u'_3, \ldots, u'_{2n-1}, u_{2n})$. Hence, G^k is also semi-sun-free and so is sun-free chordal. $\square$

Theorem 6.14. *If G is a sun-free chordal graph, then $\theta(G^{2k}) = \gamma_k(G)$.*

Proof. By Theorem 6.13, G^k is sun-free chordal. Then by Theorem 6.8, $\theta((G^k)^2) = \gamma(G^k)$. Therefore, $\theta(G^{2k}) = \theta((G^k)^2) = \gamma(G^k) = \gamma_k(G)$. $\square$

6.2 Strongly chordal graphs and strong elimination orderings

Transforming the k-domination problem in sun-free chordal graphs to the clique cover problem in chordal graphs in the previous section has several drawbacks.

(1) To find a k-dominating set of G from a clique cover of G^{2k}, it needs to find a vertex to k-dominate a clique of G^{2k}. It is time consuming.

(2) The transformation from G to G^{2k} is not linear in $|V(G)| + |E(G)|$.

(3) The method does not work for the weighted case.

From this point of view, a direct approach for the domination problem in sun-free chordal graphs is desirable. This was done by Farber (1984) and Hoffman *et al.* (1985). Farber used a method similar to tree orderings or perfect elimination orderings. Hoffman *et al.* (1985) treated the problem as a general covering problem in terms of matrices. They all used linear programming methods.

Recall that the primal-dual approach for the domination problem in trees relies on the fact that for a leaf v, its neighbor u is more powerful than v in the sense that u dominates all vertices dominated by v. This cannot be extended to a simplicial vertex v of a chordal graph, as v may have many neighbors whose closed neighborhoods may not be comparable. In that case, it is not clear which neighbor of v is most powerful to dominate it and other vertices. This is also reflected by the fact that the domination problem is NP-complete for chordal graphs.

However, if a simplicial vertex v has the additional condition that the closed neighborhoods of its two neighbors are comparable, then an algorithm similar to that for trees is possible.

More precisely, a *simple vertex* of a graph G is a vertex v such that for every two $u, u' \in N[v]$ either $N[u] \subseteq N[u']$ or $N[u'] \subseteq N[u]$. A *maximal neighbor* of a simple vertex v is a vertex $u' \in N[v]$ such that $N[u] \subseteq N[u']$ for all $u \in N[v]$.

Note that a leaf is a simple vertex. Also, a simple vertex is a simplicial vertex. To see this, suppose v is a simple vertex and $u, u' \in N[v]$. Then either $N[u] \subseteq N[u']$ or $N[u'] \subseteq N[u]$. For either case, $u \sim u'$. Thus v is a simplicial vertex.

To see an argument parallel to trees, consider the following condition:

every induced subgraph G has a simple vertex.

If G has this property, its vertex set $V(G)$ can be ordered as $[v_1, v_2, \ldots, v_n]$ such that

for $1 \le i \le n$, vertex v_i is a simple vertex of $G[\{v_i, v_{i+1}, \ldots, v_n\}]$.

This is called a *simple elimination ordering*. There is then an algorithm parallel to Algorithm DomTreePR as follows.

Algorithm DomSimplePD: Find a minimum dominating set D and a maximum 2-stable set S of a graph G with a simple elimination ordering $[v_1, v_2, \ldots, v_n]$.

$D \leftarrow \emptyset; \quad S \leftarrow \emptyset;$
for $i = 1$ **to** n **do**
if $N[v_i] \cap D = \emptyset$ **then**
$\{ \qquad$ find a maximal neighbor v_{m_i} of v_i in $G[\{v_i, v_{i+1}, \ldots, v_n\}]$;
$\qquad\quad D \leftarrow D \cup \{v_{m_i}\};$
$\qquad\quad S \leftarrow S \cup \{v_i\};$
$\}$

To verify the algorithm, it is sufficient to prove that D is a dominating set, S is a 2-stable set, and $|D| \leq |S|$.

First, D is clearly a dominating set which is guaranteed by the if-then statement.

Next, suppose S is not a 2-stable set, i.e., there exist v_i and v_j in S such that $i < j$ but $d_G(v_i, v_j) \leq 2$. Let $G_i = G[\{v_i, v_{i+1}, \ldots, v_n\}]$. Since $d(v_i, v_j) \leq 2$, a shortest v_i-v_j path P is either (v_i, v_j) or (v_i, v_k, v_j).

If $P = (v_i, v_j)$, then $v_j \in N_{G_i}[v_i] \subseteq N_{G_i}[m_i]$ and so $m_i \in N[v_j]$.

If $P = (v_i, v_k, v_j)$ and $k < i$, then $v_i, v_j \in N_{G_k}[v_k]$. Since v_k is a simple vertex and so a simplicial vertex of G_k, vertex v_i is adjacent to v_j, contradicting the assumption that P is shortest.

If $P = (v_i, v_k, v_j)$ and $i \leq k$, then $v_j \in N_{G_i}[v_k] \subseteq N_{G_i}[m_i]$ and so $m_i \in N[v_j]$.

In either case, $m_i \in N[v_j]$. Thus, at end of iteration i, D contains m_i. When the algorithm processes v_j later, $N[v_j] \cap D \neq \emptyset$ which causes that S does not contain v_j. This proves that S is a 2-stable set.

Finally, the inequality $|D| \leq |S|$ follows from that when m_i is added into D, which may or may not already be in D, a new vertex v_i is always added to S.

In conclusion, $|S| \leq \alpha_2(G) \leq \gamma(G) \leq |D| \leq |S|$ and so all inequalities are in fact equalities. Consequently, the following theorem holds.

Theorem 6.15. *Algorithm DomSimplePD gives a minimum dominating set D and a maximum 2-stable set S of a graph G with a simple elimination ordering. Moreover, the strong duality equality $\alpha_2(G) = \gamma(G)$ holds.*

The arguments above leave two important issues to be resolved:

(1) an efficient way to find a simple elimination ordering,
(2) an efficient way to find a maximal neighbor of every vertex in the graph.

In the following, a vertex ordering with a stronger property than a simple elimination ordering is investigated. More precisely, a *strong elimination ordering* is a vertex ordering $[v_1, v_2, \ldots, v_n]$ satisfying the following condition.

$$\text{If } i \leq j \leq k \text{ and } v_j, v_k \in N_i[v_i], \text{ then } N_i[v_j] \subseteq N_i[v_k], \qquad \text{(SEO)}$$

where $N_i[v_j] = \{v_r : r \geq i, v_r \sim v_j\}$.

Note that if $G_i := G[\{v_i, v_{i+1}, \ldots, v_n\}]$, then $N_i[v_j] = N_{G_i}[v_j]$. By the definition, a strong elimination ordering is a simple elimination ordering. And the vertex $v_{m_i} \in N_i[v_i]$ with a maximum index m_i is a maximal neighbor of v_i in G_i.

Farber (1984) defined *strongly chordal graphs* as those admitting a strong elimination ordering, and used strong elimination orderings to develop an efficient algorithm for the weighted domination problem. In fact his definition of the property for a strongly chordal graph looks slightly different from (SEO), but it is not hard to verify that they are equivalent. His definition of a strong elimination ordering is a vertex ordering $[v_1, v_2, \ldots, v_n]$ such that the following two conditions hold.

(SEO-a) If $i < j < k$ and $v_i v_j, v_i v_k \in E(G)$, then $v_j v_k \in E(G)$.
(SEO-b) If $i < j < k < \ell$ and $v_i v_k, v_i v_\ell, v_j v_k \in E(G)$, then $v_j v_\ell \in E(G)$.

He proved that the two conditions (SEO-a) and (SEO-b) are equivalent to a single condition (SEO-c), which is easier to use.

(SEO-c) If $i \leq j$, $k \leq \ell$, $i \sim k$, $i \sim \ell$ and $j \sim k$, then $j \sim \ell$.

(SEO-c) is equivalent to (SEO-c$'$), as (SEO-c) holds for $i = j$ or $k = \ell$.

(SEO-c$'$) If $i < j$, $k < \ell$, $i \sim k$, $i \sim \ell$ and $j \sim k$, then $j \sim \ell$.

Strong elimination ordering can also be interpreted from the matrix point of view. For a graph G with a vertex ordering $[v_1, v_2, \ldots, v_n]$, consider its *closed neighborhood matrix* $M = [a_{i,j}]_{n \times n}$ defined by

$$a_{i,j} = \begin{cases} 1, & \text{if } v_i \sim v_j; \\ 0, & \text{if } v_i \not\sim v_j. \end{cases}$$

For any 2×2 submatrix

$$M' = \begin{bmatrix} a_{i,k} & a_{i,\ell} \\ a_{j,k} & a_{j,\ell} \end{bmatrix}$$

of M obtained from rows $i < j$ and columns $k < \ell$, condition (SEO-c$'$) is the same as that if $a_{i,k} = a_{i,\ell} = a_{j,k} = 1$ then $a_{j,\ell} = 1$. In other words, M has no submatrix of the form

$$\begin{bmatrix} 1 & 1 \\ 1 & 0 \end{bmatrix},$$

which resembles the Greek letter Γ if 0 is ignored. Hoffman *et al.* (1985) called it a *Gamma matrix*. They used a linear programming method to solve covering problem whose corresponding matrix is Γ-free.

In the following, it is shown that a graph being sun-free chordal, having a simple vertex ordering, or having a strong elimination ordering are the same. First, a useful lemma.

Lemma 6.16. *Any graph G has a vertex ordering $[v_1, v_2, \ldots, v_n]$ satisfying the following condition.*

If $i < j$ and $v_k \in N[v_i] \backslash N[v_j]$, then $v_\ell \in N[v_j] \backslash N[v_i]$ for some $\ell > k$. (SLex)

Proof. For any ordering $\sigma = [v_1, v_2, \ldots, v_n]$, define the vector

$$A(\sigma) = (a_{n,n}, a_{n,n-1}, a_{n-1,n}, a_{n,n-2}, a_{n-1,n-1}, a_{n-2,n}, \ldots, a_{2,1}, a_{1,2}, a_{1,1})$$

of length n^2, where $a_{r,s}$ is before $a_{r',s'}$ when $r+s > r'+s'$ or $r+s = r'+s'$ with $r > r'$, and $a_{r,s} = 1$ when $v_r \in N[v_s]$ and $a_{r,s} = 0$ otherwise. Choose an ordering σ such that $A(\sigma)$ is the lexicographically largest.

Suppose $i < j$ and $v_k \in N[v_i] \backslash N[v_j]$. Choose the largest index $m \geq k$ such that $v_m \in N[v_i] \backslash N[v_j]$. Then $a_{m,i} = a_{i,m} = 1$ and $a_{m,j} = a_{j,m} = 0$. If there is some $\ell > m$ with $v_\ell \in N[v_j] \backslash N[v_i]$, then the lemma holds.

Otherwise, by the maximality of m,

$$a_{i,r} = a_{j,r} \quad \text{for all } r > m. \tag{6.2}$$

Then consider the ordering σ' obtained from σ by interchanging v_i and v_j. It follows that

$$a'_{r,s} = a_{r,s} \quad \text{for all } r, s \notin \{i, j\}; \quad a'_{i,r} = a_{j,r} \quad \text{and} \quad a'_{j,r} = a_{i,r} \quad \text{for all } r. \tag{6.3}$$

By (6.2) and (6.3), $a'_{r,s} = a_{r,s}$ for all $a_{r,s}$ before $a_{j,m}$ and $a_{m,j}$; while $a'_{j,m} = a'_{m,j} = 1 > 0 = a_{j,m} = a_{m,j}$, a contradiction to the lexicographical maximality of σ. $\qquad\qquad\square$

Theorem 6.17. *The following are equivalent for every graph G.*

(1) *The graph G has a strong elimination ordering.*
(2) *The graph G has a simple elimination ordering.*
(3) *Every induced subgraph of G has a simple vertex.*
(4) *The graph G is sun-free chordal.*

Proof. (1) $\Rightarrow$ (2). This follows from that a strong elimination ordering is a simple elimination ordering.

(2) $\Rightarrow$ (3). Suppose G has a simple elimination ordering $[v_1, v_2, \ldots, v_n]$. For any induced subgraph H of G, choose the vertex v_i of H with the minimum index i. Then for $v_j, v_k \in N_H(v_i)$, where $j \geq i$ and $k \geq i$, either $N_{G_i}[v_j] \subseteq N_{G_i}[v_k]$ or else $N_{G_i}[v_k] \subseteq N_{G_i}[v_j]$. Consequently, either $N_H[v_j] = N_{G_i}[v_j] \cap V(H) \subseteq N_{G_i}[v_k] \cap V(H) = N_H[v_k]$ or else $N_H[v_k] = N_{G_i}[v_k] \cap V(H) \subseteq N_{G_i}[v_j] \cap V(H) = N_H[v_j]$. This gives that v_i is a simple vertex of H.

(3) $\Rightarrow$ (4). This follows from that a chordless cycle of length at least 4 and an n-sun have no simple vertices.

(4) $\Rightarrow$ (1). Suppose G is sun-free chordal. By Lemma 6.16, G has a vertex ordering $\sigma = [v_1, v_2, \ldots, v_n]$ satisfying (SLex). We first prove the following claim.

Claim. G has no semi-walk $W = (v_{i_0}, v_{i_1}, v_{i_2}, v_{i_3}, \ldots, v_{i_{2n-2}}, v_{i_{2n-1}})$ with $n \geq 2$ such that the following conditions hold.

(a) $i_0 > i_j$ for all even $j \neq 0$.
(b) $i_{2n-1} > i_j$ for all odd $j \neq 2n - 1$.
(c) W has no even-odd semi-chord $v_{i_j} v_{i_k}$ with $|j - k| \geq 3$.

Proof of the claim. Suppose to the contrary that such a semi-walk exists. Choose one with largest $i_0 + i_{2n-1}$.

Since $i_{2n-2} < i_0$ and $v_{i_{2n-1}} \in N[v_{i_{2n-2}}] \backslash N[v_{i_0}]$, by (SLex), there exists some $i_{-1} > i_{2n-1}$ such that $v_{i_{-1}} \in N[v_{i_0}] \backslash N[v_{i_{2n-2}}]$.

Similarly, since $i_1 < i_{2n-1}$ and $v_{i_0} \in N[v_{i_1}] \backslash N[v_{i_{2n-1}}]$, by (SLex), there exists some $i_{2n} > i_0$ such that $v_{i_{2n}} \in N[v_{i_{2n-1}}] \backslash N[v_{i_1}]$.

Consider the semi-walk $W' = (v_{i_{-1}}, v_{i_0}, v_{i_1}, v_{i_2}, v_{i_3}, \ldots, v_{i_{2n-2}}, v_{i_{2n-1}}, v_{i_{2n}})$.

If W' has an even-odd chord $v_{i_{-1}} v_{i_{2j}}$ for some even $2j$ such that $0 < 2j < 2n - 2$, then choose the maximum such $2j$ and consider $W'' = (v_{i_{-1}}, v_{i_{2j}}, v_{i_{2j+1}}, \ldots, v_{i_{2n-2}}, v_{i_{2n-1}}, v_{i_{2n}})$.

If W'' has an odd-even chord $v_{i_{2k+1}} v_{i_{2n}}$ for some odd $2k + 1$ such that $2j < 2k + 1 < 2n - 1$, then choose the minimum such $2k + 1$ and consider $W''' = (v_{i_{-1}}, v_{i_{2j}}, v_{i_{2j+1}}, \ldots, v_{i_{2k}}, v_{i_{2k+1}}, v_{i_{2n}})$.

Then W''' is a semi-walk satisfying the condition in the Claim. But $i_{-1} + i_{2n} > i_0 + i_{2n-1}$, contradicting the maximality of $i_0 + i_{2n-1}$. This proves the Claim. $\square$

By the Claim, σ is a strong elimination ordering of G. For otherwise, by (SEO-c), there exist $i \le j \le k$ and $i \le \ell$ such that $v_j, v_k \in N_i[v_i]$ and $v_\ell \in N_i[v_j] \backslash N_i[v_k]$. Then (v_k, v_i, v_j, v_ℓ) is a semi-walk satisfying the conditions (a), (b), (c) in the Claim, a contradiction. $\square$

By repeatedly searching for and successively removing simple vertices, Farber (1983) gave a polynomial-time algorithm for recognizing strongly chordal graphs. Later, improved algorithms for a graph with n vertices and m edges were given by Lubiw (1987) in $O((n + m) \log^2 n)$ time, by Paige and Tarjan in $O((n + m) \log n)$ time and by Spinrad (1993) in $O(n^2)$ time.

6.3 Weighted domination in strongly chordal graphs

There are quite a few papers designing algorithms for variations of domination in strongly chordal graphs, including Farber (1984), Chang and Nemhauser (1984, 1985), Hoffman *et al.* (1985), White *et al.* (1985), Chang (1988/89) and Kratsch (1990), etc.

In particular, Fraber gave a linear-time algorithm for the weighted domination problem in strongly chordal graphs providing strong elimination orderings are given. His approach is by means of linear programming. The following method does not use linear programming.

Suppose G is a strongly chordal graph with a strong elimination ordering $[v_1, v_2, \ldots, v_n]$ and vertex weights $[w_1, w_2, \ldots, w_n]$ consisting of real numbers. According to Lemma 1.3, we may assume that these vertex weights are nonnegative.

A w-2-*stable vector* is a nonnegative vector $y = (y_1, y_2, \ldots, y_n)$ with

$$h(i) := w_i - \sum_{j \sim i} y_j \geq 0 \quad \text{for each } i.$$

The *value* of y is $\mathrm{val}(y) = \sum_{j=1}^{n} y_j$. The w-*weight* of a set D is $w(D) = \sum_{v_i \in D} w_i$. Also, let $g(j)$ counts the number of closed neighbors of v_j in D, that is

$$g(j) = |\{v_i \in D : i \sim j\}|.$$

Lemma 6.18. *Suppose G is a w-weighted strongly chordal graph with a strong elimination ordering $[v_1, v_2, \ldots, v_n]$ and nonnegative vertex weights $[w_1, w_2, \ldots, w_n]$. If y is a w-2-stable vector and D is a dominating set, then* $\mathrm{val}(y) \leq w(D)$. *Moreover,* $\mathrm{val}(y) = w(D)$ *if and only if the following complementary slackness conditions hold.*

(CS1D) *If $v_i \in D$, then $h(i) = 0$.*
(CS2D) *If $y_j > 0$, then there exists exactly one $v_i \in D$ such that $i \sim j$, meaning $g(j) = 1$.*

Proof.   By the definition of domination, $\sum_{v_i \in D, i \sim j} 1 \geq 1$ for all j. Then

$$w(D) - \mathrm{val}(y) = \sum_{v_i \in D} w_i - \sum_{j=1}^{n} y_j$$

$$\geq \sum_{v_i \in D} w_i - \sum_{j=1}^{n} \left(y_j \left(\sum_{v_i \in D, i \sim j} 1 \right) \right)$$

$$= \sum_{v_i \in D} w_i - \sum_{v_i \in D} \sum_{j \sim i} y_j$$

$$= \sum_{v_i \in D} \left(w_i - \sum_{j \sim i} y_j \right)$$

$$= \sum_{v_i \in D} h(i)$$

$$\geq 0.$$

This gives val(y) $\leq w(D)$. Moreover, val(y) $= w(D)$ if and only if all inequalities in the above are equalities.

First, $\sum_{v_i \in D} h(i) = 0$ if and only if (CS1D) holds, since $h(i) \geq 0$ for all i.

Next, for all j, $y_j = y_j \sum_{v_i \in D, i \sim j} 1$ if and only if $y_j > 0$ implying $1 = \sum_{v_i \in D, i \sim j} 1$. This gives (CS2D). □

The algorithm for the weighted domination in strongly chordal graphs has two stages.

Initially, $y_j = 0$ for all j and so $h(i) = w_i$ for all i. Hence, y is a w-2-stable vector, since all weights w_i are nonnegative. However, val(y) usually is not maximum initially. In stage one, at iteration j for j from 1 to n, reassigns y_j to the largest possible such that $h(i) \geq 0$ for all $i \sim j$. So, y remains a w-2-stable vector and val(y) increases.

In stage two, at iteration i, for i from n down to 1, if $h(i) = 0$ and $g(j) = 0$ for all $j \sim i$ with $y_j > 0$, then put v_i into D and increase $g(j)$ by 1 for all $j \sim i$.

More precisely, the algorithm is as follows.

Algorithm wDomStrongPD: Find a maximum valued w-2-stable vector y and a minimum weighted dominating set D of a strongly chordal graph G with a strong elimination ordering $[v_1, v_2, \ldots, v_n]$ and nonnegative vertex weights $[w_1, w_2, \ldots, w_n]$.

each $y_j \leftarrow 0$;   each $h(i) \leftarrow w_i$; $D \leftarrow \emptyset$;   each $g(j) \leftarrow 0$;
(One)   **for** $j = 1$ **to** n **do**
            { $y_j \leftarrow \min\{h(i) : i \sim j\}$; $h(i) \leftarrow h(i) - y_j$ for all $i \sim j$; }
(Two)   **for** $i = n$ **down to** 1 **do**
            if $h(i) = 0$ and $g(j) = 0$ for all $j \sim i$ with $y_j > 0$
            then { $D \leftarrow D \cup \{v_i\}$; $g(j) \leftarrow g(j) + 1$ for all $j \sim i$; }

Theorem 6.19. *Suppose G is a strongly chordal graph with a strong elimination ordering $[v_1, v_2, \ldots, v_n]$ and nonnegative vertex weights $[w_1, w_2, \ldots, w_n]$. Algorithm wDomStrongPD finds a maximum valued w-2-stable vector y and a minimum weighted dominating set D of G in linear time. Moreover, val(y) $= w(D)$.*

Proof.   **(i) Feasibility of the dual solution:** From the instructions in stage one, each y_j starts from 0 and remains unchanged or becomes positive.

Also, each $h(i)$ is decreasing but keeps nonnegative all times. Hence, at all times, y is a w-2-stable vector.

(ii) Feasibility of the primal solution: It needs to claim that every vertex v_j is dominated by D at end of the algorithm as follows.

By the choice of y_j in stage one, there is some $i \sim j$ such that $h(i) = 0$; also, if $j' > j$ and $j' \sim i$, then $y_{j'} = 0$. If $v_i \in D$, then the claim holds. Otherwise, by iteration i of stage two, $g(j') \geq 1$ for some $j' \sim i$ with $y_{j'} > 0$; and so $j' \leq j$. The reason for $g(j') \geq 1$ is because at some iteration i' in stage two, the condition holds and $v_{i'}$ is put into D, $j' \sim i'$ and $g(j')$ is changed from 0 to 1. Then $i < i'$, since in stage two the vertices are processed in the order $v_n, v_{n-1}, \ldots, v_1$.

In summary, $j' \leq j$, $i < i'$, $j' \sim i$, $j' \sim i'$ and $j \sim i$. By condition (SEO-c) for a strongly chordal graph, $j \sim i'$ and so v_j is dominated by $v_{i'} \in D$ as desired. Consequently, the primal solution is feasible.

(iii) Complementary slackness: If $v_i \in D$, then by the condition in stage two, $h(i) = 0$ and so (CS1D) holds.

To check (CS2D), suppose $y_j > 0$. By the instructions in stage two, $g(j)$ is changed from 0 to 1 at some iteration and then keep the same after that time. $\qquad\square$

Farber also presented an algorithm for the weighted independent domination problem with arbitrary real weights for strongly chordal graphs. The approach is similar to that for chordal graphs (with restricted weights) in Section 4.4.

6.4   Domatic numbers in strongly chordal graphs

Peng and Chang (1992) presented an elegant algorithm for the domatic number problem in strongly chordal graphs. Their algorithm uses a primal-dual approach.

Recall that the domatic number $d(G)$ of a graph G is the maximum number of pairwise disjoint dominating sets $D_1, D_2, \ldots, D_{d(G)}$ that G can have. The definition is equivalent to saying that these dominating sets form a partition of $V(G)$.

Choose a vertex x of degree $\delta(G)$. As any dominating set D_i contains at least one vertex v_i in $N[x]$ and two distinct D_i have different corresponding v_i, it is easy to see the following inequality.

Weak duality inequality: $d(G) \leq \delta(G) + 1$ for any graph G.

Now consider a strongly chordal graph G with a strong elimination ordering $[v_1, v_2, \ldots, v_n]$. Peng and Chang's algorithm maintains $\delta(G) + 1$ disjoint sets. Initially, these sets are empty. The algorithm processes the vertices in the reverse order of the strong elimination ordering. A vertex is included in a set when it is processed. When the algorithm terminates, these $\delta(G) + 1$ sets are dominating sets of G.

A vertex v is *completely dominated* if v is dominated by all of these $\delta(G) + 1$ dominating sets.

Algorithm DomaticSC: Find $\delta(G) + 1$ pairwise disjoint dominating sets $D_1, D_2, \ldots, D_{\delta(G)+1}$ of a strongly chordal graph G with a strongly elimination ordering $[v_1, v_2, \ldots, v_n]$.

$D_i \leftarrow \emptyset$ for $1 \leq i \leq \delta(G) + 1$;
for $i = n$ **to** 1 **step** -1 **do**
$\{$ find the largest $k \sim i$ such that v_k is not completely dominated;
      $D_\ell \leftarrow D_\ell \cup \{v_i\}$, where D_ℓ is a set that does not dominate v_k;
      if there is no such k, then put v_i into any D_ℓ;
$\}$

Note that deleting the last line also gives a valid algorithm, except it may not generate a full partition of $V(G)$.

Before proving the correctness of the algorithm, two lemmas are needed.

Lemma 6.20. *Assume* $D_\ell \subseteq \{v_{i+1}, v_{i+2}, \ldots, v_n\}$ *and* $k \sim i$, *where* $1 \leq i \leq n$. *If* D_ℓ *does not dominate* v_k, *then* D_ℓ *does not dominate* v_j *for all* $j \leq k$ *with* $j \sim i$.

Proof. Suppose to the contrary that D_ℓ has a vertex v_p dominating v_j for some $j \leq k$ with $j \sim i$. Then $i < p$, $j \leq k$, $i \sim j$, $i \sim k$ and $p \sim j$ imply $p \sim k$ by (SEO-c), which contradicts that S_ℓ does not dominate v_k. $\square$

Let $r(v_j) = |\{x \in N[v_j] : x$ is not in any of the $\delta(G) + 1$ sets $D_k\}|$ and $\text{ndom}(v_j)$ be the number of sets D_k that do not dominate v_j during the execution of Algorithm DomaticSC.

Lemma 6.21. *Algorithm DomaticSC maintains the following invariant:*

$$r(v_j) \geq \text{ndom}(v_j) \text{ for } 1 \leq j \leq n.$$

Proof.   The lemma is proved by induction. Initially,

$$r(v_j) = \deg(v_j) + 1 \geq \delta(G) + 1 = \text{ndom}(v_j)$$

for every vertex $v_j \in V(G)$. During iteration i, only values of $r(v_j)$ and $\text{ndom}(v_j)$, where $j \sim i$, may be altered when v_i is included in a set D_ℓ.

Note that the algorithm determines the largest index $k \sim i$ such that v_k is not completely dominated. If there is no such k, then for all $j \sim i$ vertex v_j is completely dominated, i.e., $\text{ndom}(v_j) = 0$, and so $r(v_j) \geq \text{ndom}(v_j)$. Now such k exists and then the algorithm finds a set D_ℓ that does not dominate v_k.

For any $j \sim i$ with $j \leq k$, by Lemma 6.20, v_j was not dominated by D_ℓ. Therefore, $r(v_j)$ and $\text{ndom}(v_j)$ are decremented by one after v_i is included in S_ℓ.

On the other hand, for any $j \sim i$ with $j > k$, by the choice of the vertex v_k in the algorithm, vertex v_j is completely dominated, i.e., $\text{ndom}(v_j) = 0$. Thus the invariant is maintained.   □

Theorem 6.22. *Algorithm DomaticSC finds $d(G) = \delta(G) + 1$ disjoint dominating sets of a strongly chordal graph G of n vertices and m edges in $O(n + m)$ time provided that a strong elimination ordering is given. Moreover, the strong duality equality $d(G) = \delta(G) + 1$ holds.*

Proof.   Upon termination of the algorithm, every vertex of the graph is in some D_i and so $r(v_j) = 0$ for all $v_j \in V(G)$. According to Lemma 6.21, $\text{ndom}(v_j) = 0$ for all $v_j \in V(G)$. That is, these $\delta(G) + 1$ sets are dominating sets of G. The strong duality equality

$$d(G) = \delta(G) + 1$$

then follows from the weak duality inequality.

To implement the algorithm efficiently, each vertex v_j is associated with the number $\text{ndom}(v_j)$ and the numbers $L_j(i) = 1$ or 0 indicating vertex v_j is or is not dominated by the dominating set D_i, where there are a total of $O(n(\delta(G) + 1)) = O(m)$ numbers $L_j(i)$. Initially, $\text{ndom}(v_j) = \delta(G) + 1$ and

$L_j(i) = 0$ for all v_j and D_i. Note that in the algorithm:

$$v_k \text{ is not completely dominated} \iff \text{ndom}(v_k) > 0;$$
$$D_\ell \text{ does not dominate } v_k \iff L_k(\ell) = 0.$$

After putting v_i into D_ℓ:

for each $j \sim i$, $\text{ndom}(v_j)$ is decreased by 1 when $L_j(\ell) = 0$;

for each $j \sim i$, $L_j(\ell)$ is reset to be 1 no matter which value it was.

Thus, for each vertex it takes $O(\deg(v_j))$ time to test $\text{ndom}(v_j)$ to determine v_k. It then takes $O(\delta(G)+1)$ time to decide which set v_i should go. Finally, for each $v_j \in N[v_i]$, it takes $O(1)$ time to update $\text{ndom}(j)$ and $L_j(\ell)$. Therefore, the algorithm takes

$$O\left(\sum_{j=1}^{n}(\deg(v_j) + \delta(G) + 1) \right) = O(n + m) \text{ time.}$$

□

6.5 Variations of domination in strongly chordal graphs

This section discusses several variations of domination in strongly chordal graphs, including distance k-domination, connected k-domination, total domination and clique domination.

Recall that if v is a simplicial vertex of a graph G, then $G-v$ is a distance invariant subgraph of G, i.e., $d_{G-v}(x,y) = d_G(x,y)$ for $x, y \in V(G - v)$. In particular, if v is a simplicial vertex of a connected graph G, then $G - v$ is also connected.

Also note that a simple vertex is a simplicial vertex.

6.5.1 *Distance k-domination in strongly chordal graphs*

Chang and Nemhauser (1985) solved the distance k-domination problem in a strongly chordal graph G by transforming it to the clique covering problem in the chordal graph G^{2k}. However, the transformation is not linear. Chang (1988/89) gave a linear-time algorithm for the distance k-domination in strongly chordal graphs using the method by Slater (1976) for the problem in trees for solving a slightly more general problem called the R-domination problem, see Section 3.2.

Now, each vertex v of the graph G is associated with an ordered pair $R_v = (a_v, b_v)$, where $a_v \geq 0$ and $b_v \geq 1$ are integers. An R-*dominating set* of G is a vertex subset D such that for any vertex v in $V(G)$, at least one of the following two conditions holds.

(RD1) $d(u, v) \leq a_v$ for some $u \in D$.

(RD2) $b_u + d(u, v) \leq a_v$ for some $u \in V(G)$.

The R-*domination number* $\gamma^R(G)$ of G is the minimum size of an R-dominating set. Note that R-domination with each $R_v = (k, k+1)$ is the distance k-domination.

Chang's algorithm is based on the following lemmas.

Lemma 6.23. *If v is an isolated vertex and D a minimum R-dominating set of a graph G, then $v \in D$ if and only if $a_v < b_v$.*

Proof.   Since v is an isolated vertex, the only possible vertex u with $d(u, v) < \infty$ is $u = v$. Hence, $v \in D$ if and only if condition (RD2) for v does not hold, which is the same as $a_v < b_v$. $\square$

Lemma 6.24. *In a graph G of at least two vertices, suppose x is a simplicial vertex with $a_x = 0$. If D' is a minimum R'-dominating set of $G-x$ with R' obtained from R by resetting b'_u to 1 for all $u \in N_G(x)$, then $D' \cup \{x\}$ is a minimum R-dominating set of G.*

Proof.   In G, consider the set $D' \cup \{x\}$. Since $x \in D' \cup \{x\}$, Condition (RD1) holds for $v = x$.

Note that $G - x$ is a distance invariant subgraph of G, since x is a simplicial vertex of G. If $v \neq x$, then either $d_{G-x}(u, v) \leq a_v$ for some $u \in D'$ (condition (RD1) for $G - x$) or else $b'_u + d_{G-x}(u, v) \leq a_v$ for some $u \in V(G-x)$ (condition (RD2) for $G-x$). For the former case, $d_G(u, v) \leq a_v$ for $u \in D' \cup \{x\}$ (condition (RD1) for G).

For the latter case with $u \notin N_G(x)$, $b_u = b'_u$ and so $b_u + d_G(u, v) \leq a_v$ for $u \in V(G)$ (condition (RD2) for G). For the latter case with $u \in N_G(x)$, $b_u = 1$ and so $d_G(x, v) = 1 + d_{G-x}(u, v) \leq a_v$ for $x \in D' \cup \{x\}$ (condition (RD1) for G). Therefore, $D' \cup \{x\}$ is an R-dominating set of G.

On the other hand, suppose D is a minimum R-dominating set of G. Then $x \in D$, since (RD2) is not true and (RD1) implies that the chosen u being x. By similar arguments as above, $D \backslash \{x\}$ is an R'-dominating set of

$G - x$. Hence,

$$|D'| \leq |D\backslash\{x\}| = |D| - 1 \leq |D' \cup \{x\}| - 1 = |D'|.$$

Consequently, all inequalities are equalities and so $D' \cup \{x\}$ is a minimum R-dominating set of G as desired. $\qquad \square$

Lemma 6.25. *In a graph G of at least two vertices, suppose x is a non-isolated simplicial vertex with $a_x \geq 1$. Assume $b_x \leq a_x$ or ($b_{u'} + 1 \leq a_x$ or $a_{u'} = 0$) for some $u' \in N_G(x)$. If D' is a minimum R'-dominating set of $G - x$ with R' obtained from R by resetting b'_u to $\min\{b_u, b_x + 1\}$ for all $u \in N_G(x)$, then D' is a minimum R-dominating set of G.*

Proof. In G, consider the set D'. For the case of $b_x \leq a_x$, condition (RD2) for $v = x$ holds by choosing $u = x$. For the case of $b_{u'} + 1 \leq a_x$ for some $u' \in N_G(x)$, condition (RD2) for x holds by choosing $u = u'$. For the case of $a_{u'} = 0$, we have $u' \in D'$ and so condition (RD1) for $v = x$ holds by choosing $u = u'$.

Note that $G - x$ is a distance invariant subgraph of G, since x is a simplicial vertex of G. If $v \neq x$, then either $d_{G-x}(u, v) \leq a_v$ for some $u \in D'$ (condition (RD1) for $G - x$) or else $b'_u + d_{G-x}(u, v) \leq a_v$ for some $u \in V(G-x)$ (condition (RD2) for $G-x$). For the former case, $d_G(u, v) \leq a_v$ for $u \in D'$ (condition (RD1) for G).

For the latter case, $b'_u = b_u$ or $b'_u = b_x + 1$ with $u \in N_G(x)$. If $b'_u = b_u$, then $b_u + d_G(u, v) \leq a_v$ and so condition (RD2) for G holds. If $b'_u = b_x + 1$ with $u \in N_G(x)$, then $b_x + d_G(x, v) = b_x + 1 + d_G(u, v) = b'_u + d_{G-x}(u, v) \leq a_v$ and so condition (RD2) for G holds. Therefore, D' is an R-dominating set of G.

On the other hand, suppose D is a minimum R-dominating set of G. If $x \in D$, then $(D\backslash\{x\}) \cup \{u'\}$ is also a minimum R-dominating set of G for any vertex $u' \in N_G(x)$. So, we may in fact assume that $x \notin D$. By similar arguments as above, D is an R'-dominating set of $G - x$. Hence,

$$|D'| \leq |D| \leq |D'|.$$

Consequently, all inequalities are equalities and so D' is a minimum R-dominating set of G as desired. $\qquad \square$

By similar arguments, the following lemma holds. Note that x only needs to be a simplicial vertex in Lemmas 6.24 and 6.25, while it needs to be a simple vertex in Lemma 6.26.

Lemma 6.26. *In a graph G of at least two vertices, suppose x is a non-isolated simple vertex with $a_x \geq 1$ and m a maximal neighbor of x. Assume $b_x > a_x$, $b_u + 1 > a_x$ and $a_u > 0$ for all $u \in N_G(x)$. If D' is a minimum R'-dominating set of $G - x$ with R' obtained from R by resetting a'_m to $\min\{a_m, a_x - 1\}$, then D' is a minimum R-dominating set of G.*

The lemmas then give the following linear-time algorithm. Recall that $N_i(v_i) = \{v_j \in N(v_i) : j > i\}$.

Algorithm RDomStronglyL: For a strongly chordal graph G in which every vertex v has a label $R_v = (a_v, b_v)$, where $a_v \geq 0$ and $b_v \geq 1$ are integers, find a minimum R-dominating set of G, providing a strong elimination ordering $[v_1, v_2, \ldots, v_n]$ is given.

$D \leftarrow \emptyset$;
for $i = 1$ **to** n **do**
$\{$　　**if** v_i is an isolated vertex in $G[\{v_i, v_{i+1}, \ldots, v_n\}]$
　　　　then $\{$ **if** $a_{v_i} < b_{v_i}$ **then** $D \leftarrow D \cup \{v_i\};\}$
　　　　else if $a_{v_i} = 0$ **then** $\{ b_{v_j} \leftarrow 1$ for all $v_j \in N_i(v_i);$　$D \leftarrow D \cup \{v_i\};\}$
　　　　else if $b_{v_i} \leq a_{v_i}$ or ($b_{v_j} + 1 \leq a_{v_i}$ or $a_{v_j} = 0$) for some $v_j \in N_i(v_i)$
　　　　then $\{ b_{v_j} \leftarrow \min\{b_{v_j}, b_{v_i} + 1\}$ for all $v_j \in N_i(v_i);\}$
　　　　else $\{ a_{v_m} \leftarrow \min\{a_{v_m}, a_{v_i} - 1\}$ where $v_m \in N_i(v_i)$ with m max; $\}$
$\}$

6.5.2　Connected domination in strongly chordal graphs

Connected domination in strongly chordal graphs was first studied by White *et al.* (1985). They transformed the problem to the *Steiner tree problem*. Given a graph G and a subset $S \subseteq V(G)$ of *terminal vertices*, a *Steiner tree* of G (with respect to S) is a tree that is a subgraph of G containing S. White *et al.* (1985) gave a beautiful min-max theorem for the Steiner tree problem in strongly chordal graphs. The proof of the theorem also provides a polynomial algorithm for finding a minimum Steiner tree.

They then reduced the connected domination problem to the Steiner tree problem. Suppose G is a connected chordal graph and $S \subseteq V(G)$ consists of a vertex from each component of the subgraph induced by the set of all simplicial vertices. They proved that a nonempty subset $D \subseteq V(G)$ is a connected dominating set of G if and only if $D \cup S$ is connected. Hence, the algorithm for the Steiner tree problem can be used to solve the connected domination problem in strongly chordal graphs.

In fact, a direct algorithm for the connected domination problem in strongly chordal graphs is possible.

Suppose $S \subseteq V(G)$ for a connected graph G. A *connected S-dominating set* of G is a connected dominating set $D \supseteq S$. The *connected S-domination number* $\gamma_c^S(G)$ of G is the minimum size of a connected S-dominating set. Note that connected $\emptyset$-domination is just the usual connected domination.

The following lemmas are useful for an efficient algorithm for connected S-domination in strongly chordal graphs.

Lemma 6.27. *Suppose $S \subseteq V(G)$ for a connected graph G. If x is a vertex such that $S \cap N_G(x) = \emptyset$ and $N_G[x] = V(G)$, then $\{x\}$ is a minimum connected S-dominating set.*

Lemma 6.28. *Suppose $S \subseteq V(G)$ for a connected graph G of at least two vertices. Suppose x is a simple vertex with m as a maximal neighbor such that $S \cap N_G(x) \neq \emptyset$ or $N_G[x] \neq V(G)$. Let $S' = S$ when $S \cap N_G(x) \neq \emptyset$ and $S' = S \cup \{m\}$ otherwise; and $S'' = S' \backslash \{x\}$. If D'' is a minimum connected S''-dominating set of $G - x$, then D'' (respectively, $D'' \cup \{x\}$) is a minimum connected S-dominating set of G when $x \notin S$ (respectively, $x \in S$).*

Proof. In the graph G, consider the set $\widehat{D''} := D''$ when $x \notin S$ and $\widehat{D''} := D'' \cup \{x\}$ otherwise. Since D'' is a connected S''-dominating set of $G - x$ with $N_G(x) \cap D'' \neq \emptyset$, $\widehat{D''}$ is a connected S-dominating set of G.

On the other hand, suppose D is a minimum connected S-dominating set of G. Since $S \cap N_G(x) \neq \emptyset$ or $N_G[v] \neq V(G)$, $D \cap N_G(x)$ is not empty, and then contains some u. For the case of $S \cap N_G(x) \neq \emptyset$, $D' := D$ is a connected S'-dominating set of G. For the case of $S \cap N_G(x) = \emptyset$, $u \notin S$ and $D' := (D \backslash \{u\}) \cup \{m\}$ is a connected S'-dominating set of G, since $N_G[u] \subseteq N_G[m]$. Then $\widehat{D'} := D' \backslash \{x\}$ is a connected S''-dominating set of $G - x$. Hence,

$$|D| \leq |\widehat{D''}| = \begin{cases} |D''| \leq |\widehat{D'}| = |D'| = |D|, & \text{if } x \notin S; \\ |D''| + 1 \leq |\widehat{D'}| + 1 = |D'| = |D|, & \text{if } x \in S. \end{cases}$$

Consequently, all inequalities are equalities and so $\widehat{D''}$ is a minimum connected S-dominating set of G as desired. $\square$

A linear-time algorithm for the connected S-domination problem in strongly chordal graphs then follows.

Algorithm Conn$S$DomStrongly: For a connected strongly chordal graph G and $S \subseteq V(G)$, find a minimum connected S-dominating set D of G, providing a strong elimination ordering $[v_1, v_2, \ldots, v_n]$ is given.

$D \leftarrow \emptyset$;
for $i = 1$ **to** n **do**
$\{$ **if** $v_{i+1}, v_{i+2}, \ldots, v_n$ are all adjacent to v_i and not in S;
    then $D \leftarrow D \cup \{v_i\}$ **else**
    $\{$ **if** $S \cap N_i(v_i) = \emptyset$
        then $S \leftarrow S \cup \{v_j\}$ where $v_j \in N[v_i]$ with j maximum;
        if $v_i \in S$ **then** $D \leftarrow D \cup \{v_i\}$;
    $\}$
$\}$

This method was extended to connected distance k-domination in strongly chordal graphs by Chang (1988/89). The method is similar to R-domination in the previous section.

Now, each vertex v of the graph G is associated with a nonnegative integer L_v. An *L-dominating set* of G is a vertex subset D such that for every vertex $v \in V(G)$ there exists some vertex $u \in D$ with $d(u, v) \leq L_v$. The *L-domination number* $\gamma^L(G)$ of G is the minimum size of a L-dominating set.

A *connected L-dominating set* is an L-dominating set D such that $G[D]$ is connected. The *connected L-domination number* $\gamma_c^L(G)$ of G is the minimum size of a connected L-dominating set. Note that connected L-domination with all $L_v = k$ is the original connected distance k-domination.

Chang's algorithm is based on the following lemmas.

Suppose G is a connected strongly chordal graph of at least two vertices, in which s is a simple vertex with m as a maximal neighbor. Note that $G - s$ is a connected and distance invariant subgraph of G.

Lemma 6.29. *Suppose $L_s > 0$. When $L_{u'} > 0$ for all $u' \in N_G(s)$, setting L'_m for $G - s$ to be $\min\{L_m, L_s - 1\}$. Then a minimum connected L'-dominating set D' of $G - s$ is a minimum connected L-dominating set of G.*

Proof.   Since $L'_v \leq L_v$ for every vertex $v \in V(G-s)$, D' being a connected L'-dominating set of $G - s$ implies that $d_G(u,v) = d_{G-s}(u,v) \leq L'_v \leq L_v$ for some vertex $u \in V(G - s)$. Furthermore, $d_G(u,s) = 1 + d_{G-s}(u,m) \leq 1 + L'_m \leq 1 + (L_s - 1) = L_s$ for some $u \in V(G - s)$. Thus, D' is a connected L-dominating set of G. Then

$$|D| \leq |D'|,$$

where D is a minimum connected L-dominating set of G.

Let $D_1 = (D \backslash \{s\}) \cup \{m\}$ when $s \in D$ and $D_1 = D$ otherwise. Then $|D_1| \leq |D|$, $D_1 \subseteq V(G - s)$ and $G[D_1]$ is connected, since $N_G[s] \subseteq N_G[m]$. In fact D_1 is a connected L-dominating set of G by checking the following conditions for the case when $D_1 = (D \backslash \{s\}) \cup \{m\}$.

- For $v = s$, $d_G(m,s) = 1 \leq L_s$ for $m \in D_1$.
- For $v \neq s$, note that $d_G(u,v) \leq L_v$ for some $u \in D$. If $u \neq s$, then $u \in D_1$ and so $d_G(u,v) \leq L_v$ for $u \in D_1$. If $u = s$, then since $m \in D_1$, we have $d_G(m,v) \leq d_G(s,v) \leq L_v$ for $m \in D_1$.

If there is some $u' \in N_G(s) \cap D_1$ with $L_{u'} > 0$, then let $D_2 = (D_1 \backslash A) \cup \{m\}$ with $A = \{u' \in N_G(s) \cap D_1 : L_{u'} > 0\}$ and $D_2 = D_1$ otherwise. Then $|D_2| \leq |D_1|$ and D_2 is connected since $N_G[u'] \subseteq N[m]$ for $u' \in A$. In fact D_2 is a connected L-dominating set of G by checking the following conditions for the case when $D_2 = (D_1 \backslash A) \cup \{m\}$.

- For $v = s$, $d_G(m,s) = 1 \leq L_s$ for $m \in D_2$.
- For $v \neq s$, note that $d_G(u,v) \leq L_v$ for some $u \in D_1$. If $u \notin A$, then $u \in D_2$ and so $d_G(u,v) \leq L_v$ for $u \in D_2$. If $u \in A$, then $m \in D_2$. For the case of $u = v$, $L_v = L_u > 0$ and so $d_G(m,v) \leq 1 \leq L_v$ for $m \in D_2$. For the case of $u \neq v$, $d_G(m,v) \leq d_G(u,v) \leq L_v$ for $m \in D_2$.

In summary, $D_2 \subseteq V(G - s)$ is a connected L-dominating set of G such that $u' \in N_G(s) \cap D_2$ with $L_{u'} > 0$ only possibly for $u' = m$ and

$$|D_2| \leq |D_1| \leq |D|.$$

Finally, D_2 is a connected L'-dominating set of $G - s$ by checking the following conditions. Note that for $v \in V(G)$, $d_G(u,v) \leq L_v$ for some $u \in D_2$ and so $u \neq s$. Now consider $v \in V(G - s)$.

- If $L_v = L'_v$, then $d_{G-s}(u, v) = d_G(u, v) \le L_v = L'_v$ for $u \in D_2$.
- If $L_v \ne L'_v$, then $v = m$, $L'_m = L_s - 1$ and $L_{u'} > 0$ for all $u' \in N_G(s)$. Consider the condition $d_G(u', s) \le L_s$ for some $u' \in D_2$. If $u' \in N_G(s)$, then $u' = m$ and so $d_{G-s}(u', v) = d_{G-s}(m, m) = 0 \le L'_m$. If $u' \notin N_G(s)$, then $d_{G-s}(u', v) = d_{G-s}(u', m) = d_G(u', s) - 1 \le L_s - 1 = L'_m = L'_v$.

These give that D_2 is a connected L'-dominating set of $G - s$ and so

$$|D'| \le |D_2|.$$

Therefore, $|D'| \le |D_2| \le |D_1| \le |D| \le |D'|$. Consequently, all inequalities are equalities and so D' is a minimum L-dominating set of G as desired.

$\square$

Lemma 6.30. *Suppose $L_s = 0$ and $L_{s'} = 0$ for another vertex s'. When $L_{u'} > 0$ for all $u' \in N_G(s)$, setting L'_m for $G-s$ to be 0. If D' is a minimum connected L'-dominating set of $G-s$, then $D' \cup \{s\}$ is a minimum connected L-dominating set of G.*

Proof.   Since $L'_v \le L_v$ for every vertex $v \in V(G-s)$, D' being a connected L'-dominating set of $G - s$ implies that $d_G(u, v) = d_{G-s}(u, v) \le L'_v \le L_v$ for some vertex $u \in D' \subseteq D' \cup \{s\}$. Furthermore, $d_G(s, s) = 0 \le L_s$ for $s \in D' \cup \{s\}$. By the definition of L'_m, $L'_{u''} = 0$ for some $u'' \in N_G(s)$, which implies that $u'' \in D'$ and so $G[D' \cup \{s\}]$ is connected. Thus, $D' \cup \{s\}$ is a connected L-dominating set of G. Then

$$|D| \le |D' \cup \{s\}| = |D'| + 1,$$

where D is a minimum connected L-dominating set of G.

If there is some $u' \in N_G(s) \cap D$ with $L_{u'} > 0$, then let $D_1 = (D \backslash A) \cup \{m\}$ with $A = \{u' \in N_G(s) \cap D : L_{u'} > 0\}$ and $D_1 = D$ otherwise. Then $|D_1| \le |D|$ and D_1 is connected since $N_G[u'] \subseteq N[m]$ for $u' \in A$. In fact D_1 is a connected L-dominating set of G by checking the following conditions for the case when $D_1 = (D \backslash A) \cup \{m\}$. Note that for every $v \in V(G)$ there is some $u \in D$ such that $d_G(u, v) \le L_v$.

- If $u \notin A$, then $u \in D_1$ and so $d_G(u, v) \le L_v$ for $u \in D_1$.
- If $u \in A$, then $m \in D_2$. For the case of $u = v$, $L_v = L_u > 0$ and so $d_G(m, v) \le 1 \le L_v$ for $m \in D_1$. For the case of $u \ne v$, $d_G(m, v) \le d_G(u, v) \le L_v$ for $m \in D_1$.

In summary, D_1 is a connected L-dominating set of G such that $u' \in N(v) \cap D_1$ with $L_{u'} > 0$ only possibly for $u' = m$ and

$$|D_1| \leq |D|.$$

Let $D_2 = D_1 \backslash \{s\}$. Then $|D_2| = |D_1| - 1$ and $G[D_2]$ is connected.

Finally, D_2 is a connected L'-dominating set of $G - s$ by checking the following conditions. Note that for $v \in V(G)$, $d_G(u,v) \leq L_v$ for some $u \in D_1$. Now consider $v \in V(G - s)$.

- If $L_v = L'_v$ and $u \neq s$, then $u \in D_2$ and so $d_{G-s}(u,v) = d_G(u,v) \leq L_v = L'_v$ for $u \in D_2$.
- If $L_v = L'_v$ and $u = s$, then $d_{G-s}(u'',v) \leq d_G(s,v) = d_G(u,v) \leq L_v = L'_v$ for some $u'' \in D_2 \cap N_G(s)$.
- If $L_v \neq L'_v$, then $v = m$, $L'_m = 0$ and $L_{u'} > 0$ for all $u' \in N_G(s)$. By the definitions of A, D_1 and D_2, $m \in D_2$ and so $d_{G-s}(m,v) = d_{G-s}(m,m) = 0 \leq L'_m$.

These give that D_2 is a connected L'-dominating set of $G - s$ and so

$$|D'| \leq |D_2| = |D_1| - 1.$$

Therefore, $|D'| + 1 \leq |D_2| + 1 \leq |D_1| \leq |D| \leq |D' \cup \{s\}| = |D'| + 1$. Consequently, all inequalities are equalities and so $D' \cup \{s\}$ is a minimum L-dominating set of G as desired. □

The lemmas then give the following linear-time algorithm.

Algorithm ConnLDomStronglyL: For a strongly chordal graph G in which every vertex v is associated with a nonnegative integer L_v, find a minimum connected L-dominating set D of G.

$D \leftarrow \emptyset$;
if $L_{v_i} > 0$ for all vertices $v_i \in V(G)$ **then**
{   find a strong elimination ordering $[v_1, v_2, \ldots, v_n]$ of G;
    for $i = 1$ **to** $n - 1$ **do**
    {   let v_j be the maximum neighbor of v_i;
        $L_{v_j} \leftarrow \min\{L_{v_j}, L_{v_i} - 1\}$;
        $G \leftarrow G - v_i$;
        if $L_{v_j} = 0$ **then goto** outloop;
    }
}

$D \leftarrow \{v_n\}$ and **stop**;
outloop: find a strong elimination ordering $[v_1, v_2, \ldots, v_q]$ of G with $L_{v_q} = 0$;
for $i = 1$ **to** q **do**
{   let v_j be the maximum neighbor of v_i;
         if $L_{v_i} = 0$ **then** $D \leftarrow D \cup \{v_i\}$;
         if $L_{v_k} > 0$ for all $v_k \in N(v_i)$ **then**
                  if $L_{v_i} = 0$ **then** $L_{v_j} \leftarrow 0$ **else** $L_{v_j} \leftarrow \min\{L_{v_j}, L_{v_i} - 1\}$;
}

6.5.3   *Clique domination in strongly chordal graphs*

Recall that a dominating clique of a graph is a dominating set that forms a clique. For a graph G to have a dominating clique D it is necessary that the graph has diameter at most three. To see this, choose vertices x and y such that $d(x, y) = \mathrm{diam}(G)$. Let x be dominated by $x' \in D$ and y is dominated by $y' \in D$. By the triangle inequality for the distance function,

$$d(x, y) \leq d(x, x') + d(x', y') + d(y', y) \leq 1 + 1 + 1 = 3$$

and so $\mathrm{diam}(G) \leq 3$.

However, not every graph of diameter at most three has a dominating clique. Examples include C_5, C_6, C_7 and $K_{4,4} - 4K_2$ etc. It is interesting that Kratsch *et al.* (1994) showed that a chordal graph of diameter at most three does have a dominating clique, even the problem of finding a minimum dominating clique is NP-complete in split graphs, which are special chordal graphs. To prove this result, recall that Lemma 6.1 gives that

> *for a closed semi-walk $(v_0, v_1, \ldots, v_r)$ with $r \geq 4$ in a chordal graph, for any i there exists some $j \neq i, i+1$ such that $v_j \sim v_i$ and $v_j \sim v_{i+1}$.*

Theorem 6.31 (Kratsch *et al.*, 1994). *A chordal graph has a dominating clique if and only if it has diameter at most three.*

Proof.   The necessity was showed above.

Now suppose G is a chordal graph of diameter at most three. Choose a clique K such that $|N[K]|$ is as large as possible. If $N[K] = V(G)$, then K is a dominating clique as desired. Suppose $N[K] \neq V(G)$. Choose a vertex w such that $d(w, K) = 2$. Then $K' = \{y \in K : d(w, y) = 2\}$ is not empty.

Then there exists some vertex x' adjacent to w and all vertices in K'. For otherwise, there exist vertices x_1, y_1, x_2, y_2 such that $y_i \in K'$ and x_i is adjacent to both w and y_i but not to y_{3-i} for $i = 1, 2$. Hence, $(w, x_1, y_1, y_2, x_2, w)$ is a cycle in which the only possible chord is $x_1 x_2$. So, there is either a chordless 4-cycle or 5-cycle, a contradiction.

Since $w \in N[K' \cup \{x'\}] \backslash N[K]$, by the maximality of $|N[K]|$, the set $L = N[K] \backslash N[K' \cup \{x'\}]$ is not empty. Any vertex $x \in L$ is not in K and is adjacent to some $y \in K \backslash K'$. Also $y \notin K'$ implies $d(w, y) \geq 3$ and so $d(w, x) \geq 2$. Consider a shortest w-x path $P = (w = w_0, w_1, \ldots, w_s = x)$, where $2 \leq s \leq 3$.

For any $y' \in K'$, the path $P' = (w, x', y', y)$ is chordless. In the closed semi-walk $(w = w_0, w_1, \ldots, w_s = x, y, y', x', w_0)$, there exists a vertex x^* other than y' and y but $x^* \sim y'$ and $x^* \sim y$. Note that x^* is not in P', since it is chordless. Then $x^* = w_i$ for some $0 < i \leq s$. But $i \neq 1$, for otherwise $d(w, y) \leq 2$ is impossible. Also, $i \neq s$, since x is not adjacent to y'. Therefore, $i = 2$ and $s = 3$.

In conclusion, for every $x \in L$ adjacent to $y \in K \backslash K'$, $d(w, x) = d(w, y) = 3$, and there exists some x^* adjacent to all vertices in $\{x, y\} \cup K'$ and $d(w, x^*) = 2$.

For any two distinct vertices $x_1, x_2 \in L$ adjacent to $y_1, y_2 \in K \backslash K'$, respectively, in the closed semi-walk $(w, *, x_1^*, y_1, y_2, x_2^*, *, w)$ there exists a vertex z other than x_1^* and y_1 but $z \sim x_1^*$ and $z \sim y_1$. It is only possible that $z = x_2^*$ or $z = y_2$. For the former case, x_1^* is adjacent to x_2^*. For the latter case, in the closed semi-walk $(w, *, x_1^*, y_2, x_2^*, *, w)$ there exists a vertex z' other than x_1^* and y_2 but $z' \sim x_1^*$ and $z' \sim y_2$. It is only possible that $z' = x_2^*$, and so again x_1^* is adjacent to x_2^*.

Therefore, $K'' = K' \cup \{x^* : x \in L\}$ is a clique. Moreover, $N[K] \subseteq N[K']$ and $d(w, y'') = 2$ for all $y'' \in K''$. By the same arguments as in the third paragraph of the proof, there exists some x'' adjacent to w and all vertices in K''. Then $K'' \cup \{x''\}$ is a clique such that $N[K] \subseteq N[K''] \subsetneq N[K'' \cup \{x''\}]$, since the last set contains w that is not in the previous two sets. This contradicts the maximality of $|N[K]|$. The theorem then follows. $\square$

A more general result in terms of clique radius was established by Chang (1991). Recall that $\text{rad}(G) \leq \text{diam}(G) \leq 2\text{rad}(G)$ for any connected graph G. And in general, for any positive integers r and d with $r \leq d \leq 2r$, there is always a graph G with radius $\text{rad}(G) = r$ and diameter

$\operatorname{diam}(G) = d$. However, relations between radii and diameters are restricted for special graphs. An oldest such kind of results is for trees.

Theorem 6.32 (Jordan, 1869). *For any tree T, either* $\operatorname{diam}(T) = 2\operatorname{rad}(T) - 1$ *with T having exactly two adjacent centers, or else* $\operatorname{diam}(T) = 2\operatorname{rad}(T)$ *with T having just one center.*

Chang and Nemhauser (1984) extended the result to chordal graphs. While $\operatorname{diam}(T)$ has two possible values $2\operatorname{rad}(T) - 1$ and $2\operatorname{rad}(T)$ for a tree T, $\operatorname{diam}(G)$ has three possible values for a chordal graph G.

Theorem 6.33 (Chang and Nemhauser, 1984). *If G is a connected chordal graph, then $2\operatorname{rad}(G) - 2 \le \operatorname{diam}(G) \le 2\operatorname{rad}(G)$. If G is a 3-sun-free connected chordal graph, then $2\operatorname{rad}(G) - 1 \le \operatorname{diam}(G) \le 2\operatorname{rad}(G)$.*

Chang (1991) introduced the concept the clique radius. The *eccentricity* of a vertex subset S is

$$e(S) = \max\{d(v, S) : v \in V(G)\}.$$

The *clique radius* $\operatorname{cr}(G)$ of a graph is the minimum eccentricity of a clique; and a clique K with $e(K) = \operatorname{cr}(G)$ is called a *clique center*. Note that a graph G has a dominating clique if and only if $\operatorname{cr}(G) \le 1$. Chang (1991) established the following result, which implies Theorem 6.31.

Theorem 6.34 (Chang, 1991). *If G is a connected chordal graph, then $2\operatorname{cr}(G) \le \operatorname{diam}(G) \le 2\operatorname{cr}(G) + 1$.*

For an interval graph G of diameter at most three, it not only has a clique dominating set but also has one of size at most two. The following shows two proofs of this fact.

The first proof. Suppose $\{I_i = [a_i, b_i] : 1 \le i \le n\}$ is an interval representation of G, where interval I_i represents vertex v_i. Let D be a dominating clique of G. Then $\cap_{v_i \in D} I_i$ is not empty. Choose a point p in this intersection. Choose $v_j \in D$ such that $a_j = \min_{v_i \in D} a_i$ and $v_k \in D$ such that $b_k = \max_{v_i \in D} b_k$. Since p is in I_i for all $v_i \in D$, $I_j \cup I_k = \cup_{v_i \in D} I_i$ and so $\{v_j, v_k\}$ is a dominating clique of size at most two. □

The second proof. Suppose $[v_1, v_2, \ldots, v_n]$ is an interval ordering of G. Let D be a dominating clique of G. Let v_1 be dominated by $v_r \in D$ and s is the maximum index for which $v_s \in D$. First, for $1 \le i \le r$, v_i is dominated by v_r, since $v_1 \in N[v_r]$. For $r \le i \le s$, v_i is dominated by v_s, since $v_r \in N[v_s]$. For $i > s$, v_i is dominated by some $v_j \in D$ with $j \leq s < i$ and so is dominated by v_s. In summary, $\{v_r, v_s\}$ is a dominating clique of size at most two. $\qquad\square$

According to the second proof, if the interval graph has a dominating clique of size two, then it can be chosen to be the set $\{v_i, v_j\}$, where v_i is the maximal neighbor of v_1 and v_j is the maximal neighbor of v_i.

Kratsch (1990) gave a linear-time algorithm for determining if a connected strongly chordal graph has a dominating clique, and in case of a positive answer it also gives a minimum one.

Note that the algorithm is a modification of Algorithm DomSimplePD in Section 6.2 by modifying "add to D the largest indexed neighbor of an undominated vertex" with "add to D the largest indexed neighbor of an undominated vertex which is adjacent to all vertices of D." The following is a simplification of the algorithm by Kratsch.

Algorithm DomCliqueStrongly: Determine if a connected strongly chordal graph G has a dominating clique. In case of positive answer, give a minimum dominating clique D and a maximum 2-stable set S, providing a strong elimination ordering $[v_1, v_2, \ldots, v_n]$ is given.

$D \leftarrow \emptyset; \quad S \leftarrow \emptyset;$
$C \leftarrow V(G); \quad //*$ The set of candidate vertices of D. $*//$
for $i = 1$ **to** n **do**
    if $N[v_i] \cap D = \emptyset$ **then**
        if $\quad N[v_i] \cap C = \emptyset$
        then report "G has no dominating clique" and **stop**;
        else $\{ \ m_i \leftarrow \max\{j : v_j \in N[v_i] \cap C\};$
                $S \leftarrow S \cup \{v_i\};$
                $D \leftarrow D \cup \{v_{m_i}\};$
                $C \leftarrow C \cap N(v_{m_i});$
         $\}$

Theorem 6.35. *Algorithm DomCliqueStrongly correctly determines if a connected strongly chordal graph G has a dominating clique in linear time providing a strong elimination ordering is provided. In case when the answer is positive, a minimum dominating clique D and a maximum 2-stable set S are given; and moreover the strong duality equalities $\alpha_2(G) = \gamma(G) = \gamma_t(G) = \gamma_c(G) = \gamma_{cl}(G)$ hold, where $\gamma_t(G)$ appears only when $\gamma(G) \geq 2$.*

Proof.   First consider the case when the algorithm exits normally. Let the final $S = \{v_{i_1}, v_{i_2}, \ldots, v_{i_r}\}$ and $D = \{v_{m_{i_1}}, v_{m_{i_2}}, \ldots, v_{m_{i_r}}\}$, where $i_1 < i_2 < \cdots < i_r$. It is clear that $|S| = |D|$ and D is a dominating clique by the checking of $N[v_i] \cap D = \emptyset$ and $N[v_i] \cap C = \emptyset$.

The following proves that S is a 2-stable set.

First, $v_{i_1} = v_1$ and $v_{m_{i_1}} \in D$ is the maximum neighbor of v_1. For $1 < a \leq r$, $d(v_1, v_{i_a}) \geq 3$ for otherwise suppose $v_1, v_{i_a} \in N[u]$ for some vertex u. Then $v_{i_a} \in N[u] \subseteq N[v_{m_{i_1}}]$, violating the condition $N[v_i] \cap D = \emptyset$ at iteration i_a.

Next to prove that $d(v_{i_a}, v_{i_b}) \geq 3$ for $1 < a < b \leq r$. Suppose to the contrary that $d(v_{i_a}, v_{i_b}) \leq 2$ for some $1 < a < b \leq r$. By the algorithm, $v_{i_a} v_{m_{i_1}}, v_{i_b} v_{m_{i_1}}, v_{i_b} v_{m_{i_a}} \notin E(G)$. Consider the following two cases, both give a contradiction and so S is a 2-stable set.

Case 1. $v_{i_a} v_{m_{i_b}} \in E(G)$. In this case, by the maximality of m_{i_a} at iteration i_a, it is the case that $m_{i_b} < m_{i_a}$. Then $i_b < m_{i_b}$, for otherwise $m_{i_b} < i_b$ together with $m_{i_b} < m_{i_a}$ and $v_{i_b}, v_{m_{i_a}} \in N(v_{m_{i_b}})$ imply that $v_{i_b} v_{m_{i_a}} \in E(G)$, a contradiction. Then $i_a < i_b < m_{i_b} < m_{i_a}$ imply that $v_{i_b} \in N_{i_a}[v_{m_{i_b}}] \subseteq N_{i_a}[v_{m_{i_a}}]$, a contradiction again.

Case 2. $v_{i_a} v_{m_{i_b}} \notin E(G)$. In this case, $v_{i_a} v_{i_b} \notin E(G)$ for otherwise $(v_{i_a}, v_{v_{i_b}}, v_{m_{i_b}}, v_{m_{i_a}}, v_{i_a})$ is a chordless 4-cycle, a contradiction. Hence, $d(v_{i_a}, v_{i_b}) = 2$, say v_{i_a} and v_{i_b} are both adjacent to v_c. The 5-cycle $(v_{i_a}, v_c, v_{i_b}, v_{m_{i_b}}, v_{m_{i_a}}, v_{i_a})$ then has exactly two chords $v_c v_{m_{i_a}}$ and $v_c v_{m_{i_b}}$. For the six vertices $\{v_{i_a}, v_c, v_{i_b}, v_{m_{i_b}}, v_{m_{i_1}}, v_{m_{i_a}}\}$ not inducing a 3-sun, $v_c v_{m_{i_1}} \in E(G)$. By the maximality of m_{i_a} at iteration i_a, it is the case that $c < m_{i_a}$. Then $i_b < c$, for otherwise $c < i_b$ together with $c < m_{i_a}$ and $v_{i_b}, v_{m_{i_a}} \in N(v_c)$ imply that $v_{i_b} v_{m_{i_a}} \in E(G)$, a contradiction. Then $i_a < i_b < c < m_{i_a}$ imply that $v_{i_b} \in N_c[v_c] \subseteq N_c[v_{m_{i_a}}]$, a contradiction.

Hence,

$$|S| \leq \alpha_2(G) \leq \gamma(G) \leq \gamma_t(G) \leq \gamma_c(G) \leq \gamma_{cl}(G) \leq |D| = |S|$$

and then all inequalities are equalities as desired, where $\gamma_t(G)$ appears only when $\gamma(G) \geq 2$.

Now consider the case when the algorithm reports "G has no dominating clique" and stops at iteration i. Suppose at that time $S = \{v_{i_1}, v_{i_2}, \ldots, v_{i_s}\}$ and $D = \{v_{m_{i_1}}, v_{m_{i_2}}, \ldots, v_{m_{i_s}}\}$. Then $1 = i_1 < i_2 < \ldots < i_s < i$.

We shall claim that G has no dominating clique. Suppose to the contrary that G has a dominating clique D'. Let t be the maximum index such that $D_t := \{v_{m_{i_1}}, v_{m_{i_2}}, \ldots, v_{m_{i_t}}\} \subseteq D \cap D'$. We may assume that D' is chosen so that t is as large as possible.

Suppose $t < s$. Then D'' dominates $\{v_1, v_2, \ldots, v_{i_{t+1}-1}\}$ but not $v_{i_{t+1}}$. Let $v_{i_{t+1}}$ be dominated by $v_a \in D'$. By the maximality of m_{t+1} at iteration i_{t+1}, it is the case that $a < m_{t+1}$. For all $v_j \in N[v_a]$, consider three cases.

If $j < i_{t+1}$, then v_j is dominated by D_t.

If $j = i_{t+1}$, then v_j is dominated by $v_{m_{t+1}}$.

If $j > i_{t+1}$, then together with $a < m_{t+1}$ and $v_{i_{t+1}} v_a, v_{i_{t+1}} v_{m_{t+1}}, v_j v_a \in E(G)$ imply $v_j v_{m_{t+1}}$ by condition (SEO-c) for strongly chordal graphs.

In conclusion, $D'' = (D' \backslash \{v_a\}) \cup \{m_{t+1}\}$ is a new dominating clique such that $D_{t+1} := \{v_{m_{i_1}}, v_{m_{i_2}}, \ldots, v_{m_{i_{t+1}}}\} \subseteq D \cap D''$. This contradicts the maximality of t; therefore, $t = s$ and $D \subseteq D'$.

Suppose v_i is dominated by $v_b \in D' \supseteq D$. Then at iteration i, it is the case that $b \in C$ and so it is not the case that $N[v_i] \cap C = \emptyset$. This confirms that G has no dominating clique. $\qquad\square$

Note that the strong duality equality in fact was shown by Kratsch *et al.* (1994), who gave an alternative modification of Farber's method without using the set C as above.

There are chordal graphs G for which $\gamma(G)$ and $\gamma_{cl}(G)$ have a gap, as shown below. For $k \geq 2$, let G_k be the chordal graph with

$$\text{vertex set } V(G_k) = \{a_i, a, b_i, c, c_i : 1 \leq i \leq k\}$$

$$\text{and edge set } E(G_k) = \{a_i a, a b_i, b_i c, c c_i, a_i b_i : 1 \leq i \leq k\} \cup$$

$$\{b_i b_j : 1 \leq i < j \leq k\}.$$

Note that $\text{diam}(G_k) = 3$, $\gamma(G_k) = 2 < k + 1 = \gamma_{cl}(G_k)$. For $k \geq 3$, let H_k be the chordal graph with

$$\text{vertex set } V(H_k) = V(G_k)$$

$$\text{and edge set } E(H_k) = E(G_k) \cup \{b_i c_i : 1 \leq i \leq k\}.$$

Note that $\text{diam}(H_k) = 2$ and $\gamma(H_k) = 2 < k = \gamma_{cl}(H_k)$.

6.6   Exercises

(6.1) For every positive integer n, construct a graph G with $\theta(G^2) = n$ and $\gamma(G) = 2n$.

(6.2) For every positive integer n, construct a chordal graph G with $\theta(G^2) = 1$ and $\gamma(G) = n$.

(6.3) Does there exist any graph G for which G^2 is chordal but G^3 is not?

(6.4) Let a_n be the number of n-suns, where isomorphic n-suns are counted as one. Determine a_n for $3 \le n \le 6$.

(6.5) Show that an n-sun contains a complete m-sun as an induced subgraph for some $m \le n$.

(6.6) Determine all n-suns that are interval graphs.

(6.7) An n-cycle matrix is the vertex-edge incident matrix of an n-cycle $(v_0, e_1, v_1, e_2, v_2, \ldots, e_{n-1}, v_{n-1}, e_n, v_n)$, where $v_n = v_0$.
A totally balanced matrix (respectively, balanced matrix) is a 0–1 matrix having no n-cycle matrices (respectively, odd n-cycle matrices) as submatrices. Note that the property of being (totally) balanced of a matrix is invariant under row/column permutations.

   (a) Show that G is a sun-free chordal graph if and only if the closed neighborhood matrix $M(G)$ of G is totally balanced.

   (b) Show that G is an odd-sun-free chordal graph if and only if $M(G)$ is balanced.

(6.8) (a) Show that an $r \times s$ 0-1 matrix M has an n-cycle submatrix if and only if the matrix $M^* = \begin{pmatrix} I & M^T \\ M & J \end{pmatrix}$ has an n-cycle submatrix, where J is the $r \times r$ matrix of all ones.

   (b) Show that M is a totally balanced matrix if and only if M^* is the closed neighborhood matrix of some sun-free chordal graph.

(6.9) Show that if M is totally balanced, then so is $M^T M$, which is calculated using boolean products and sums. (Hint: Consider the square of the graph corresponding M^*.)

(6.10) Show that a totally balanced matrix is Γ-freeable.

(6.11) Determine the domatic number $d(C_n)$ of the n-cycle C_n for $n \ge 3$.

(6.12) Determine the domatic number $d(S_n)$ of the complete n-sun S_n for $n \ge 3$.

(6.13) Prove that domatic number $d(G) \ge 2$ for any graph G of $\delta(G) \ge 1$. Consequently, if $\delta(G) \le 1$, then $d(G) = \delta(G) + 1$.

(6.14) Provide examples of chordal graphs G for which $d(G) < \delta(G) + 1$.

(6.15) For integers $r > s \geq 2$, provide an example of chordal graph G for which $d(G) = s$ and $\delta(G) = r - 1$.

(6.16) Provide examples of chordal graphs G for which the independent domatic number $d_i(G) < d(G)$.

(6.17) Prove or disprove that $d_i(G) = \delta(G) + 1$ for every strongly chordal graph G.

(6.18) Prove Lemma 6.26.

(6.19) Suppose G is a graph in which every vertex v has a nonnegative integer L_v. An L-*stable set* of G is a vertex subset $S \subseteq V(G)$ such that $d(u,v) > L_u + L_v$ for every two distinct vertices u and v in S. The L-*stability number* $\alpha_L(G)$ of G is the maximum size of an L-stability set of G.

(a) Show that $\alpha_L(G) \leq \gamma_L(G)$ for every graph G.

(b) Provide examples of graphs G with L for which $\alpha_L(G) < \gamma_L(G)$.

(c) Determine $\alpha_L(C_n)$ and $\gamma_L(C_n)$ for every nonnegative integer-valued L and every n-cycle C_n.

(d) Show that $\alpha_L(G) = \gamma_L(G)$ for every strongly chordal graph G.

(6.20) (a) Show that if v is a simplicial vertex of a connected graph G, then $G - v$ is a distance invariant subgraph of G.

(b) Suppose G is a noncomplete connected chordal graph and $S(G)$ is the set of all simplicial vertices of G. Show that $S(G) \neq V(G)$ and $D := V(G) - S(G)$ is a connected dominating set of G.

(6.21) Show that $cr(G) \leq r(G) \leq cr(G) + 1$ for every connected graph G.

(6.22) Determine all trees T for which $cr(T) = r(T)$.

(6.23) Determine all connected chordal graphs G for which $cr(T) = r(G)$.

(6.24) Prove Theorem 6.32.

(6.25) Prove Theorem 6.33.

(6.26) Prove Theorem 6.34.

(6.27) Design an efficient algorithm for the total domination problem in strongly chordal graphs.

Chapter 7

Cocomparability Graphs and Asteroidal Triple-Free Graphs

7.1 Comparability graphs as classical perfect graphs

Chordal graphs and comparability graphs are two classical graph classes in the perfect graph theory. The stability numbers and chromatic numbers of comparability graphs can be calculated in polynomial time. On the other hand, the domination problem and most of its variants are NP-complete for chordal graphs and bipartite graphs, see Chapter 4, where bipartite graphs are special comparability graphs.

Comparability graphs came very early in terms of partially ordered sets. A *partial order* on a set P is a binary relation $\preceq$ on P such that for $a, b, c \in P$, the following three conditions hold.

(Reflexivity) $\quad a \preceq a$.
(Antisymmetry) If $a \preceq b$ and $b \preceq a$, then $a = b$.
(Transitivity) $\quad$ If $a \preceq b$ and $b \preceq c$, then $a \preceq c$.

A *partially ordered set* (or *poset* for short) is an ordered pair $(P, \preceq)$ consisting of a set P and a partial order $\preceq$ on P. The notion $a \prec b$ stands for $\preceq b$ but $a \neq b$; $a \succeq b$ stands for $b \preceq a$; and $a \succ b$ stands for $b \prec a$.

The following are some examples of posets. The set of all real numbers is a poset under the standard less-than-or-equal-to relation. The set of all natural numbers is a poset under the relation of divisibility. The set of all subsets of a set is a poset under the inclusion relation. The set of all subspaces of a vector space is a poset under the inclusion relation.

The set of all vertices of a graph is a poset under the relation of being in the same component.

In a poset $(P, \preceq)$, two elements a and b are *comparable* if $a \preceq b$ or $b \preceq a$; otherwise they are *incomparable*. A *totally ordered set* (or *linearly ordered set*) is a poset in which every two elements are comparable. In the above examples, only the first one is totally ordered. A *chain* is a subset of P in which every two elements are comparable.

An *antichain* is a subset of P in which every two distinct elements are incomparable.

There are two well-known decomposition theorems on partially ordered sets as follows.

Theorem 7.1 (Dilworth, 1950). *In a finite poset, the maximum size of an antichain is equal to the minimum number of disjoint chains whose union is the poset.*

Another one is the dual version of Dilworth's theorem.

Theorem 7.2 (Mirsky, 1971). *In a finite poset, the maximum size of a chain is equal to the minimum number of disjoint antichains whose union is the poset.*

These two theorems can be presented in terms of graphs. The *comparability graph* of a finite partially ordered set $(P, \preceq)$ is the graph G with vertex set $V(G) = P$ and edge set

$$E(G) = \{uv : u \text{ and } v \text{ are two distinct comparable elements}\}.$$

In this way, a chain in P is a clique in G, and an antichain in P is a stable set in G. Therefore,

Dilworth's theorem is equivalent to the duality equality $\alpha(G) = \theta(G)$;

Mirsky's theorem is equivalent to the duality equality $\omega(G) = \chi(G)$.

Note that a graph G is a comparability graph (of some finite partially ordered set) if and only if it has a *transitive orientation* $\vec{G}$, that is, $uv, vw \in E(\vec{G})$ imply $uw \in E(\vec{G})$; or equivalently, G has a *comparability ordering* which is an ordering $[v_1, v_2, \ldots, v_n]$ of $V(G)$ satisfying

$$i < j < k \text{ and } v_i v_j, v_j v_k \in E(G) \text{ imply } v_i v_k \in E(G). \tag{CO}$$

A comparability ordering of a comparability graph can be obtained by performing a topological sort on its transitive orientation. Comparability graphs include bipartite graphs and trapezoid graphs, which further include interval graphs and permutation graphs.

A graph G is a *cocomparability graph* if the complement of G is a comparability graph; or equivalently, G has a *cocomparability ordering* which is an ordering $[v_1, v_2, \ldots, v_n]$ of $V(G)$ satisfying

$$i < j < k \text{ and } v_i v_k \in E(G) \text{ imply } v_i v_j \in E(G) \text{ or } v_j v_k \in E(G). \quad \text{(CCO)}$$

Cocomparability graphs include trapezoid graphs.

While the domination problem and most of its variations are NP-complete in comparability graphs, there are polynomial algorithms for most of these problems in cocomparability graphs, as to be discussed in the latter sections of this chapter.

Recognizing comparability graphs was first studied by Gilmore and Hoffman (1964) on the way to study interval graphs. A key is the concept of *forcing relation* Γ on $E(G)$ defined by: for $xy, yz \in E(G)$ with $x \neq z$,

$$xy \, \Gamma yz \text{ if and only if } xz \notin E(G).$$

In a comparability graph G, if $xy \, \Gamma yz$ then in a transitive orientation $\vec{G}$:

$$xy \in E(\vec{G}) \text{ implies } zy \in E(\vec{G}),$$
$$yx \in E(\vec{G}) \text{ implies } yz \in E(\vec{G}).$$

These are because $xy, yz \in E(\vec{G})$ imply $xz \in E(\vec{G})$ and so $xz \in E(G)$; and $zy, yx \in E(\vec{G})$ imply $zx \in E(\vec{G})$ and so $zx \in E(G)$; both contradict the definition of $xy \, \Gamma yz$.

The *closure* $\tilde{\Gamma}$ of the relation Γ on $E(G)$ is defined by

$$e \, \tilde{\Gamma} e' \text{ if } e = e_1 \, \Gamma e_2 \, \Gamma e_3 \cdots \Gamma e_r = e' \text{ for some } e_1, e_2, \ldots, e_r \text{ with } r \geq 1.$$

This yields an equivalence relation on $E(G)$, that is, the following conditions hold for $e_1, e_2, e_3 \in E(G)$.

(Reflexivity) $e_1 \, \tilde{\Gamma} e_1$.
(Symmetry) If $e_1 \, \tilde{\Gamma} e_2$, then $e_2 \, \tilde{\Gamma} e_1$.
(Transitivity) If $e_1 \, \tilde{\Gamma} e_2$ and $e_2 \, \tilde{\Gamma} e_3$, then $e_1 \, \tilde{\Gamma} e_3$.

For each $e \in E(G)$, let $\tilde{e} = \{f : f \, \tilde{\Gamma} e\}$, which is called an *equivalence class* of the equivalence relation $\tilde{\Gamma}$ or a $\tilde{\Gamma}$-*class* containing e. Then for $e, f \in E(G)$, either $\tilde{e} = \tilde{f}$ or $\tilde{e} \cap \tilde{f} = \emptyset$. Hence, the relation $\tilde{\Gamma}$ induces a partition of $E(G)$ into equivalence classes.

Here is the characterization of comparability graphs by Gilmore and Hoffman, who used the term "cycle" for "closed walk."

Theorem 7.3 (Gilmore and Hoffman, 1964). *A graph G is a comparability graph if and only if it has no closed walk $v_0, v_1, v_2, \ldots, v_r$ of odd length $r \geq 5$ with $v_r = v_0$ and $v_i v_{i+2} \notin E(G)$ for $0 \leq i \leq r - 1$, where indices are taken modulo r.*

Proof.   ($\Rightarrow$) Suppose G has a transitive orientation $\vec{G}$ but G has an odd walk as described in the theorem. Without loss of generality, assume that $v_0 v_1 \in E(\vec{G})$ (the arguments for $v_1 v_0 \in E(\vec{G})$ is similar). By the transitivity, $v_0 v_2 \notin E(G)$ implies $v_2 v_1 \in E(\vec{G})$, and in turn, $v_1 v_3 \notin E(G)$ implies $v_2 v_3 \in E(\vec{G})$. Continue this process, $E(\vec{G})$ contains the following directed edges:

$$v_0 v_1, v_2 v_1, v_2 v_3, v_4 v_3, \ldots, v_{r-1} v_r, v_{r+1} v_r.$$

This is a contradiction since $v_0 v_1$ ends up directed both as $v_0 \to v_1$ and $v_{r+1} \to v_r$.

($\Leftarrow$) The proof of this direction is quite long in the paper by Gilmore and Hoffman and hence is not presented here. In fact, their proof is equivalent to the proof of the correctness of the algorithm for determining whether a graph has a transitive orientation to be described as follows.   □

The proof of the theorem by Gilmore and Hoffman contains an algorithm of recognizing comparability graphs implicitly without clear explanations. Later, Pnueli *et al.* (1971) presented a precise algorithm for recognizing comparability graphs on the way to study permutation graphs.

Suppose $G = (V, E)$ is a graph. A *partially oriented graph* $G^* = (V, E^*)$ of G is a "mixed" graph obtained from G by orienting some but may not all edges of $E(G)$. An edge of $E(G^*)$ is an *implicant* if it is directed and is Γ-related to at least one undirected edge of $E(G^*)$. A partially oriented graph G^* is *stable* if it contains no implicants; otherwise it is *unstable*.

To test if a given graph G has a transitive orientation, start by arbitrarily choosing and directing one edge e of $E(G)$. If the resulting partially oriented graph G^* is unstable, proceed according to the following Γ-orientation rule as long as it is applicable.

The Γ-orientation rule.
Select a directed edge xy that is an implicant of G^*.
To every undirected edge $x'x$ with $x'x \, \Gamma xy$ assign the direction $x \to x'$.
To every undirected edge yy' with $xy \, \Gamma yy'$ assign the direction $y' \to y$.

It is clear that repeated applications of the above rule will result, eventually, in a stable partially oriented graph G^* which either is fully oriented or has none of its undirected edges Γ-relating to a directed edge. In fact, all edges which are assigned direction in the rule form the $\tilde{\Gamma}$-class $\tilde{e}$.

Next, consider separately the subgraphs G_d^* and G_u^* of G^*, which are induced by the directed edges and the undirected edges of $E(G^*)$, respectively. First apply the following test procedure to determine whether G_d^* is transitive.

Test for transitivity.
Suppose $\vec{H}$ is a directed graph.
For every $x \in V(\vec{H})$, let $V(x) := \{y \in V(\vec{H}) : xy \in E(\vec{H})\}$.
Then $\vec{H}$ is transitive if and only if

$$V(x) \supseteq W(x) := \cup_{y \in V(x)} V(y) \quad \text{for all } x \in V(\vec{H}).$$

The validity of test for transitivity is evident directly from the definition of transitive graphs.

If G_d^* is not transitive, then the given graph G is not transitive; this is to be proved later. If G_d^* is transitive, then repeat the whole procedure for the remaining undirected graph G_u^*. The cycle described above is referred as a *phase* of the algorithm. The following is the summary of the algorithm.

Algorithm transOrientation: For a graph G, test if G has a transitive orientation. If the answer is positive, a transitive orientation is given.

$G' \leftarrow G$;
while G' has at least one edge e **do**
{   apply the Γ-orientation rule until no implicants are left;
       apply test for transitivity to $G_d'^*$;
       if the answer is positive **then** $G' \leftarrow G_u'^*$
       else G has no transitive orientation and **stop**;
}
G has a transitive orientation, as constructed;

For any subset $C \subseteq E(G)$, let $V(C) := \{u, v : uv \in C\}$.

Recall that the subgraph of G induced by a nonempty subset S of $V(G)$ is the graph $G[S] = (S, E[S])$, where $E[S] = \{xy \in E(G) : x, y \in S\}$.

Theorem 7.4. *Algorithm transOrientation determines if a graph G has a transitive orientation in $O(\Delta m)$-time, where Δ is the maximum degree of a vertex in G and $m = |E(G)|$.*

Proof.   Since each edge in $E(G)$ is oriented once and all edges adjacent to an oriented edge are checked once, the time complexity is

$$\sum_{xy \in E(G)} (\deg(x) + \deg(y)) = \sum_{x \in V(G)} \deg(x)^2 \leq \Delta \sum_{x \in V(G)} \deg(x) = 2\Delta m.$$

To prove that the algorithm terminates in a positive answer if and only if G has a transitive orientation, according to induction, it is only necessary to prove that G_d^* is transitive and G_u^* has a transitive orientation if and only if G has a transitive orientation.

($\Rightarrow$) Suppose G_d^* is transitive and G_u^* has a transitive orientation $\vec{G}_u^*$. We shall prove that G has a transitive orientation. Let $\vec{G}$ be the orientation of G with $E(\vec{G}) = E(G_d^*) \cup E(\vec{G}_u^*)$. For any $xy, yz \in E(\vec{G})$, it is desired to prove that $xz \in E(\vec{G})$.

If $xy, yz \in E(G_d^*)$, then $xz \in E(G_d^*) \subseteq E(\vec{G})$ by the transitivity of G_d^*.

If $xy, yz \in E(\vec{G}_u^*)$, then $xz \in E(\vec{G}_u^*) \subseteq E(\vec{G})$ by the transitivity of $\vec{G}_u^*$.

If $xy \in E(G_d^*)$ and $yz \in E(\vec{G}_u^*)$, then $xz \in E(G)$, for otherwise $xz \notin E(G)$ and $xy \in E(G_d^*)$ would imply that $zy \in E(G_d^*)$ by the Γ-orientation rule, a contradiction. If $zx \in E(G_d^*)$, then $zy \in E(G_d^*)$ by the transitivity of G_d^*, a contradiction to $yz \in E(G_u^*)$. If $zx \in E(\vec{G}_u^*)$, then $yx \in E(\vec{G}_u^*)$ by the transitivity of $\vec{G}_u^*$, a contradiction to $xy \in E(G_d^*)$. Hence, $xz \in E(\vec{G})$.

Similarly, if $xy \in E(\vec{G}_u^*)$ and $yz \in E(G_d^*)$, then $xz \in E(\vec{G})$.

($\Leftarrow$) Suppose G has a transitive orientation $\vec{G}$. We shall prove that $(V(A), A)$ and $(V(B), B)$ both have transitive orientations, where $A = \tilde{e}$ is the ground edges of G_d^* and $B = E(G) \backslash A$ which is precisely $E(G_u^*)$.

For this purpose, the following claim plays an important role.

Claim 1. *Suppose graph H has a transitive orientation $\vec{H}$. If $C \subseteq E(H)$ satisfies condition $(*)$, then $(V(D), D)$ has a transitive orientation $(V(D), \vec{D})$, where $D = E(H) \backslash C$.*

(∗) If $x'y \in E(H)$ for some $x' \in V(C)$ and $y \in V(D)\backslash V(C)$, then $xy \in E(H)$ for all $x \in V(C)$.

Proof of Claim 1. Let $Y = \{y \in V(D)\backslash V(C) : x'y \in E(\vec{H})$ for some $x' \in V(C)$ and $yx'' \in E(\vec{H})$ for some $x'' \in V(C)\}$. Reversing all and only those edges $xy \in E(\vec{H})$ with $x \in V(C)$ and $y \in Y$, and then deleting those directed edges corresponding to edges in C and all isolated vertices resulted, to get an orientation $(V(D), \vec{D})$ of $(V(D), D)$. Note that after the edge reversing process, there is no edge $xy \in \vec{D}$ with $x \in V(C)$ and $y \in Y$.

Next prove that $(V(D), \vec{D})$ is transitive. Suppose to the contrary that $(V(D), D)$ is not transitive. There are two cases to be considered.

Case 1. There are edges $yx, xz, zy \in \vec{D}$. By the transitivity of $\vec{H}$, one of $yx, xz, zy \in \vec{D}$ was obtained from an edge in $E(\vec{H})$ by reversing.

By symmetry, assume $yx \in \vec{D}\backslash E(\vec{H})$, and so $x \in V(C)$, $y \in Y$ and $xy \in E(\vec{H})$. Then $z \notin V(C)$ as $zy \in \vec{D}$ and $y \in Y$; also $z \notin Y$ as $xz \in \vec{D}$ and $x \in V(C)$. Hence, $xz, zy \in E(\vec{H})$.

By the definition of $y \in Y$, there is some $x' \in V(C)$ such that $yx' \in E(\vec{H})$. Then $zx' \in E(\vec{H})$ by the transitivity of $\vec{H}$ and $zy, yx' \in E(\vec{H})$. But then $xz \in E(\vec{H})$ and $zx' \in E(\vec{H})$ imply that $z \in Y$, a contradiction.

Case 2. There are edges $yx, xz \in \vec{D}$ but $zy \notin D$. By the transitivity of $\vec{H}$, one of $yx, xz \in \vec{D}$ was obtained from an edge in $\vec{H}$ by reversing.

For the case of $yx \in \vec{D}\backslash E(\vec{H})$, then $x \in V(C)$, $y \in Y$ and $xy \in E(\vec{H})$. Then $z \notin V(C)$, otherwise $xy \in E(G)$, $z \in V(C)$ and $y \in V(D)\backslash V(C)$ imply $zy \in E(H)$ by condition (∗) and so $zy \in D$, a contradiction. Also $z \notin Y$ as $xz \in \vec{D}$ and $x \in V(C)$. So, $xz \in E(\vec{H})$. By the definition of $y \in Y$, there is some $x' \in V(C)$ such that $yx' \in E(\vec{H})$. Also, $xz \in E(G)$, $x' \in V(C)$ and $z \in V(D)\backslash V(C)$ imply $x'z \in E(G)$ by condition (∗). If $x'z \in E(\vec{H})$, then $yz \in E(\vec{H})$ by the transitivity of $\vec{H}$ and $yx', x'z \in E(\vec{H})$. Then $zy \in D$, a contradiction. If $zx' \in E(\vec{H})$, then $z \in Y$, a contradiction.

For the case of $xz \in \vec{D}\backslash E(\vec{H})$, then $z \in V(C)$, $x \in Y$ and $zx \in E(\vec{H})$. Then $y \notin V(C)$, as $yx \in E(\vec{D})$ and $x \in Y$. By the definition of $x \in Y$, there is some $z' \in V(C)$ such that $xz' \in E(\vec{H})$. Then $yz' \in E(\vec{H})$ by the transitivity of $\vec{H}$ and $yx, xz' \in E(\vec{H})$. By condition (∗) and $z'y \in E(H)$, we have $zy \in E(H)$, a contradiction. □

Having Claim 1 in mind, first to prove that $(V(B), B)$ has a transitive orientation. This follows from Claim 1 by letting $H = G$, $C = A$ and checking condition (∗) as in Claim 2.

Claim 2. *If $x'y \in E(G)$ for some $x' \in V(A)$ and $y \in V(B)\backslash V(A)$, then $xy \in E(G)$ for all $x \in V(A)$.*

Proof of Claim 2. First, $(V(A), A)$ is connected, since an edge only $\tilde{\Gamma}$-related to edges adjacent to it. Then for any $x \in V(A)$, there is an x'–x path $(x' = x_1, x_2, \dots, x_r = x)$ in $(V(A), A)$. If to the contrary that $xy \notin E(G)$, then there is some index j such that $x_j y \in E(G)$ but $x_{j+1} y \notin E(G)$. Therefore, $x_i x_{j+1} \in A$, $x_j y \in E(G)$ and $x_{j+1} y \notin E(G)$ will force $x_j y \in A$ by the Γ-orientation rule. Then $y \in V(A)$, violating the assumption that $y \in V(B)\backslash V(A)$. Hence, the claim holds. $\qquad \square$

Next to prove that $(V(A), A)$ has a transitive orientation.

Since G has a transitive orientation $\vec{G}$, $G_0 := G[V(A)]$ has a transitive orientation $\vec{G}_0 := \vec{G}[V(A)]$. Suppose $G[V(A)] - A$ has exactly p components (X_i, E_i), $i = 1, 2, \dots, p$. Inductively define $G_i = G_{i-1} - E_i$ for $i = 1, 2, \dots, p$. Then $G_p = (V(A), A)$.

To prove that G_p has a transitive orientation, it is only necessary to prove that if G_{i-1} has a transitive orientation then so is G_i for $2 \le i \le p$. This follows from Claim 1 by letting $H = G_{i-1}$, $C = E_i$ and checking condition (∗) as in Claim 3. Note that $V(C) = X_i$, $D = A \cup (\cup_{i < j \le p} E_j)$ and $V(D) = V(A)$.

Claim 3. *If $x'y \in E(G_{i-1})$ for some $x' \in X_i$ and $y \in V(A)\backslash X_i$, then $xy \in E(G_{i-1})$ for all $x \in X_i$.*

Proof of Claim 3. Since (X_i, E_i) is connected, for any $x \in X_i$, there is an x'–x path $(x' = x_1, x_2, \dots, x_r = x)$ in (X_i, E_i). If to the contrary that $xy \notin E(G_{i-1})$, then there is some index j such that $x_j y \in E(G_{i-1})$ but $x_{j+1} y \notin E(G_{i-1})$. Then $x_j, x_{j+1} \in X_i$ and $y \notin X_i$ imply $x_j y \in A$ and $x_{j+1} y \notin E(G)$ as (X_i, E_i) is a component of $G[V(A)] - A$. Therefore, $x_i y \in A$ and $x_{i+1} y \notin E(G)$ will force $x_i x_{i+1} \in A$ by the Γ-orientation rule, a contradiction to $x_i x_{i+1} \in E_i$. Hence, the claim holds. $\square$ $\qquad \square$

Spinrad (1983, 1985) presented an (n^2)-time algorithm and McConnell and Spinrad (1999) presented an $O(n + m)$-time algorithm for constructing an orientation of any given graph G such that the orientation is a transitive orientation of G if and only if G has a transitive orientation. However, in order to determine if G has a transitive it still needs to check if the given orientation is transitive. This can be done by using matrix multiplication, which takes $O(n^3)$ time. Theoretically, the algorithm by Strassen (1969) takes $O(n^{\log_2 7} = n^{2.8074})$ time. Recently, Duan *et al.* (2023) presented an $O(n^{2.371866})$-time algorithm and Williams, Xu *et al.* (2023) presented an $O(n^{2.371552})$-time algorithm for matrix multiplication. Note that these algorithms are not used in practices, since they are *galactic algorithms*: the coefficient hidden by the big O notation is so large that they are only worthwhile for matrices that are too large to handle on present-day computers.

7.2 Weighted independent domination in cocomparability graphs

Kratsch and Stewart (1993) presented an $O(n^3)$-time algorithm for the weighted independent domination problem in cocomparability graphs. In fact, their algorithm was described for the cardinality case. However, it is easy to be rewritten for the weighted case.

Suppose G is a cocomparability graph with a cocomparability ordering $[v_1, v_2, \ldots, v_n]$ and vertex v_i has a weight w_i of real number for $1 \leq i \leq n$. For technical reasons, let G' be the graph obtained from G by adding two new isolated vertices v_0 and v_{n+1} with weights $w_0 = w_{n+1} = 0$. Then G' is a cocomparability graph with a cocomparability ordering $[v_0, v_1, v_2, \ldots, v_n, v_{n+1}]$.

It is evident that any independent dominating set of G' must contain both v_0 and v_{n+1}; and D is an independent dominating set of G if and only if $D \cup \{v_0, v_{n+1}\}$ is an independent dominating set of G'.

The algorithm for the weighted independent domination in cocomparability graphs relies on the following lemma.

Lemma 7.5. *Suppose G' is a cocomparability graph with a cocomparability ordering $[v_0, v_1, \ldots, v_n, v_{n+1}]$, where v_0 and v_{n+1} are isolated vertices. An independent set $D = \{v_{i_0}, v_{i_1}, \ldots, v_{i_r}, v_{i_{r+1}}\}$, where $0 = i_0 < i_1 < \cdots < i_r < i_{r+1} = n+1$, is an independent dominating set of G' if and only if for $0 \leq j \leq r$ and $i_j < k < i_{j+1}$ the vertex v_k is adjacent to either v_{i_j} or $v_{i_{j+1}}$.*

Proof. ($\Leftarrow$) This is evident.

($\Rightarrow$) Suppose D is an independent dominating set of G'. For $0 \leq j \leq r$ and $i_j < k < i_{j+1}$, there is some j' such that v_k is adjacent to $v_{i_{j'}}$.

If $j' = j$ or $j' = j+1$, then v_k is adjacent to either v_{i_j} or $v_{i_{j+1}}$.

If $j' < j$, then by (CCO), $i_{j'} < i_j < k$ and $v_{i_j}, v_k \in E(G')$ imply that v_{i_j} is adjacent to either $v_{i_{j'}}$ or v_k. But D is independent and so v_{i_j} is not adjacent to $v_{i_{j'}}$, and then is adjacent to v_k.

If $j' > j+1$, then similarly $v_{i_{j+1}}$ is adjacent to v_k.   □

An $O(n^3)$-time dynamic programming algorithm for weighted independent domination in cocomparability graphs then follows.

Let D_i be a minimum w-weighted independent dominating set of $G'[\{v_0, v_1, \ldots, v_i\}]$ that contains v_i; and $d_i = w(D_i)$. Then $\gamma_i(G, w) = d_{n+1}$. The algorithm for the weighted independent domination in cocomparability graphs is as follows.

Algorithm wIndDomCocomparability: Suppose G is a cocomparability graph with a cocomparability ordering $[v_1, v_2, \ldots, v_n]$ and each vertex v_i has a weight w_i for $1 \leq i \leq n$. Find a minimum w-weighted independent dominating set D of G and $\gamma_i(G, w)$.

let $w_0 \leftarrow w_{n+1} \leftarrow 0$; $d_0 \leftarrow 0$;
for $i = 1$ **to** $n+1$ **do**
{ $t \leftarrow \infty$;
    for $j = i - 1$ **to** 0 **step** -1 **do**;
        if $(v_j v_i \notin E(G)$, v_k is adjacent to v_i or v_j for $j < k < i$, $d_j < t)$
        then { $t \leftarrow d_j$; $p_i \leftarrow j$ };
    $d_i \leftarrow t + w_i$;
}
$D \leftarrow \emptyset$;
$i \leftarrow p_{n+1}$;
while $i > 0$ **do** { $D \leftarrow D \cup \{v_i\}$; $i \leftarrow p_i$; }
$\gamma_i(G, w) \leftarrow d_{n+1}$;

Theorem 7.6. *Algorithm wIndDomCocomparability gives a minimum weighted independent dominating set of a cocomparability graph of n vertices in $O(n^3)$ time.*

It is possible to implement the algorithm in the following way to reduce the time complexity to $O(\Delta n^2)$, where n is the number of vertices and Δ is the maximum degree of G.

Algorithm wIndDomCocomparabilityRev: Suppose G is a cocomparability graph with an cocomparability ordering $[v_1, v_2, \ldots, v_n]$ and each vertex v_i has a weight w_i for $1 \le i \le n$. Find a minimum w-weighted independent dominating set D of G and $\gamma_i(G, w)$.

let $w_0 \leftarrow w_{n+1} \leftarrow 0;$ $d_0 \leftarrow 0;$
for $i = 1$ **to** $n + 1$ **do**
{   find the maximum index j^* such that $v_{j^*} v_i \notin E(G);$
     $t \leftarrow w_{j^*};$
     for all $j < j^*$ such that $v_j v_{j^*} \in E(G)$ but $v_j v_i \notin E(G)$ & $d_j < t$ **do**
        if (v_k is adjacent to v_i or v_j for $j < k < j^*$ with $v_k v_{j^*} \in E(G)$)
        then { $t \leftarrow d_j;$ $p_i \leftarrow j$ };
     $d_i \leftarrow t + w_i;$
}
$D \leftarrow \emptyset;$
$i \leftarrow p_{n+1};$
while $i > 0$ **do** { $D \leftarrow D \cup \{v_i\};$ $i \leftarrow p_i;$ }
$\gamma_i(G, w) \leftarrow d_{n+1};$

In two manuscripts, Breu and Kirkpatrick (1993) and Arvind *et al.* (1996) (see Kratsch (1998)) presented an $O(n^{2.376})$-time algorithm for the weighted independent domination problem in cocomparability graphs. It is curious if more efficient algorithms are possible for the (weighted) independent domination problem in cocomparability graphs.

Question 4. *Is it possible to design an $O(n^2)$-time algorithm for the independent domination problem in cocomparability graphs?*

Question 5. *Is it possible to design an $O(n^2)$-time algorithm for the weighted independent domination problem in cocomparability graphs?*

7.3   Connected domination in cocomparability graphs

While the weighted independent domination problem is polynomially solvable for cocomparability graphs, Chang (1995, 1997) proved that the weighted domination, the weighted total domination and the weighted connected domination problems are NP-complete for cocomparability graphs.

On the other hand, Kratsch and Stewart (1993) first presented an $O(n^3)$-time algorithm for the connected domination problem in cocomparability graphs.

Suppose G is a cocomparability graph with a cocomparability ordering $[v_1, v_2, \ldots, v_n]$. According to this vertex ordering, $V(G)$ can be linearly ordered as follows. A vertex u is *smaller than* (or *less than* or *before*) another vertex v, denoted by $u < v$, if $u = v_i$ and $v = v_j$ with $i < j$. Also, $u \le v$ means $u < v$ or $u = v$; vertex u is *larger than* (or *greater than* or *after*) vertex v, denoted by $u > v$, if $v < u$; and $u \ge v$ means $v \le u$. For a subset S of $V(G)$, the smallest element of S is denoted by $\min(S)$, and the largest element by $\max(S)$.

For vertices $u \le v$, the *interval* $[u, v]$ is the set $\{x : u \le x \le v\}$. A *source* is a vertex u which dominates $[v_1, u]$, and a *sink* is a vertex v which dominates $[v, v_n]$. The following lemma is a generalization of (CCO): if $uv \in E(G)$ with $u < v$, then $\{u, v\}$ dominates $[u, v]$.

Lemma 7.7. *If S is a vertex subset of a comparability graph G such that $G[S]$ is connected, then S dominates $[\min(S), \max(S)]$.*

Proof.   Choose a $\min(S)$–$\max(S)$ path $(x_1, x_2, \ldots, x_r)$ in $G[S]$, where $x_1 = \min(S)$ and $x_r = \max(S)$. If $y \in [\min(S), \max(S)] \backslash S$, then $x_i < y < x_{i+1}$ for some $1 \le i \le r - 1$. By (CCO), y is dominated by $\{x_i, x_{i+1}\}$. This proves the lemma.   □

The following are key lemmas for the algorithm on connected domination in cocomparability graphs by Kratsch and Stewart.

Lemma 7.8. *Any connected cocomparability graph G has a minimum connected dominating set that induces a chordless path.*

Proof.   Choose a minimum connected dominating set S of G with a cocomparability ordering $[v_1, v_2, \ldots, v_n]$. If $|S| \le 2$, then S induces a chordless path as desired. By now suppose $|S| \ge 3$.

The set $S' = S \cup \{v_1, v_n\}$ is also a connected dominating set. Choose a shortest v_1–v_n path $(x_1, x_2, \ldots, x_r)$ in $G[S']$, where $x_1 = v_1$ and $x_r = v_n$. By Lemma 7.7, $S'' := \{x_1, x_2, \ldots, x_r\}$ is a connected dominating set of G. Therefore, $|S| \le |S''| \le |S'| \le |S| + 2$.

For the case of $|S''| = |S|$, the set S'' is a minimum connected dominating set as desired.

For the case of $|S| + 1 \leq |S''| \leq |S'| \leq |S| + 2$, either $S \subseteq S''$ or there is exactly one vertex $y \in S \backslash S''$. For the former case, the connected dominating set S is as desired. For the latter case, $|S''| = |S| + 1$ and $y \in S \backslash S''$ and $x_1, x_n \notin S$. Further consider some subcases.

If $S'' \backslash \{x_1\}$ dominates all vertices smaller than $\min(S'' \backslash \{x_1\})$, then it is a connected dominating set as desired. Otherwise, there is a smallest vertex $z < \min(S'' \backslash \{x_1\})$ that is not dominated by $S'' \backslash \{x_1\} \supseteq S \backslash \{y\}$. Therefore, $zy \in E(G)$. For any vertex $z' < z$, z' is dominated by some x_i with $i \geq 2$. So $z' < z < x_i$, $z'x_i \in E(G)$ and $zx_i \notin E(G)$ imply $z'z \in E(G)$. Thus, z is a source adjacent to y.

Similarly, there is a sink w adjacent to y.

In conclusion, $\{z, y, w\}$ is a minimum connected dominating set inducing a chordless path as desired. $\qquad\square$

Lemma 7.9. *If* $P = (x_1, x_2, \ldots, x_r)$ *with* $x_1 < x_r$ *is a chordless path in a cocomparability graph* G, *then* $x_i < x_j$ *for* $1 \leq i < i + 1 < j \leq r$. *Consequently, if* $S = \{x_1, x_2, \ldots, x_r\}$ *with* $r \geq 2$, *then* $\min(S) = \min(\{x_1, x_2\})$ *and* $\max(S) = \max(\{x_{r-1}, x_r\})$.

Proof. Suppose to the contrary that $x_j < x_i$ for some $1 \leq i < i + 1 < j \leq r$.

If $x_1 < x_j$, then $x_1 < x_j < x_i$ and so there exists an edge $x_k x_{k+1}$ in P for some $k \leq i$ such that $x_{k-1} < x_j < x_k$, forcing a chord $x_j x_{k-1}$ or $x_j x_k$ by (CCO). This contradicts the fact that P is chordless.

If $x_j < x_1$, then $x_j < x_1 < x_r$ and so there exists an edge $x_k x_{k+1}$ in P for some $k \geq j$ such that $x_k < x_1 < x_{k+1}$, forcing a chord $x_1 x_k$ or $x_1 x_{k+1}$ by (CCO). This contradicts the fact that P is chordless.

Hence, the lemma holds. $\qquad\square$

Lemma 7.10. *If* S *is a connected dominating set of a connected cocomparability graph* G *with* $|S| \geq 2$ *that induces a chordless path* $(x_1, x_2, \ldots, x_r)$ *with* $x_1 < x_r$, *then every vertex* $y < \min(S)$ *is dominated by* $\{x_1, x_2\}$ *and every vertex* $z > \max(S)$ *is dominated by* $\{x_{r-1}, x_r\}$.

Proof. As S is a dominating set of G, every vertex $y < \min(S)$ is dominated by some x_i. If $i = 1$ or $i = 2$, then the lemma holds. If $i \geq 3$, then $y < \min(S) \leq x_1 < x_i$ by Lemma 7.9. This together with $yx_i \in E(G)$ but $x_1 x_i \notin E(G)$ implies $yx_1 \in E(G)$, as desired.

Similarly, every vertex $z > \max(S)$ is dominated by $\{x_{r-1}, x_r\}$. $\qquad\square$

An $O(n^3)$-time algorithm for the connected cocomparability graphs then follows, as described below. In fact this is an $O(\Delta m)$-time algorithm, where Δ is the maximum degree of G, since the running time is bounded by

$$\sum_{uv \in E(G)} (\deg(u) + \deg(v)) = \sum_{v \in V(G)} \deg(v)^2 \leq \Delta \sum_{v \in V(G)} \deg(v) = 2\Delta m.$$

Algorithm ConnDomCocomparability. Suppose G is a cocomparability graph with a cocomparability ordering $[v_1, v_2, \ldots, v_n]$. Find a minimum connected dominating set D of G and $\gamma_c(G)$.

if there is vertex v that dominates $V(G)$ **then**
    $\{\ D \leftarrow \{v\}; \quad \gamma_c(G) \leftarrow 1; \quad$ **stop**; $\}$
for each edge $uv \in E(G)$ **do**
$\{$ **if** $\{u, v\}$ dominates $[v_1, \min\{u, v\}]$ **then** label u, v, uv by L;
    if $\{u, v\}$ dominates $[\max\{u, v\}, v_n]$ **then** label u, v, uv by R;
    if uv is labeled both L and R **then**
        $\{\ D \leftarrow \{u, v\}; \quad \gamma_c(G) \leftarrow 2; \quad$ **stop**; $\}$
$\}$
add a new vertex s adjacent to all vertices labeled by L;
add a new vertex t adjacent to all vertices labeled by R;
find a shortest s–t path $(s, x_1, x_2, \ldots, x_r, t)$ in the new graph;
$D \leftarrow \{x_1', x_1, x_2, \ldots, x_{r-1}, x_r, x_r'\}$, where $x_1' x_1$ is labeled by L
                                  and $x_r' x_r$ is labeled by R;
$\gamma_c(G) \leftarrow r + 2;$

For the connected domination problem in cocomparability graphs, Kratsch (1998) presented a more efficient algorithm, which was originally given in two manuscripts by Breu and Kirkpatrick (1993) and Arvind *et al.* (1996) (see Kratsch (1998)).

For this purpose, it was first pointed out that a cocomparability graph has a *canonical cocomparability ordering* $[v_1, v_2, \ldots, v_n]$, which is one with two extra conditions:

$$v_1, v_2, \ldots, v_r \text{ are all sources, where } 1 \leq r < n;$$

$$v_s, v_{s+1}, \ldots, v_n \text{ are all sinks, where } 1 < s \leq n.$$

However, this is not always possible. For instance, in the star $K_{1,r}$ with $r \geq 3$, the root is a source and a sink in any cocomparability ordering of $K_{1,r}$. If the ordering is canonical, then every vertex in $K_{1,r}$ is either a source or a sink. Note that all sources form a clique and all sinks form another clique. But, any two clique of $K_{1,r}$ cannot cover all vertices of $K_{1,r}$, a contradiction.

In fact, the algorithm does not require the ordering to be canonical. This can be seen in the following proof of their result.

Theorem 7.11. *Every connected cocomparability graph G satisfying $\gamma_c(G) \geq 3$ has a minimum connected dominating set which is the vertex set of a shortest path between a source and a sink of G.*

Proof. Let $[v_1, v_2, \ldots, v_n]$ be a cocomparability ordering of G. By Lemma 7.8, G has a minimum connected dominating set S inducing a chordless path $P = (x_1, x_2, \ldots, x_r)$, where $r \geq 3$. Assuming $x_1 < x_r$, then by Lemma 7.10, every vertex $y < \min(S)$ is dominated by $\{x_1, x_2\}$ and every vertex $z > \max(S)$ is dominated by $\{x_{r-1}, x_r\}$. If x_1 is a source and x_r is a sink, then we may replace P by a shortest x_1–x_r path to conclude the theorem.

If x_1 is not a source, then there is a minimum $x' < x_1$ that is not dominated by x_1. For every $y < x'$, by the minimality, $yx_1 \in E(G)$ and so $yx' \in E(G)$. This proves that x' is a source. In case of $x_2 < x' < x_1$, by $x_2 x_1 \in E(G)$ but $x' x_1 \notin E(G)$, we have $x' x_2 \in E(G)$ by (CCO). In case of $x' < x_2$, by Lemma 7.10, $x' x_2 \in E(G)$. In any case, replacing x_1 in P by x' resulting a dominating path from a source x' to x_r of the same length.

Similarly, one may replace x_r by a sink x'' to result a dominating path of the same length from a source to a sink.

These reduce to the same situation as in the first paragraph. □

Note that the condition $\gamma_c(G) \geq 3$ is necessary in the proof of the above theorem. In case of $\gamma_c(G) = 1$, the theorem is true. But if $\gamma_c(G) = 2$ and x_1 is not a source and x_2 is not a sink, we may replace x_1 by x' so that $\{x', x\}$ dominates all vertices before for this set, but now it is not necessary for this set to dominate all vertices after the set, as $\{x_{r-1}, x_r\} = \{x_1, x_2\}$ but now x_1 is removed.

The following is a slightly simplified algorithm of theirs. Note that the only dominating step is the case of $\gamma_c(G) = 2$, which takes $O(\Delta m)$ time; all other places only need $O(n + m)$ time.

Algorithm ConnDomCocomparabilityRev: Suppose G is a
cocomparability graph with an cocomparability ordering $[v_1, v_2, \ldots, v_n]$.
Find a minimum connected dominating set D of G and $\gamma_c(G)$.

if there is a vertex v that dominates $V(G)$ **then**
    $\{ D \leftarrow \{v\}; \quad \gamma_c(G) \leftarrow 1; \quad$ **stop;** $\}$
for each edge $uv \in E(G)$ **do**
    if $\{u, v\}$ dominates G **then** $\{ D \leftarrow \{u, v\}; \quad \gamma_c(G) \leftarrow 2; \quad$ **stop;** $\}$
identify all sources and add a new vertex s adjacent to all of them;
identify all sinks and add a new vertex t adjacent to all of them;
find a shortest s–t path $(s, x_1, x_2, \ldots, x_r, t)$ in the new graph;
$D \leftarrow \{x_1, x_2, \ldots, x_{r-1}, x_r\}; \quad \gamma_c(G) \leftarrow r;$

The performance bottleneck lies in the test for $\gamma_c(G) = 2$. The crux
is that there are permutation graphs for which each minimum connected
dominating set of size two contains neither a source nor a sink vertex, see
the papers by Ibarra and Zheng (1994) and Köhler (1996, 2000). Minimum
dominating sets of this type cannot be found by a shortest path approach.

Question 6. *Whether lines 3 and 4 of Algorithm ConDomComparabilityRev can be implemented in a more efficient way?*

7.4   (Total, clique) domination in cocomparability graphs

Kratsch and Stewart (1993) presented an $O(n^6)$-time algorithm for the
domination problem in cocomparability graphs by combining the ideas
for independent domination and connected domination. Namely, each connected component of the subgraph induced by a minimum domination
set behaves like that for connected domination, as shown in the following
lemma.

Lemma 7.12. *Every cocomparability graph G has a minimum dominating
set D such that each connected component of $G[D]$ is an induced path.
Furthermore, different components of a minimum dominating set cannot
overlap in the cocomparability ordering.*

Proof.   This follows from Lemma 7.8. Choose a minimum dominating
set D. If one of the connected component of $G[D]$, say C, is not an induced
path, then there is an induced path C' such that $N[C] \subseteq N[C']$ and
$|C'| \leq |C|$. Hence, $D' = (D \backslash C) \cup C'$ is also a minimum dominating set.

If $G[D']$ still has a component that is not an induced path, continue the same process. Eventually, a desired minimum dominating set is reached. The furthermore part holds, for otherwise one component would cover a vertex of the other component and therefore there is an edge between the components.   □

Consider a cocomparability graph G with a cocomparability ordering $[v_1, v_2, \ldots, v_n]$. The first step to implement the lemma is to construct the graph G' obtained from G by adding two isolated vertices v_0 and v_{n+1}. Then G' is a cocomparabilty graph for which $[v_0, v_1, v_2, \ldots, v_n, v_{n+1}]$ is a cocomparability ordering. Moreover, D' is a dominating set of G' if and only if $D' = \{v_0, v_{n+1}\} \cup D$ for some dominating set of G.

Next, the following lemma shows that we only need to pay attentions to at most four vertices.

Lemma 7.13. *Suppose $D \subseteq V(G')$ such that each component of $G'[D]$ is an induced path. Then D is a dominating set of G' if and only if for every two consecutive vertices s and t of D in the cocomparability ordering in two different components of $G'[D]$ one of the following is true:*

(i) *$\{s, t\}$ dominates $[s, t]$ when s and t are isolated vertices in $G'[D]$,*

(ii) *$\{s', s, t\}$ dominates $[s, t]$ when s has a neighbor s' and t is an isolated vertex in $G'[D]$,*

(iii) *$\{s', s, t', t\}$ dominates $[s, t]$ when s has a neighbor s' and t has a neighbor t' in $G'[D]$.*

Proof. The sufficiency is obvious. The necessity follows from the fact that any vertex $a \in [s, t]$ that is dominated by $v \in D \setminus \{s', s, t', t\}$ is also dominated by $\{s, t\}$, since the edge av must cover one of $\{s, t\}$, forcing $as \in E[G']$ or $at \in E[G']$ by the cocomparabilitry ordering.   □

To implement the lemmas, Kratsch and Stewart used five state parameters $z_1 \in \{0, 1, \ldots, n\}$ and $z_2, z_3, z_4, z_5 \in \{v_0, v_1, v_2, \ldots, v_n, v_{n+1}\}$, where

z_1 is the number of vertices in current D,

z_2 is the third last vertex of D (in the order of the components),

z_3 is the second last vertex of D,

z_4 is the last vertex of D,

z_5 is the last processed vertex.

The reason to keep three vertices z_2, z_3, z_4 rely on Lemma 7.9 that $x_i < x_{i+2}$ for a chordless path component $P = (x_1, x_2, \ldots, x_r)$ of D with $x_1 < x_r$.

To get the solution from old states to a new states, eight cases each further contains four subcases are discussed for the parameters z_2, z_3, z_4, i. At end, an $O(n^6)$-time algorithm is established.

Kratsch and Stewart also gave an $O(n^6)$-time algorithm for the total domination problem in cocomparability graphs. The method is precisely the same as that for domination, except that the induced path in Lemma 7.12 has at least two vertices.

In two manuscripts, Breu and Kirkpatrick (1993) and Arvind *et al.* (1996) (see Kratsch (1998)) gave an $O(nm^2)$-algorithm for the domination and the total domination problems in cocomparability graphs.

Kratsch and Stewart also proved that finding dominating cliques is NP-hard for cocomparability graphs, while the problem is trivial for interval graph and comparability graphs. Note that for both interval graphs and comparability graphs, if the graph has a dominating clique, then it has a dominating clique of size at most two.

Theorem 7.14. *Given a cocomparability graph G and an integer k, it is NP-complete to determine if G has a dominating clique of size at most k.*

Proof.   Given a graph H and an integer m, let $k = m + 1$ and construct a cocomparability graph G as follows:

$$V(G) = \{a, b\} \cup \{v, v' : v \in V(H)\} \quad \text{and}$$
$$E(G) = \{ab\} \cup \{bv : v \in V(H)\}$$
$$\cup \{uv, u'v' : u, v \in V(H), u \neq v\}$$
$$\cup \{uv' : u = v \text{ or } uv \in E(H)\}.$$

Note that G is a cocomparability graph with a cocomparability ordering $(a, b$, any ordering of all $v \in V(H)$, any ordering of all v' with $v \in V(H))$.

The following shows that G has a dominating clique of size at most k if and only if H has a dominating set of size at most m.

If G has a dominating clique D of size at most k, then it must contain b and so the remaining vertices are all in $V(H)$. Since every vertex $v' \in V(G)$ is dominated by some $u \in D\backslash\{b\}$, the set $D\backslash\{b\}$ is a dominating set of H with size at most $k - 1 = m$.

If H has a dominating set D of size at most m, then $D \cup \{b\}$ is clearly a dominating clique of G with size at most $m + 1 = k$.

Therefore, it is NP-complete to determine if a cocomparability graph G has a dominating clique of size at most k, since it is NP-complete to determine if a graph H has a dominating set of size at most m. □

To prove that finding dominating cliques is NP-hard for cocomparability graphs, the following NP-complete problem is used for reduction.

The Monotone 3-Satisfiability Problem (Monotone 3-SAT)
Parameters: A set $V = \{u_1, u_2, \ldots, u_n\}$ of n logical variables and a family $\{C_j = a_{j,1} \lor a_{j,2} \lor a_{j,3} : 1 \le j \le m\}$ of clauses using the variables in V, where every clause C_j is either positive (i.e., $a_{j,1}, a_{j,2}, a_{j,3} \in V$) or negative (i.e., $a_{j,1}, a_{j,2}, a_{j,3} \in \overline{V} := \{\bar{u}_1, \bar{u}_2, \ldots, \bar{u}_n\}$) and at least one clause is positive and at least one is negative.
Problem: Is it possible to assign true/false values to V such that every clause C_j is satisfiable?

Theorem 7.15. *It is NP-complete to determine if a cocomparability graph has a dominating clique.*

Proof. For an instance of Monotone 3-SAT, construct a cocomparability graph G as follows:

$$V(G) = \{C_j : 1 \le j \le m\} \cup \{u_i, \bar{u}_i : 1 \le i \le n\} \quad \text{and}$$

$$E(G) = \{C_j C_{j'} : j \ne j', \text{ both } C_j \text{ and } C_{j'} \text{ are either positive or negative}\}$$

$$\cup \{u_i u_{i'}, u_i \bar{u}_{i'}, \bar{u}_i \bar{u}_{i'} : i \ne i', 1 \le i \le n, 1 \le i' \le n\}$$

$$\cup \{u_i C_j : u_i \text{ is some } a_{j,i'} \text{ for some positive clause } C_j\}$$

$$\cup \{\bar{u}_i C_j : \bar{u}_i \text{ is some } a_{j,i'} \text{ for some negative clause } C_j\}.$$

Note that G is a cocomparability graph with a cocomparability ordering (any ordering of all positive clauses C_j, $u_1, u_2, \ldots, u_n, \bar{u}_1, \bar{u}_2, \ldots, \bar{u}_n$, any ordering of all negative clauses C_j).

Suppose it is possible to assign true/false values to V such that every clause C_j is satisfiable. Then $D = \{u_i : u_i \in V \text{ is assigned true}\} \cup \{\bar{u}_i : u_i \in V \text{ is assigned false}\}$ is a dominating clique, since every clause C_j contains a literal from D.

Now suppose D is a dominating clique of G. If D contains some negative clause C_j, then D only contains negated variables $\bar{u}_i$ but no nonnegated variable $u_{i'}$. And so D does not dominate positive clauses, a contradiction.

Hence, D cannot contain any negative clause. Similarly, D cannot contain any positive clause. In other words, $D \subseteq \{u_i, \bar{u}_i : 1 \leq i \leq n\}$.

Since D is a clique, it cannot contain both u_i and $\bar{u}_i$ for all $u_i \in V$. Then for every $u_i \in D$ assign u_i true, for every $\bar{u}_i \in D$ assign u_i false, and for all other variables u_i assign it true. Then every clause has a literal assigned true since D dominates it.

Therefore, it is NP-complete to determine if a cocomparability graph G has a dominating clique, since Monotone 3-SAT is NP-complete. $\square$

7.5   Asteroidal triple-free graphs

A stable set of three vertices is an *asteroidal triple* if between each pair in the triple there exists a path that avoids all neighborhoods of the third. For example, the outer vertices of a 3-sun form an asteroidal triple. A graph is *asteroidal-triple-free* (or *AT-free* for short) if it contains no asteroidal triples. The concept of AT-freeness was first considered by Lekkerkerker and Boland (1962). The following is an alternative proof for their characterization of interval graphs in terms of chordal graphs and AT-freeness.

Theorem 7.16 (Lekkerkerker and Boland, 1962). *A graph is an interval graph if and only if it is chordal and AT-free.*

Proof.   ($\Rightarrow$) Suppose G is an interval graph. Choose an interval ordering $[v_1, v_2, \ldots, v_n]$. For every cycle C of length at least four in G, choose a minimum indexed vertex v_i in C. For the two neighbors v_j and v_k of v_i in the cycle, either $i < j < k$ or $i < k < j$. By property (IO), $v_j v_k \in E(G)$ is a chord of C. This shows that G is chordal.

Suppose $\{v_i, v_j, v_k\}$ is a stable set with $i < j < k$. For any v_i–v_k path $P = (v_i = v_{r_1}, v_{r_2}, \ldots, v_{r_s} = v_k)$, since $r_1 = i < j < k = r_s$, there is an index t such that $r_{t-1} \leq j < r_t$. By property (IO), v_j is adjacent to v_{r_t} which is a vertex in P. Hence, $\{v_i, v_j, v_k\}$ is not an asteroidal triple.

($\Leftarrow$) Suppose G is chordal and AT-free. We shall prove by induction on $n := |V(G)|$ that G is an interval graph. The claim is clear for $n = 1$. Suppose $n \geq 2$ and the claim holds for $n' < n$.

First, we may assume that G is connected.

If G has two adjacent simplicial vertices x and y, then $N[x] = N[y]$. Choose an interval model $\mathcal{F} = \{I_v : v \in V(G - x)\}$ of $G - x$ and let $I_x = I_y$. Then G is an interval graph with $\mathcal{F} \cup \{I_x\}$ as an interval model.

Hence, we may assume that every two simplicial vertices are not adjacent. In particular, G is not a complete graph and $n \geq 3$.

Claim 1. If x is a simplicial vertex of a connected interval graph H with $H - N[x]$ connected, then H has an interval model $\mathcal{F} = \{I_v = [a_v, b_v] : v \in V(H)\}$ such that $\cup_{v \in V(H)} I_v$ is an interval $[a, b]$ and the following hold.

(1) $C = V(H) \backslash N[x] = \{z_1, z_2, \dots, z_m\}$ such that

$$b_{z_1} < b_{z_2} < \dots < b_{z_m} < a_x < b_x = b_y = b \quad \text{for } y \in N(x) \quad \text{and}$$

$$N(x) \cap N(z_1) \subseteq N(x) \cap N(z_2) \subseteq \dots \subseteq N(x) \cap N(z_m).$$

(2) $a = a_y < b_{z_1} < a_{y'}$ for $y \in N(x) \cap N(z_1)$ and $y' \in N(x) \backslash N(z_1)$.

Proof of Claim 1. Choose an interval model $\mathcal{F} = \{I_v = [a_v, b_v] : v \in V(H)\}$ of H. Since H is connected, $\cup_{v \in V(H)} I_v$ is an interval $[a, b]$.

As $H - N[x]$ is connected, $\cup_{z \in C} I_z$ is an interval $[a', b']$ disjoint from $I_x = [a_x, b_x]$. By symmetry, assume that $b' < a_x$ and so one may indexing vertices in C such that $b_{z_1} < b_{z_2} < \dots < b_{z_m} < a_x$. Since $N[x]$ is a clique, we may replace b_x and b_y by b for all $y \in N(x)$ keeping $\mathcal{F}$ still being an interval model of H. The last statement of (1) then also follows.

If $y \in N(x) \cap N(z_1)$, then $y \in N(x) \cap N(z_i)$ for all $z_i \in C$ and so $I_y \cap I_{z_i} \neq \emptyset$ for all $z_i \in C$. One may then replace a_y by a for all $y \in N(x) \cap N(z_1)$ keeping $\mathcal{F}$ still being an interval model of H. If $y' \in N(x) \backslash N(z_1)$, then $I_{y'} \cap I_{z_1} = \emptyset$. This together with $b_{z_1} < b = b_{y'}$ give that $b_{z_1} < a_{y'}$. $\qquad\square$

We now consider the following two cases.

Case 1. $G - N[x]$ is connected for every simplicial vertex x of G.

Choose a simplicial vertex x_1 of G and two simplicial vertices x_2 and x_3 of $G_1 := G - x_1$ (either G_1 is complete, or is not complete and then has two nonadjacent simplicial vertices).

If $x_2, x_3 \in V(G) \backslash N[x_1]$, then x_2 and x_3 are also simplicial vertices of G, since $N_{G_1}[x_i] = N_G[x_i]$ for $i = 2, 3$. Hence, $\{x_1, x_2, x_3\}$ is stable. In fact it is an asteroidal triple, since $G - N[x_i]$ is connected and so for $j, k \neq i$ there is an x_j–x_k path avoiding neighbors of x_i for $i = 1, 2, 3$. Hence, one may assume that $x_2 \in N_G(x_1)$.

Next prove that $G_1 - N_{G_1}[x_2]$ is connected. Suppose to the contrary that $G_2 := G_1 - N_{G_1}[x_2]$ has at least two components, say H_1 and H_2. Then for $i = 1, 2$ the graph $G_1[N_{G_1}[x_2] \cup V(H_i)]$ has at least two simplicial vertices and one of them, say y_i, is in $V(H_i)$. Then y_1 and y_2 are also simplicial vertices of G, since $N_{G_1}[y_i] = N_G[y_i]$ for $i = 1, 2$. Again, $\{x_2, y_1, y_2\}$ is an asteroidal triple by the same arguments as above, a contradiction.

Apply Claim 1 to $H = G_1$ and $x = x_2$ to get an interval model $\mathcal{F}$ of G_1 as described in the claim. For every $y \in N_G(x_1) \subseteq N_{G_1}[x_2]$, we may extend the right endpoint $b_y = b$ of I_y to $b + 2$ keeping $\mathcal{F}$ still being an interval model of G_1. In this way, we may let $I_{x_1} = [b + 1, b + 2]$ only intersecting those I_y for $y \in N_G(x_1)$. Therefore, $\mathcal{F} \cup \{I_{x_1}\}$ is an interval model of G and then G is an interval graph.

Case 2. $G - N[x]$ is not connected for some simplicial vertex x of G.

Choose a simplicial vertex x_1 of G for which $G - N[x_1]$ has $k \geq 2$ components $C_1, C_2, \ldots, C_k$. Let $G_1 = G - x_1$ and choose an interval model $\mathcal{F}' = \{I_v' : v \in V(G_1)\}$ for G_1. For $1 \leq i \leq k$, since C_i is connected, $\cup_{v \in V(C_i)} I_v'$ is an interval $[a_i', b_i']$. Without loss of generality, assume

$$a_1' < b_1' < a_2' < b_2' < \cdots < a_k' < b_k'.$$

Since $N(x_1)$ is a clique, $\cap_{y \in N(x)} I_y'$ is an interval $[c, d]$.

For the case when $N(x_1) \not\subseteq N(y)$ for every $y \in V(G) \backslash N[x_1] = \cup_{i=1}^k V(C_i)$, it is the case that I_y' is disjoint from $[c, d]$. Then one may let $I_{x_1} = [c, d]$ so that $\mathcal{F}' \cup \{I_{x_1}\}$ is an interval model of G. This gives that G is an interval graph.

By now there exists some C_j such that $A = \{y \in V(C_j) : N(x_1) \subseteq N(y)\} \neq \emptyset$. Apply Claim 1 to $H = G[V(C_j) \cup N[x_1]])$ and $x = x_1$ to get an interval model $\mathcal{F}$ of H as described in the claim. Note that $V(C_j) = \{z_1, z_2, \ldots, z_m\}$ as in (1) of Claim 1.

Note that in the interval model $\mathcal{F}$, the interval I_{z_1} is disjoint and on the left of the intervals $\cap_{y \in N(x)} I_y$ and I_{x_1}. In the interval model $\mathcal{F}'$, the interval I_{z_1}' is disjoint from $\cap_{y \in N(x)} I_y'$. If necessary, one may reflect intervals of $\mathcal{F}'$ and assume that I_{z_1}' is on the left of $\cap_{y \in N(x)} I_y'$.

Also, I_y being on the right of I_{z_1} is the same as I_y' being on the right of I_{z_1}' for $y \in N(x) \backslash N(z_1)$. By shifting and scaling, one may change $\cup_{v \in H} I_v$ from $[a, b]$ to $[a_j', b_j']$. Then change the part in $[a_j', b_j']$ of the model $\mathcal{F}'$ by $\mathcal{F}$, which also includes I_{x_1}, results an interval model of G. This proves that G is an interval graph. $\square$

A similar proof for the ($\Rightarrow$) part of the above theorem also gives that a cocomparability graph is AT-free.

Theorem 7.17. *A cocomparability graph is AT-free.*

Proof. Choose a cocomparability ordering $[v_1, v_2, \ldots, v_n]$ of the graph. Suppose G has an asteroidal triple v_i, v_j, v_k. Assume $i < j < k$. For any v_i–v_k path $P = (v_i = v_{r_1}, v_{r_2}, \ldots, v_{r_s} = v_k)$, since $r_1 = i < j < k = r_s$, there is an index t such that $r_{t-1} \leq j < r_t$. By property (CO), either $v_j = v_{t-1}$ or else v_j is adjacent to v_{t-1} or v_t which is a vertex P, contradicting that v_i, v_j, v_k is an asteroidal triple. $\qquad\square$

We say that a vertex v *intercepts* a path P if $N[v] \cap V(P) \neq \emptyset$ or equivalently v is dominated by some vertex in P, otherwise v is said to *miss* P. For a pair of vertices u and v in a connected graph G,

$$D(u, v) = \{z : z \text{ intercepts all } u\text{–}v \text{ paths.}\}$$

Note that $N[u] \cup N[v] \subseteq D(u, v) = D(v, u)$. Also, $\{u, v, w\}$ is an asteroidal triple if and only if $u \notin D(v, w)$, $v \notin D(u, w)$ and $w \notin D(u, v)$. A *dominating pair* of G is a pair (u, v) such that $D(u, v) = V(G)$.

Balakrishnan *et al.* (1993) gave an $O(n^3)$-time algorithm for the connected domination problem in AT-free graphs. The paper used the following interesting theorem quoted from a technical report by Corneil *et al.* (1992), which I cannot find. Corneil *et al.* (1997) gave a non-algorithmic proof of the theorem in 1997 and an algorithmic one in 1999 by using the LexBFS procedure. The following is the first proof.

Theorem 7.18 (Corneil et al., 1997). *Every connected AT-free graph G contains a dominating pair.*

Proof. Choose a vertex x in G. We may assume that $N[x] \neq V(G)$ for otherwise (x, x) is a dominating pair. Let $G' = G - N[x]$. First, we have several claims, where C is a component of G'.

Claim 1. If $u, v \in V(C)$, then $D(u, x) \backslash V(C) = D(v, x) \backslash V(C)$.

Proof of Claim 1. Choose a u–v path P in C. Suppose $w \in D(u, x) \backslash V(C)$. If $w \in N[x]$, then $w \in D(v, x) \backslash V(C)$. Now suppose $w \in V(G') \backslash V(C)$. For any v–x path P', PP' is a u–x path and so dominates w. As w is not dominated by P, it is dominated by P'. Then $w \in D(v, x)$ and so $D(u, x) \backslash V(C) \subseteq D(v, x) \backslash V(C)$.

Similarly, $D(v, x) \backslash V(C) \subseteq D(u, x) \backslash V(C)$ and so the claim holds. $\qquad\square$

Claim 2. If $u \in D(v, x)$, $w \in D(u, x)$ but $w \notin N[u]$, then $w \in D(v, x)$.

Proof of Claim 2. For any v–x path $P = (v_0, v_1, \ldots, v_r)$, since $u \in D(v, x)$, vertex u is dominated by some v_j in P. For the u–x path $P' = (u, v_j, v_{j+1}, \ldots, v_r)$, since $w \in D(u, x)$, vertex w is dominated by some v_k in P'. But $v_k \neq u$, so w is dominated by P and then $w \in D(v, x)$. $\square$

Claim 3. If $u, v \in V(C)$, then $u \in D(v, x)$ or $v \in D(u, x)$.

Proof of Claim 3. Suppose to the contrary that $u \notin D(v, x)$ and $v \notin D(u, x)$. Then $\{u, v, x\}$ is an asteroidal triple, since there exists a u–v path in C that does not dominate x. The claim then follows. $\square$

Claim 4. If $u, v \in V(C)$ and $v \notin D(u, x)$, then $D(u, x) \subsetneq D(v, x)$.

Proof of Claim 4. By Claim 3, $u \in D(v, x)$. Suppose $w \in D(u, x)$. If $w \in N[x]$, then $w \in D(v, x)$. If $w \notin N[u]$, then $u \in D(v, x)$ and $w \in D(u, x)$ imply $w \in D(v, x)$ by Claim 2.

If $w \notin N[x]$ and $w \in N[u]$, then $w \in V(C)$. Since $v \notin D(u, x)$, some u–x path P does not dominate v and so the w–x path wP does not dominate v. This gives $v \notin D(w, x)$. However, by Claim 3, $w \in D(v, x)$. In any case, $D(u, x) \subseteq D(v, x)$. The claim then follows from that $v \in D(v, x) \backslash D(u, x)$.

$\square$

Now choose a set $S = \{v_1, v_2, \ldots, v_n\} \subseteq V(G')$ such that $U = V(G)$, where $U = \cup_{v_i \in S} D(v_i, x)$. Note that such S exists, for instance we may choose $S = V(G')$. We further assume that S is chosen so that n and m are minimum, where $m = |\{(p, q) : p \neq q, v_p \in D(v_q, x)\}|$.

Suppose C_i is the component of G' containing v_i for $1 \leq i \leq n$.

We first claim that $C_i \neq C_j$ for $v_i \neq v_j$. Suppose to the contrary that $C_i = C_j$ for some $i \neq j$. Choose a vertex $u \in C_i$ such that $D(u, x) \subsetneq D(u', x)$ for no vertex $u' \in C_i$. Then $C_i \subseteq D(u, x)$, for otherwise any vertex $v \in C_i$ but $v \notin D(u, x)$ would imply that $D(u, x) \subsetneq D(v, x)$ by Claim 4, a contradiction. By Claim 1, $D(v_i, x) \cup D(v_j, x) \subseteq D(u, x)$. Replacing v_i and v_j by u in S reduces the size of S, contradicting the minimality of n.

Next, we claim that $m = 0$, i.e., $v_i \notin D(v_j, x)$ for $i \neq j$. Suppose to the contrary that $v_p \in D(v_q, x)$ for some $p \neq q$. Let $U' = \cup_{v_i \in S, v_i \neq v_p} D(v_i, x)$ and $\overline{U'} = U \backslash U'$. By the minimality of n, $\overline{U'} \neq \emptyset$. Choose a vertex $v'_p \in \overline{U'}$ such that $D(v'_p, x) \subsetneq D(u', x)$ for no vertex $u' \in \overline{U'}$. Then $\overline{U'} \subseteq D(v'_p, x)$, for otherwise any vertex $v \in \overline{U'}$ but $v \notin D(v'_p, x)$ would imply that $D(v'_p, x) \subsetneq D(v, x)$, a contradiction. Note that for $j \neq p$: $v'_p \notin D(v_j, x)$, and by Claim 1,

$v_j \in D(v_p, x)$ iff $v_j \in \cup D(v'_p, x)$. Replacing v_p by v'_p in S decreases the value of m, contradicting the minimality of m.

Having these two properties in mind, we now consider three cases.

Case 1. $n = 1$.

In this case, $D(v_1, x) = V(G)$ and so (v_1, x) is a dominating pair.

Case 2. $n = 2$.

In this case (v_1, v_2) is a dominating pair. For otherwise, there is a v_1–v_2 path $(u_0, u_2, \ldots, u_r, \ldots, u_s, u_{s+1}, \ldots, u_t)$ that does not dominate some $u \in V(G)$, where r (respectively, s) is the first (respectively, last) index such that $u_r \in N(x)$ (respectively, $u_s \in N(x)$). Then $u \in N(x)$. This implies that $v_2 \notin D(v_1, u)$, since the path $(u_0, u_2, \ldots, u_r, x, u)$ does not dominate v_2. Similarly, $v_1 \notin D(v_2, u)$, since the path $(u, x, u_s, u_{s+1}, \ldots, u_t)$ does not dominate v_1. Then $\{v_1, v_2, u\}$ is an asteroidal triple, a contradiction.

Case 3. $n \geq 3$.

In this case, $\{v_1, v_2, v_3\}$ is an asteroidal triple, a contradiction. $\quad\square$

Note that the converse of the theorem is not true, i.e., not every graph with dominating pairs is AT-free. For instance, C_6 has exactly three dominating pairs, but any stable set of size three is an asteroidal triple. Also, for any graph G the graph obtained from G by adding a new vertex adjacent to all old vertices has dominating pairs, but it may have asteroidal triples.

In fact the 1997 paper also proved a structure theorem for all dominating pairs if the AT-free graph has diameter at least four.

Theorem 7.19 (Corneil *et al.*, 1997). *A connected AT-free graph G of diameter at least four has two nonempty and disjoint sets X and Y such that (x, y) is a dominating pair if and only if $x \in X$ and $y \in Y$.*

Note that the theorem is best possible in the sense that, for AT-free graphs of diameter less than four, the sets X and Y are not guaranteed to exist. To wit, C_5 and $P_2 \,\square\, P_3$ provide counterexamples of diameter two and three, respectively.

7.6 Connected domination in AT-free graphs

Balakrishnan *et al.* (1993) gave an $O(n^3)$-time algorithm for the connected domination problem in AT-free graphs. They also extend the method to design an $O(n^3)$-time algorithm for the Steiner set problem in AT-free graphs.

They started with an $O(n^3)$-time algorithm finding all dominating pairs of an arbitrary graph. It bases on the fact that (u, v) is a dominating pair of a graph G if and only if for every vertex $x \in V(G) \setminus (N[u] \cup N[v])$; u and v are in different components of $G - N[x]$. (Otherwise, there exists a u–v path that does not contain a neighbor of x.)

Algorithm DomPairG: Find all dominating pairs of a connected graph G with $V(G) = \{v_1, v_2, \ldots, v_n\}$, representing by $A[i, j] = true$ if (v_i, v_j) is a dominating pair and $A[i, j] = false$ otherwise.

$A[i, j] \leftarrow true$ for $i = 1, 2, \ldots, n$ and $j = 1, 2, \ldots, n$;
for all $x \in V(G)$ **do**
    $G' \leftarrow G - N[x]$;
    do a DFS to find all components of G';
    for $i = 1$ **to** n **do**
    for $j = 1$ **to** n **do**
        if v_i and v_j are in a same component of G'
        then $A[i, j] \leftarrow false$;

Suppose (u, v) is a dominating pair of a connected AT-free graph G. For each $x \in N(u)$, let $A_x \subseteq V(G)$ be the set of all vertices in $V(G) \setminus \{u\}$ such that if $a \in A_x$, then $\{a, x\}$ dominates all vertices in $N(u)$. Similarly, for each $y \in N(v)$, let $B_y \subseteq V(G)$ be the set of all vertices in $V(G) \setminus \{v\}$ such that if $b \in B_y$, then $\{b, y\}$ dominates all vertices in $N(v)$. Define Γ to be the following set of paths:

$\{P : P = (u, \ldots, v)$, or

    $P = (x, a, \ldots, v)$ for $x \in N(u)$ and $a \in A_x$, or

    $P = (u, \ldots, b, y)$ for $y \in N(v)$ and $b \in B_y$, or

    $P = (x, a, \ldots, b, y)$ for $x \in N(u)$, $a \in A_x$, $y \in N(v)$ and $b \in B_y\}$.

Theorem 7.20. *Every path $P' \in \Gamma$ with $|P'| = \min_{P \in \Gamma} |P|$ is a minimum connected dominating set of the AT-free graph G.*

Proof. We first show that every path $P \in \Gamma$ is a connected dominating set of G. Clearly, P is connected. To see that P is a dominating set, we consider four cases.

Case 1. $P = (u, \ldots, v)$.

In this case, P is a dominating set, since (u, v) is a dominating pair.

Case 2. $P = (x, a, \ldots, v)$ for $x \in N(u)$ and $a \in A_x$.

In this case, uP is a dominating set. Since $\{x, a\}$ dominates all vertices dominated by u, then P is also a dominating set.

Case 3. $P = (u, \ldots, b, y)$ for $y \in N(v)$ and $b \in B_y$.

In this case, Pv is a dominating set. Since $\{b, y\}$ dominates all vertices dominated by v, then P is also a dominating set.

Case 4. $P = (x, a, \ldots, b, y)$ for $x \in N(u)$, $a \in A_x$, $y \in N(v)$ and $b \in B_y$.

In this case, uPv is a dominating set. Since $\{x, a, b, y\}$ dominates all vertices dominated by $\{u, v\}$, then P is also a dominating set.

Next, we show that P' is a minimum connected dominating set. Choose a minimum connected dominating set D of G.

Case a. $u, v \in D$.

In this case, choose a u–v path P in $G[D]$. Then $|P'| \leq |P| \leq |D|$ and so P' is a minimum connected dominating set.

Case b. $u \notin D$ and $v \in D$.

In this case, u is a dominated by some $x \in D \cap N(u)$. Choose an x–v path $P = (x, a, \ldots, v)$ in $G[D]$.

If $P \subsetneq D$, then $P \cup \{u\}$ is connected and so has a u–v path P''. Hence, $|P'| \leq |P''| \leq |P| + 1 \leq |D|$ and so P' is a minimum connected dominating set.

Now suppose $P = D$. If $a \in A_x$, then $P \in \Gamma$ and so $|P'| \leq |P| = |D|$. Thus, P' is a minimum connected dominating set. If $a \notin A_x$, then $N(u)$ has some vertex x' not dominated by $\{x, a\}$. So, x' is dominated by some a' in P, say $P = (x, a, \ldots, a', \ldots, v)$. Replacing the part of P before a' by $\{u, x'\}$ to get a u–v path $P'' = (u, x', a', \ldots, v) \in \Gamma$. Then $|P'| \leq |P''| \leq |P| = |D|$ and so P' is a minimum connected dominating set.

Case c. $u \in D$ and $v \notin D$.

This case is symmetric to Case b.

Case d. $u \notin D$ and $v \notin D$.

In this case, consider x, x', a, a' as in the otherwise subcase in Case b, and replace $\{x, a\}$ by $\{u, x'\}$ to get D' which satisfies the condition of Case c. Using the same technique to conclude the theorem. $\qquad \square$

The following implementation of the theorem gives an $O(n^3)$-time algorithm for finding a connected dominating set of a connected AT-free graph.

Algorithm ConnDomATfree: Find a minimum connected dominating set of a connected AT-free graph G.

find a dominating pair (u, v) of G;
construct A_x for all $x \in N(u)$;   let $A \leftarrow \cup_{x \in N(u)} A_x$;
construct B_y for all $y \in N(v)$;   let $B \leftarrow \cup_{y \in N(v)} B_y$;
compute all-pairs shortest paths on G and store the result in array D,
i.e., $D[i, j]$ is the length of a shortest i–j path;
$\ell \leftarrow \min\{\ell_1, \ell_2, \ell_3, \ell_4\}$, where
$\qquad \ell_1 \leftarrow D[u, v]$,
$\qquad \ell_2 \leftarrow \min_{a \in A} D[a, v] + 1$,
$\qquad \ell_3 \leftarrow \min_{b \in B} D[u, b] + 1$,
$\qquad \ell_4 \leftarrow \min_{a \in A, b \in B} D[a, b] + 2$;
ℓ gives the size of a minimum connected dominating set,
and such a set can be constructed from the above information;

The reason for Algorithm ConnDomATfree to be $O(n^3)$ is because the first line uses Algorithm DomPairG to find dominating pairs of G. But actually it only needs one dominating pair. Corneil *et al.* (1999) gave a linear-time algorithm for finding a dominating pair of a connected AT-free graph. To do this, they first gave an algorithmic proof for a connected AT-free graph to have a dominating path. The proof uses the procedure LexBFS(G, x) introduced in Section 4.3 for testing if a graph is chordal.

Recall that LexBFS(G, x) produces an ordering $[v_1, v_2, \ldots, v_n]$ of vertices of an arbitrary graph G of n vertices with $v_n = x$. For convenience, the procedure is repeated as follows.

LexBFS(G, x):
label$(y) \leftarrow \emptyset$ for all $y \in V(G)$;   label$(x) \leftarrow \{1\}$;
for $i = |V(G)|$ **down to** 1 **do**
{   choose a vertex y with the largest label;
$\qquad v_i \leftarrow y$;   //* Assign number i to y. *//
$\qquad$ **for** each unnumbered neighbor z of y **do** add i to label(z);
}

Suppose $[v_1, v_2, \ldots, v_n]$ is the vertex ordering obtained by LexBFS(G, x) for a general graph G. Vertex a is *smaller than* vertex b (or b is *larger than* a), denoted by $a \prec b$, if $a = v_i$ and $b = v_j$ for some $i < j$. And $a \preceq b$ means $a \prec b$ or $a = b$. For $a \preceq b$, let $\lambda(a, b)$ be the label of a when b about to be numbered. For a vertex y in G, let G_y be the subgraph induced by all vertices z with $y \preceq z$.

Label set $L = \{i_1 > i_2 > \cdots > i_r\}$ is *smaller than* label set $M = \{j_1 > j_2 > \cdots > j_s\}$, denoted by $L \prec M$, if there is an index $t \leq \min\{r, s\}$ such that $i_k = j_k$ for $1 \leq k \leq t$ but either $t = r < s$ or $t < \min\{r, s\}$ with $i_{t+1} < j_{t+1}$. And $L \preceq M$ means $L \prec M$ or $L = M$.

Lemma 7.21 (Monotonicity property).

(1) *If $a \preceq c$, $b \preceq c \preceq d$ and $\lambda(a, d) \prec \lambda(b, d)$, then $\lambda(a, c) \prec \lambda(b, c)$.*
(2) *If $a \prec b \preceq c$, then $\lambda(a, c) \preceq \lambda(b, c)$.*

Proof. (1) This follows from the fact that adding smaller elements to $\lambda(a, d)$ and $\lambda(b, d)$ does not change the monotonicity property.

(2) Suppose to the contrary that $\lambda(b, c) \prec \lambda(a, c)$. By (1), $\lambda(b, b) \prec \lambda(a, b)$ and so a must be chosen before b by LexBF, contradicting $a \prec b$. $\square$

Lemma 7.22. *If $a \prec b \prec c$, $ac \in E(G)$ and $bc \notin E(G)$, then there is some vertex d such that $c \prec d$, $ad \notin E(G)$ and $bd \in E(G)$.*

Proof. By Lemma 7.21 (2), $a \prec b$ implies $\lambda(a, b) \preceq \lambda(b, b)$. Since $a \prec b \prec c$, $ac \in E(G)$ and $bc \notin E(G)$, the number i assigned to c is in $\lambda(a, b)$ but not in $\lambda(b, b)$. Then there is some $j > i$ in $\lambda(b, b)$ but not in $\lambda(a, b)$. The vertex d assigned j then satisfies the desired property. $\square$

Lemma 7.23. *If $y \prec a \prec b \prec c$, $ac \in E(G)$ and $\{y, b, c\}$ is stable, then y misses some b–c path in G_y.*

Proof. Suppose $\{y, a, b, c\}$ is a maximal counterexample to the lemma.

By Lemma 7.21 (2), $y \prec a \prec b$ imply $\lambda(y, b) \preceq \lambda(a, b) \preceq (b, b)$. Secondly, $a \prec b \prec c$, $ac \in E(G)$ and $bc \notin E(G)$ imply $\lambda(a, b) \neq \lambda(b, b)$ and so $\lambda(y, b) \prec \lambda(b, b)$. Let $j \in \lambda(b, b) \backslash \lambda(y, b)$. Let vertex d be assigned label j. Then $db \in E(G)$ and $dy \notin E(G)$.

If i is the number assigned to c, then $i \notin \lambda(b, b)$ and so $i \neq j$, i.e., $c \neq d$. If $cd \in E(G)$, then y misses path (b, d, c) in G_y as desired. Now suppose $cd \notin E(G)$.

If $c \prec d$, then $y \prec b \prec c \prec d$, $bd \in E(G)$ and $\{y, c, d\}$ is stable. By the maximality of $\{y, a, b, c\}$, for $\{y, b, c, d\}$ vertex y misses a c–d path P in G_y. Hence, y misses the c–b path Pb as desired.

If $d \prec c$, then $y \prec a \prec d \prec c$, $ac \in E(G)$ and $\{y, d, c\}$ is stable. By the maximality of $\{y, a, b, c\}$, for $\{y, a, d, c\}$ vertex y misses a d–c path P in G_y. Hence, y misses the b–c path bP as desired.   □

The following are two basic results for finding a dominating pair of a connected AT-free graph G after performing LexBF(G, x).

Theorem 7.24. *For any vertex y and two vertices u and v in G_y, either $u \in D(v, y)$ or $v \in D(u, y)$ in G_y.*

Proof.   Suppose to the contrary that there exist vertices u and v such that $u \notin D(v, y)$ and $v \notin D(u, y)$ in G_y. Then u misses some chordless y–v path $P = (y = v_1, v_2, \dots, v_p = v)$ in G_y with $p \geq 3$, and v misses some chordless y–u path $P = (y = u_1, u_2, \dots, u_q = u)$ in G_y with $q \geq 3$.

Without loss of generality, assume that $v_p \prec u_q$.

Case 1. $u_2 \prec v_p \prec u_q$, and so $u_{i-1} \prec v_p \prec u_i$ for some $3 \leq i \leq q$.

Since $y \prec u_{i-1} \prec v_p \prec u_i$, $u_{i-1}u_i \in E(G)$ and $\{y, v_p, u_i\}$ is stable, by Lemma 7.23, y misses some v_p–u_i path P in G_y and so y misses the v_p–u_q path $(P, u_{i+1}, u_{i+2}, \dots, u_q)$ in G_y.

Case 2. $u_1 \prec v_p \prec u_2$.

Since $u_1 \prec v_p \prec u_2$, $u_1u_2 \in E(G)$ and $v_pu_2 \notin E(G)$, by Lemma 7.22, there exists v' such that $u_2 \prec v'$, $u_1v' \notin E(G)$ and $v_pv' \in E(G)$.

Then $u_iv' \notin E(G)$ for $3 \leq i \leq q$, otherwise $u_iv' \in E(G)$ implies that y misses the v_p–u_q path $(v_p, v', u_i, u_{i+1}, \dots, u_q)$ in G_y.

If $v' \prec u_3$, then since $y \prec u_2 \prec v' \prec u_3$, $u_2u_3 \in E(G)$ and $\{y, v', u_3\}$ is stable, by Lemma 7.23, y misses some v'–u_3 path P in G_y and so y misses the v_p–u_q path $(v_p, P, u_4, u_5, \dots, u_q)$ in G_y.

If $u_3 \prec v'$, then since $y \prec v_p \prec u_3 \prec v'$, $v_pv' \in E(G)$ and $\{y, u_3, v'\}$ is stable, by Lemma 7.23, y misses some u_3–v' path P in G_y and so y misses the v_p–u_q path $(v_p, P^{-1}, u_4, u_5, \dots, u_q)$ in G_y.

In any case, $y \notin D(v_p, u_q) = D(v, u)$ and so $\{v, u, y\}$ is an asteroidal triple, a contradiction. This proves the theorem.   □

Theorem 7.25. *Suppose for any two vertices u and v in G, either $u \in D(v, x)$ or $v \in D(u, x)$. If $u \prec v$, then $v \in D(u, x)$. Consequently, (x, u) is a dominating pair of G_u for all $u \in V(G)$.*

Proof. Assume the theorem is false and let v be the largest vertex such that there exists a vertex u with $u \prec v$ and $v \notin D(u, x)$. By the assumption of the theorem, $u \in D(v, x)$. Choose a u–x path $P = (u = u_0, u_1, \ldots, u_r = x)$ which is missed by v. Assume that P is chosen such that its length r is minimum.

Observe that $u = u_0 \prec v \prec u_1$, otherwise $u_1 \prec v$ would imply that the assignments $u' \leftarrow u_1$ and $P' \leftarrow (u_1, u_2, \ldots, u_r)$ contradict the minimality of the length of P.

Since $u_0 \prec v \prec u_1$, $u_0 u_1 \in E(G)$ and $v u_1 \notin E(G)$, by Lemma 7.22, there exists a vertex v_1 such that $u_1 \prec v_1$, $v_1 u_0 \notin E(G)$ and $v_1 v \in E(G)$.

Next, $v_1 u_j \notin E(G)$ for $j \geq 2$, for otherwise $v_1 u_j \in E(G)$ with $j \geq 2$ would imply that $u = u_0$ misses $(v, v_1, u_j, u_{j+1}, \ldots, u_r = x)$, contradicting that $u \in D(v, x)$.

Also, $v_1 \prec u_2$, otherwise $u_2 \prec v_1$ would imply that the assignments $u' \leftarrow u_2$, $v' \leftarrow v_1$ and $P' \leftarrow (u_2, u_3, \ldots, u_r)$ give that $v' \notin D(u', x)$, contradicting the maximality of v.

Since, $u_0 < u_1 < v_1 < u_2$, $u_1 u_2 \in E(G)$ and $\{u_0, v_1, u_2\}$ is stable, by Lemma 7.23, $u = u_0$ misses some v_1–u_2 path P and so misses the v–x path $(v, P, u_3, u_4, \ldots, u_r = x)$, contradicting that $u \in D(v, x)$.

The theorem then follows. $\qquad\square$

Based on the above two theorems, a linear-time algorithm for finding a dominating-pair of a connected AT-free graph is possible.

Algorithm DomPairAT: Find a dominating pair (x, y) of a connected AT-free graph G.

Choose an arbitrary vertex x' of G;
let x be the vertex numbered last by LexBFS(G, x');
let y be the vertex numbered last by LexBFS(G, x).

In fact the 1999 paper also provided a linear-time algorithm for finding all dominating pairs of a connected AT-free graph G of diameter at least four. That is, the algorithm finds two disjoint sets X and Y such that (x, y) is a dominating pair if and only if $x \in X$ and $y \in Y$.

7.7 Domination and total domination in AT-free graphs

Kratsch (2000) gave $O(n^6)$-time algorithms for the domination and the total domination problems in AT-free graphs.

Assume that the AT-free graph G to be considered is connected and has at least two vertices. Recall that by Theorem 7.25, there exists a vertex x such that (x, u) is a dominating pair of G_u for $u \in V(G)$ and $d(x, y) = \ell$, where $G_y = G$ and $\ell = \max\{d(x, z) : z \in V(G)\}$. Let

$$L_i = \{z : z \in V(G), d(x, z) = i\} \quad \text{for } 0 \le i \le \ell.$$

For $0 \le i \le \ell$ construct $x_0, x_1, \ldots, x_\ell$ as follows:

$$x_\ell = y \text{ and } x_i \text{ is the largest neighbor of } x_{i+1} \text{ in } L_i \text{ for } 0 \le i < \ell.$$

The following lemma appeared in the proof of Theorem 7 (respectively, 3.16) in the paper by Kloks *et al.* (1995) (respectively, (1999)).

Lemma 7.26 (Kloks *et al.*, 1995, 1999). *For every vertex $z \in L_i$, where $1 \le i \le \ell$, vertex z is dominated by $\{x_{i-1}, x_i\}$.*

Proof.   Since $\{x_0, x_1, \ldots, x_\ell = y\}$ dominates $G_y = G$ and $N[z] \subseteq L_{i-1} \cup L_i \cup L_{i+1}$, vertex z is dominated by $\{x_{i-1}, x_i, x_{i+1}\}$.

If z is dominated by $\{x_{i-1}, x_i\}$, then the lemma holds. If z is dominated by x_{i+1}, then $z \preceq x_i$ by the definition of x_i. Since $\{x_0, x_1, \ldots, x_i\}$ dominates G_{x_i}, the vertex $z \in V(G_{x_i})$ is dominated by $\{x_{i-1}, x_i\}$.   □

Theorem 7.27. *There is a minimum dominating set D of the AT-free graph G such that for $0 \le i \le \ell$ and $0 \le j \le \ell - i$,*

$$\left| D \cap \bigcup_{s=i}^{i+j} L_s \right| \le j + 3. \tag{7.1}$$

Proof.   Choose a minimum dominating set D and let

$$Q = \left\{ (i, j) : \left| D \cap \bigcup_{s=i}^{i+j} L_s \right| \ge j + 4 \right\}.$$

If $Q = \emptyset$, then the theorem holds. Assume that $Q \ne \emptyset$. Choose $(i', j') \in Q$ such that $i' = \min\{i : (i, j) \in Q\}$ and $j' = \max\{j : (i', j) \in Q\}$. We may assume that D is chosen such that i' and j' are as large as possible.

Let $A = \{x_{i'-2}, x_{i'-1}, \ldots, x_{i'+j'}, x_{i'+j'+1}\}$, where the vertices x_s with $s < 0$ or $s > \ell$ are ignored. Then $|A| \le j' + 4$. Since $\bigcup_{s=i'}^{i'+j'} L_s$ only dominates vertices in $\bigcup_{s=i'-1}^{i'+j'+1} L_s$, the set $D'' = (D \setminus \bigcup_{s=i'}^{i'+j'} L_s) \cup A$ is also a dominating set by Lemma 7.26. In fact $|D''|$ is a minimum dominating set, since $|D''| \le |D| - |\bigcup_{s=i'}^{i'+j'} L_s| + (j' + 4) \le |D|$ imply $|D''| = |D|$ and $|\bigcup_{s=i'}^{i'+j'} L_s| = j' + 4 = |A|$.

For D'', consider $Q'' = \{(i,j) : |D'' \cap \bigcup_{s=i}^{i+j} L_s| \geq j+4\}$. If $Q'' = \emptyset$, then the theorem is true. Now assume that $Q'' \neq \emptyset$ and consider $(i'', j'') \in Q''$ such that $i'' = \min\{i : (i,j) \in Q''\}$ and $j'' = \max\{j : (i'',j) \in Q''\}$.

By the choice of D and the maximality of i', we have $i'' \leq i'$. Also, $i'' + j'' \geq i' - 2$, otherwise $i'' + j'' < i' - 2$ would imply $|D \cap \bigcup_{s=i''}^{i''+j''} L_s| = |D'' \cap \bigcup_{s=i''}^{i''+j''} L_s| \geq j'' + 4$ and so $(i'', j'') \in Q$, contradicting the definition of i'. By construction, $|D'' \cap L_s| \geq 1$ for $s \in \{i' - 2, i' - 1, \ldots, i' + j', i' + j' + 1\}$. Thus, $(i'', j'') \in Q''$ with $i'' \leq i'$ and $i'' + j'' \geq i' - 2$ implies that there exists j such that $(i'', j) \in Q''$ and $i'' + j \geq i' + j' + 1$.

As $|\bigcup_{s=i'}^{i'+j'} L_s| = |A|$, we have $|D \cap \bigcup_{s=i''}^{i''+j} L_s| \geq |D'' \cap \bigcup_{s=i''}^{i''+j} L_s| \geq j+4$ and so $(i'', j) \in Q$. By the definition of i', $i'' \geq i'$ and so $i'' = i'$. Then $j > j'$, contradicting the maximality of j'. This proves the theorem. $\qquad\square$

The key idea for the algorithm on the domination problem in AT-free graphs is the dynamic programming method through the levels L_i. Note that to check if D is a dominating set if it is necessary to check if L_i is dominated by $L_{i-1} \cup L_i \cup L_{i+1}$ for $0 \leq i \leq \ell$, where $L_{-1} = L_{\ell+1} = \emptyset$. By Theorem 7.27, it is only necessary to consider those D such that $|D \cap (L_{i-1} \cup L_i \cup L_{i+1})| \leq 5$ for all i. In the following we keep S as $D \cap (L_{i-1} \cup L_i)$ and S' as $D \cap (L_0 \cup L_1 \cup \ldots \cup L_i)$.

Algorithm Dom-AT-free: Find a minimum dominating set D of a connected AT-free graph G.

find a dominating pair (x, y) of G by Algorithm DomPairAT;
$L_i \leftarrow \{z : d(x, z) = i\}$ for $0 \leq i \leq \ell$ where $\ell = d(x, y)$; $L_{\ell+1} \leftarrow \emptyset$;
let queue A_1 contain all pairs (S, S'),
    where $S = S'$ is a nonempty subset of $N[x]$ and $|S'| \leq 5$;
for $i = 2$ **to** $\ell + 1$ **do**
for all pairs (S, S') in the queue A_{i-1} **do**
for every $U \subseteq L_i$ with $|S \cup U| \leq 5$ and $N[S \cup U] \supseteq L_{i-1}$ **do**
{ $R \leftarrow (S \cup U) \backslash L_{i-2}$;
    $R' \leftarrow S' \cup U$;
    if there is no pair in A_i with first entry R
      then insert (R, R') in the queue A_i;
    if there is a pair (P, P') in A_i with $P = R$ and $|R'| < |P'|$
      then replace (P, P) in A_i by (R, R');
}
determine a pair (S, S') in A_i with minimum $|S'|$ **and** set $D \leftarrow S'$;

Theorem 7.28. *Algorithm Dom-AT-free gives a minimum dominating set of an AT-free graph in $O(n^6)$ time.*

The total dominating set problem can be solved in a similar way.

7.8   Exercises

(7.1) For positive integer n, consider the poset $(N_n, \preceq)$, where $N_n = \{1, 2, \ldots, n\}$ and $a \preceq b$ if and only if a is a factor of b. Determine the maximum size of a chain and the maximum size of an antichain.

(7.2) In a poset $(P, \preceq)$, a *minimal* (respectively, *maximal*) element is an element m for which there is no element $a \prec m$ (respectively, $a \succ m$).

  (a) Show that the set of all minimal elements is an antichain.
  (b) Show that in a finite poset, the maximum size of a chain is less than or equal to the minimum number of disjoint antichains whose union is the poset.
  (c) Use (b) and the following procedure to prove Mirsky's Theorem.

     Let A_1 be the set of all minimal elements. Delete A_1 from the poset and let A_2 be the set of all minimal elements of the new poset. Continue this process to get $A_1, A_2, \ldots, A_c$ until the poset is empty.

(7.3) Show that a graph has a transitive orientation if and only if it has a comparability ordering.

(7.4) Show that bipartite graphs are comparability graphs.

(7.5) Show that the 3-sun and its complement are not comparability graphs.

(7.6) Show that interval graphs are cocomparability graphs.

(7.7) A permutation graph is a graph G with $V(G) = \{1, 2, \ldots, n\}$ and a permutation $\pi = [\pi_1, \pi_2, \ldots, \pi_n]$ on $V(G)$ such that $ij \in E(G)$ if and only if $(i - j)(\pi_i^{-1} - \pi_j^{-1}) < 0$.
Show that a permutation graph and its complement are comparability graphs.

(7.8) A trapezoid is a quadrilateral that has a pair of parallel sides. A *trapezoid graph* is a graph such that there are two parallel lines in the plane, and each vertex of the graph corresponding to a trapezoid whose two parallel edges are in these two parallel lines, and

two vertices are adjacent if and only if their corresponding trape-
zoids intersect.

(a) Show that interval graphs are trapezoid graphs.
(b) Show that permutation graphs are trapezoid graphs.
(c) Show that trapezoid graphs are cocomparability graphs.

(7.9) Suppose G is a cocomparability graph with a cocomparability order-
ing $[v_0, v_1, \ldots, v_n, v_{n+1}]$, where v_0 and v_{n+1} are isolated vertices.
Let d_i be the minimum size of an independent dominating set of
$G[v_0, v_1, \ldots, v_i]$. Is it always true that $d_i \leq d_{i+1}$ for $0 \leq i \leq n$?

(7.10) Discuss the possibility of designing an $O(n^2)$-time algorithm for the
independent domination problem in cocomparability graphs.

(7.11) Discuss the possibility of designing an $O(n^2)$-time algorithm for
the weighted independent domination problem in cocomparability
graphs.

(7.12) Show that the weighted domination problem, the weighted total
domination problem, the weighted connected domination problem
and the weighted clique domination problem are NP-complete for
cocomparability graphs (Chang, 1997).

(7.13) Discuss the possibility of designing a more efficient method for lines
3 and 4 of Algorithm ConDomComparabilityRev.

(7.14) Recall that $\gamma(G) \leq \gamma_i(G)$ for every graph G. Provide at least two
examples of cocomparability graphs G for which $\gamma(G) < \gamma_i(G)$ but
$\gamma(H) = \gamma_i(H)$ for every proper induced subgraphs H of G.

(7.15) Recall that $\gamma(G) \leq \gamma_t(G)$ for every graph G. Provide at least two
examples of cocomparability graphs G for which $\gamma(G) < \gamma_t(G)$ but
$\gamma(H) = \gamma_t(H)$ for every proper induced subgraphs H of G.

(7.16) Recall that $\gamma_t(G) \leq \gamma_{cl}(G)$ for every graph G. Provide at least two
examples of cocomparability graphs G for which $\gamma_t(G) < \gamma_{cl}(G)$ but
$\gamma_t(H) = \gamma_{cl}(H)$ for every proper induced subgraphs H of G.

(7.17) Provide at least two examples of AT-free graphs G which are not
cocomparability graphs but their proper induced subgraphs are.

(7.18) Provide at least two examples of graphs G which are not AT-free
but their proper induced subgraphs are.

(7.19) Prove or disprove that a graph having a dominating pair is AT-free.

(7.20) Prove or disprove that a graph any of whose connected induced
subgraphs has a dominating pair is AT-free.

Chapter 8

Permutation Graphs

8.1 Permutation graphs as cocomparability graphs

As illustrated by Example 9-5 in Liu (1968, p. 245), in a design of printed circuit there are nine fixed junctions, numbered from 1 to 9, in the upper row and four fixed junctions in the lower row. The goal is to connect a wire from each upper junction to a corresponding lower junction, as specified in Figure 8.1(a). The wire connecting from upper junction i is called wire i. Some wires may inevitably cross one another, such as wires 1 and 2. Since printed wires are not insulated, no two of them should meet one another except at the junctions. If printed on multiple planes, such that the connecting wires on each plane meet only at the junctions, as shown by the dashed lines in Figure 8.1(b), then no crossing wires will result. The corresponding junctions in the planes are then tied together by conductors, as the heavy solid lines in Figure 8.1(b).

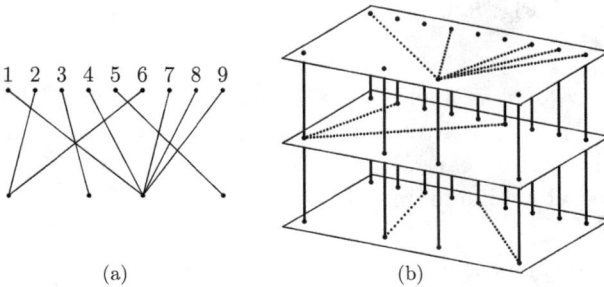

1 2 3 4 5 6 7 8 9

(a)   (b)

Fig. 8.1. An example for wire connecting.

261

To find the least number of planes needed, the intersection graph G of the nine wires in Figure 8.1(a) is constructed, see Figure 8.2, namely, $V(G) = \{1, 2, 3, 4, 5, 6, 7, 8, 9\}$ and $E(G) = \{ij : \text{wire } i \text{ crosses wire } j\}$. Finding the chromatic number $\chi(G)$ gives the answer, since each color class of an optimal coloring of G is a set of pairwise noncrossing wires.

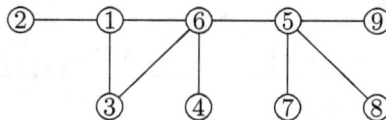

Fig. 8.2.   The intersection graph G of the nine wires in Figure 8.1(a).

Even *et al.* (1972) gave an alternative way to represent Figure 8.1(a) by "trimming" the four lower junctions into nine so that the endpoints of the wires are distinct and the wire intersections are preserved as in Figure 8.3. The endpoint of wire i in the lower row is named by i. This gives a permutation $\pi = [2, 6, 3, 1, 4, 7, 8, 9, 5]$. Note that the crossing relations of wires in Figure 8.1(a) are the same as those in Figure 8.3.

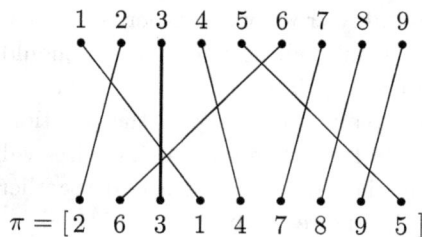

Fig. 8.3.   Lower junctions are "trimmed".

The inverse of π is another permutation $\pi^{-1} = [4, 1, 3, 5, 9, 2, 6, 7, 8]$, in which $(\pi^{-1})_i$, also written by π_i^{-1} for short, is the position of i in the permutation π. Wire i can be viewed as from head i to tail π_i^{-1}. Note that wire i crosses wire j if and only if either (a) or (b) holds.

(a) head i is on the left of head j but tail π_i^{-1} is on the right of tail π_j^{-1},
(b) head i is on the right of head j but tail π_i^{-1} is on left of the tail π_j^{-1}.

That is, either $i < j$ with $\pi_i^{-1} > \pi_j^{-1}$ or else $i > j$ with $\pi_i^{-1} < \pi_j^{-1}$; i.e.,

$$(i - j)(\pi_i^{-1} - \pi_j^{-1}) < 0.$$

The permutation π determines the intersection graph G. It is often that solving a problem by means of permutation π is more efficient than using the structure of G.

More generally, given a permutation $\pi = [\pi_1, \pi_2, \ldots, \pi_n]$ on the set $[n] = \{1, 2, \ldots, n\}$, the *permutation graph of* π is the graph G_π with vertex set $V(G_\pi) = [n]$ and edge set

$$E(G_\pi) = \{ij : (i - j)(\pi_i^{-1} - \pi_j^{-1}) < 0\}.$$

A *permutation graph* is one that is the permutation graph of some permutation. Figure 8.3 illustrates a permutation and its *line diagram*, namely, there are $n = 9$ points named by 1 to n in the upper row and n points named by π_1 to π_n in the lower row. A line is drawn from the upper point i to the lower point at position π_i^{-1}. In this way, G_π is the intersection graph of the n lines, that is, i is adjacent to j if and only if line i crosses line j, or equivalently $(i - j)(\pi_i^{-1} - \pi_j^{-1}) < 0$.

Permutation graphs were first introduced by Pnueli *et al.* (1971) and Even *et al.* (1972). They gave characterizations and applications of permutation graphs.

Theorem 8.1 (Pnueli *et al.*, 1971; Even *et al.*, 1972). *A graph G is a permutation graph if and only if both G and $\overline{G}$ are comparability graphs.*

Proof. Suppose G is the permutation graph G_π of a permutation π on $[n]$. Consider the natural ordering $1, 2, \ldots, n$ of $V(G)$ and $V(\overline{G})$. If $i < j < k$ and $ij, jk \in E(G)$, then $i < j$ gives $\pi_i^{-1} > \pi_j^{-1}$ and $j < k$ gives $\pi_j^{-1} > \pi_k^{-1}$. Both inequalities imply $\pi_i^{-1} > \pi_k^{-1}$ and so $jk \in E(G)$. This shows that G is a comparability graph. If $i < j < k$ and $ij, jk \in E(\overline{G})$, then $i < j$ gives $\pi_i^{-1} < \pi_j^{-j}$ and $j < k$ gives $\pi_j^{-1} < \pi_k^{-1}$. Both inequalities imply $\pi_i^{-1} < \pi_k^{-1}$ and so $jk \in E(\overline{G})$. This shows that $\overline{G}$ is a comparability graph.

On the other hand, suppose both $G = (V, E)$ and $\overline{G} = (V, \overline{E})$ are comparability graphs. Choose a transitive orientation (V, F_1) of G and a transitive orientation (V, F_2) of $\overline{G}$. We claim that $(V, F_1 \cup F_2)$ is an acyclic orientation of the complete graph $(V, E \cup \overline{E})$. Suppose $F_1 \cup F_2$ had a directed cycle $(v_0, v_1, \ldots, v_r = v_0)$ of the smallest possible length r. If $r > 3$, then the directed cycle can be shortened by either $v_0 v_2$ or $v_2 v_0$, contradicting the minimality of r. Now, $r = 3$. Then at least two of the edges in the directed cycle are in a same F_i, implying that F_i is not transitive. Hence, $(V, F_1 \cup F_2)$ is acyclic and so is transitive. It then induces a comparability ordering $v_1, v_2, \ldots, v_n$ of $(V, E \cup \overline{E})$. Note that F_1^{-1} is also a transitive orientation of G. Similarly, $(V, F_1^{-1} \cup F_2)$ is acyclic and so is transitive. It then induces a comparability ordering $v_1', v_2', \ldots, v_n'$ of $(V, E \cup \overline{E})$.

Identifying each vertex v_i by i and letting π be the permutation of $[n]$ such that $v_{\pi_i} = v_i'$, or equivalently $v_i = v_{\pi_i^{-1}}'$ for $i \in [n]$. If $i < j$ and $v_i v_j \in E(G)$, then $v_{\pi_i^{-1}}' v_{\pi_j^{-1}}' \in E(G)$ and so $\pi_i^{-1} > \pi_j^{-1}$, since $v_1', v_2', \ldots, v_n'$ is induced by $(V, F_1^{-1} \cup F_2)$. Hence, $(i - j)(\pi_i^{-1} - \pi_j^{-1}) < 0$. On the other hand, assume $i < j$ and $(i - j)(\pi_i^{-1} - \pi_j^{-1}) < 0$. Then $\pi_i^{-1} > \pi_j^{-1}$. Since $v_1', v_2', \ldots, v_n'$ is induced by $(V, F_1^{-1} \cup F_2)$, we have $v_i v_j \in E(G)$. In conclusion, G is the permutation graph of π as desired. $\square$

From this theorem, recognition algorithm for comparability graphs also provides recognition for permutation graphs.

8.2   Domination in permutation graphs

Farber and Keil (1985) presented $O(n^3)$-time algorithms for the weighted domination and the weighted independent domination problems in permutation graphs, and an $O(n^2)$-time algorithm for the domination problem in permutation graphs. Using a priority queue technique to refine their algorithms, Tsai and Hsu (1993) developed an $O(n^2 \log^2 n)$-time algorithm for the weighted domination problem and an $O(n \log \log n)$-time algorithm for the domination problem in permutation graphs. Chao et al. (2000) established an amortized $O(n)$-time algorithm for the domination problem in permutation graphs. Liang et al. (1991) presented a new approach to give an $O(n(m + n))$-time algorithm for the weighted domination problem in permutation graphs. By refining their method, Rhee et al. (1996) presented an $O(n + m)$-time algorithm for the weighted domination problem in permutation graphs.

8.2.1   An $O(n^2)$-time algorithm for domination

Farber and Keil (1985) presented a dynamic programming algorithm using $O(n^2)$ time for the domination problem.

Here, we use the notation by Tsai and Hsu (1993). Suppose G is a permutation graph defined by the permutation $\pi = [\pi_1, \pi_2, \ldots, \pi_n]$ on $[n]$. For $i, j \in [n]$, define

$$V_i = \{\pi_1, \pi_2, \ldots, \pi_i\} \quad \text{and} \quad V_{ij} = V_i \cap \{1, 2, \ldots, j\}.^1$$

For a subset S of $[n]$, define $\max(S) = \max\{i : i \in S\}$. For $i, j \in [n]$, construct D_{ij} satisfying the following two conditions.

(1) D_{ij} is a minimum cardinality subset of V_i dominating $V_{i,j}$.
(2) $\max(D_{ij})$ is as large as possible.

Note that D_{ij} is a subset of V_i but may not be a subset of V_{ij}. Let $d_{ij} = |D_{ij}|$ and $m_{ij} = \max(D_{ij})$. Then D_{ij} can be constructed recursively:

$$D_{ij} = \begin{cases} \emptyset, & \text{if } i = 1 \text{ and } j < \pi_1; \\ \{\pi_1\}, & \text{if } i = 1 \text{ and } j \geq \pi_1; \\ D_{i-1,j}, & \text{if } i \geq 2 \text{ and } j < \pi_i; \\ \min(A, B), & \text{if } i \geq 2 \text{ and } j \geq \pi_i, \end{cases}$$

where

$$A = \begin{cases} D_{i-1,j}, & \text{if } m_{i-1,j} > \pi_i; \\ D_{i-1,j} \cup \{\max(V_i)\}, & \text{if } m_{i-1,j} \leq \pi_i, \end{cases}$$

$$B = D_{i-1,\pi_i} \cup \{\pi_i\},$$

and $\min(A, B) = A$ if either $|A| < |B|$ or $|A| = |B|$ with $\max(A) > \max(B)$; otherwise $\min(A, B) = B$. Note that the conditions in the definition of A correct those given by Tsai and Hsu.

Next, we check the correctness of the construction of D_{ij}. The formula for $i = 1$ follows from that $V_{1j} = \emptyset$ for $j < \pi_1$ and $V_{1j} = \{\pi_1\}$ for $j \geq \pi_1$.

Suppose now $i \geq 2$. If $j < \pi_i$, then $\pi_i \notin V_{ij}$ and so $V_{ij} = V_{i-1,j}$. In this case, $D_{ij} = D_{i-1,j}$, since π_i dominates no vertex in $V_{i-1,j}$. Now suppose $j \geq \pi_i$. Then $\pi_i \in V_{ij}$ and so $V_{ij} = V_{i-1,j} \cup \{\pi_i\}$. There are two cases to be considered.

The first case is when $\pi_i \notin D_{ij}$. Then D_{ij} has some $\pi_k \in V_{i-1}$ dominates π_i. If $\pi_k \in D_{i-1,j}$, then $D_{i-1,j}$ dominates V_{ij}. This happens when $m_{i-1,j} > \pi_i$, since $m_{i-1,j}$ dominates π_i; see the first case in the definition of A. In fact

---

[1]The notation used here follows Tsai and Hsu, which differs from Farber and Keil, who defined $V_i = \{1, 2, \ldots, i\}$ and $V_{ij} = V_i \cap \{\pi_1^{-1}, \pi_2^{-1}, \ldots, \pi_j^{-1}\}$.

when $m_{i-1,j} \le \pi_i$, and so $m_{i-1,j} < \pi_i$, there is no $\pi_{k'} \in D_{i-1,j}$ dominates π_i. Then $|D_{ij}| \ge |D_{i-1,j}| + 1$, since $D_{ij} - \{\pi_k\}$ dominates $V_{i-1,j}$. However, $D_{i-1,j} \cup \{\max(V_i)\}$ dominates V_{ij}, since $\max(V_i)$ dominates π_i. Hence, in fact $D_{ij} = D_{i-1,j} \cup \{\max(V_i)\}$, see the second case in the definition of A.

The second case is when $\pi_i \in D_{ij}$. Note that π_i dominates all vertices $\pi_k \in V_{ij}$ with $\pi_k \ge \pi_i$, but does not dominate vertices $\pi_k \in V_{ij}$ with $\pi_k < \pi_i$ or equivalently any vertex $\pi_k \in V_{i-1,\pi_i}$. Hence, $D_{ij} = B$.

In summary, $D_{i,j} = \min(A, B)$ if $i \ge 2$ and $j \ge \pi_i$. This completes the check of the correctness of the construction of D_{ij} as above.

An example is shown in Figure 8.4 for the permutation graph in Figure 8.2 of the permutation in Figure 8.3. In the calculation of D_{97}, that is $i = 9$ and $j = 7$. Then $m_{i-1,j} = 7 > 5 = \pi_i$ and so $A = D_{i-1,j} = \{1, 6, 7\}$; $B = D_{85} \cup \{5\} = \{1, 5, 6\}$. Hence, $D_{97} = A = \{1, 6, 7\}$. In the calculation of D_{98}, that is $i = 9$ and $j = 8$. Then $m_{i-1,j} = 8 > 5 = \pi_i$ and so $A = D_{i-1,j} = \{1, 6, 7, 8\}$; $B = D_{85} \cup \{5\} = \{1, 5, 6\}$. Hence, $D_{98} = B = \{1, 5, 6\}$.

j	(d_{8j}, m_{8j})	D_{8j}		D_{9j}
9	$(5, 9)$	$\{1, 6, 7, 8, 9\}$		$\{1, 5, 6\}$
8	$(4, 8)$	$\{1, 6, 7, 8\}$		$\{1, 5, 6\}$
7	$(3, 7)$	$\{1, 6, 7\}$		$\{1, 6, 7\}$
6	$(2, 6)$	$\{1, 6\}$		$\{1, 6\}$
5	$(2, 6)$	$\{1, 6\}$	$\Longrightarrow$	$\{1, 6\}$
4	$(2, 6)$	$\{1, 6\}$		$\{1, 6\}$
3	$(1, 1)$	$\{1\}$		$\{1\}$
2	$(1, 2)$	$\{2\}$		$\{2\}$
1	$(1, 6)$	$\{6\}$		$\{6\}$

Fig. 8.4.   Examples of (d_{ij}, m_{ij}) and D_{ij}.

For the implementation of the algorithm, Tsai and Hsu pointed out that it only needs to compute the size d_{ij} of D_{ij} and the maximum m_{ij} of D_{ij}, according to the above construction of D_{ij}. Here are the detail formulas.

For $i = 1$,

$$d_{1j} = \begin{cases} 0, & \text{if } j < \pi_1; \\ 1, & \text{if } j \ge \pi_1, \end{cases} \qquad m_{1j} = \begin{cases} 0, & \text{if } j < \pi_1; \\ \pi_1, & \text{if } j \ge \pi_1. \end{cases}$$

For $2 \leq i \leq n$,

$$d_{ij} = \begin{cases} d_{i-1,j}, & \text{if } j < \pi_i; \\ \min\{a, b\}, & \text{if } j \geq \pi_i, \end{cases} \qquad m_{ij} = \begin{cases} m_{i-1,j}, & \text{if } j < \pi_i; \\ m_a, & \text{if } j \geq \pi_i, a < b; \\ \max\{m_a, m_b\}, & \text{if } j \geq \pi_i, a = b; \\ m_b, & \text{if } j \geq \pi_i, a > b, \end{cases}$$

where

$$a = \begin{cases} d_{i-1,j}, & \text{if } m_{i-1,j} > \pi_i; \\ d_{i-1,j} + 1, & \text{if } m_{i-1,j} \leq \pi_i, \end{cases} \qquad m_a = \begin{cases} m_{i-1,j}, & \text{if } m_{i-1,j} > \pi_i; \\ \max(V_i), & \text{if } m_{i-1,j} \leq \pi_i, \end{cases}$$

$$b = d_{i-1,\pi_i} + 1, \qquad m_b = \max\{m_{i-1,\pi_i}, \pi_i\},$$

and $\max(V_i)$'s are computed recursively by $\max(V_1) = \pi_1$ and $\max(V_i) = \max\{\max(V_{i-1}), \pi_i\}$ for $2 \leq i \leq n$.

Since the calculation of each pair (d_{ij}, m_{ij}) takes constant time, the algorithm takes $O(n^2)$ time. The value d_{nn} is $\gamma(G)$, and an optimal D_{nn} can be obtained efficiently via backtracking.

Theorem 8.2 (Farber and Keil, 1985). *There is an $O(n^2)$-time algorithm for the domination problem in permutation graphs.*

8.2.2 An $O(n \log \log n)$-time algorithm for domination

Tsai and Hsu (1993) improved the $O(n^2)$-time algorithm for the domination by Farber and Keil (1985) to an $O(n \log \log n)$-time algorithm by using an efficient priority queue.

Let M be the two-dimensional array whose (i, j)-entry is (d_{ij}, m_{ij}). For convenience, we place M in the plane such that the index i corresponds to the x-coordinate and index j corresponds to the y-coordinate. Note that this is different from the usual notation in which i denotes a row index, and j a column index.

Farber and Keil's algorithm calculates all entries of M one by one. Hence, $\Omega(n^2)$ time is required.

Tsai and Hsu's algorithm keeps track of each column in a very compact data structure and updates them from one column to the next through simple insertions or deletions in the data structure.

The following lemma provides important monotone properties of the matrix M.

Lemma 8.3. *For each i, if $j < k$, then $d_{ij} \leq d_{ik}$, i.e., if $d_{ij} < d_{ik}$, then $j < k$. Furthermore, if $j < k$ and $d_{ij} = d_{ik}$, then $m_{ij} \geq m_{ik}$, i.e., if $d_{ij} = d_{ik}$ and $m_{ij} < m_{ik}$, then $j > k$.*

Proof. Since $V_{ij} \subseteq V_{ik}$ for $j < k$, every minimum dominating set D_{ik} of V_{ik} dominates D_{ij} and so $d_{ij} \leq d_{ik}$. If $d_{ij} = d_{ik}$, then by the definition of D_{ij}, $m_{ij} = \max(D_{ij}) \geq \max(D_{ik}) = m_{ik}$. $\qquad\qquad\qquad\square$

From this lemma of monotone properties, in a column of M, all entries with the same (d, m)-value must be contiguous. Now, define a *d-block* (respectively, *(d, m)-block*) in column i to be a maximal contiguous sequence of (i, j)-entries in column i with the same d-value (respectively, (d, m)-value). Clearly, each (d, m)-block is contained in a d-block. To describe the changes of entries in M, it is sufficient to describe the change of these block structures from column to column. The block structure corresponding to Figure 8.4 is shown in Figure 8.5.

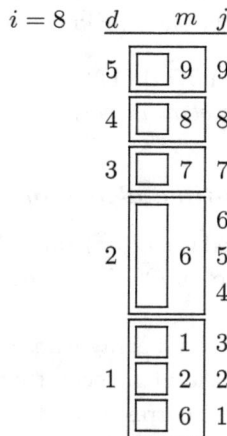

Fig. 8.5. The block structure of a column.

A (d, m)-block in a column can be denoted by an interval [low-j, high-j] associated with a pair of value (d, m), where low-j (high-j) is the j-index of the lowest (highest) entry in the block. Define the *marker* of this block to be low-j. Since no (d, m)-blocks within the same column can overlap with each other, the block structure can be maintained by keeping trace of markers. Hence, to change the block structure from the $(i - 1)$th column to the ith column is equivalent to changing the markers in the $(i-1)$th iteration to the corresponding markers in the ith iteration.

To implement the above ideas, Tsai and Hsu rewrote the above updating rules. The following formula is a simplification of their writing, where $m_b = \max\{m_{i-1,\pi_i}, \pi_i\}$.

$$(d_{ij}, m_{ij})$$
$$= \begin{cases} (d_{i-1,\pi_i}+1, m_b), & \text{if } d_{i-1,j} > d_{i-1,\pi_i}+1; \\ (d_{i-1,\pi_i}+1, m_b), & \text{if } d_{i-1,j} = d_{i-1,\pi_i}+1,\ m_{i-1,j} \le m_b; \\ (d_{i-1,j}, m_{i-1,j}), & \text{if } d_{i-1,j} = d_{i-1,\pi_i}+1,\ m_{i-1,j} > m_b; \\ (d_{i-1,j}+1, \max(V_i)), & \text{if } d_{i-1,j} = d_{i-1,\pi_i},\ j \ge \pi_i \ge m_{i-1,j}; \\ (d_{i-1,j}, m_{i-1,j}), & \text{if } d_{i-1,j} = d_{i-1,\pi_i},\ (j < \pi_i \text{ or } \pi_i < m_{i-1,j}); \\ (d_{i-1,j}, m_{i-1,j}), & \text{if } d_{i-1,j} < d_{i-1,\pi_i}. \end{cases}$$

To verify the new updating rules, consider the following three cases.

Case 1. $d_{i-1,j} \ge d_{i-1,\pi_i}+1$. In this case, $d_{i-1,j} > d_{i-1,\pi_i}$ and so $j > \pi_i$ by Lemma 8.3. By the old updating rules, $d_{ij} = \min\{a, b\}$.

If $d_{i-1,j} > d_{i-1,\pi_i}+1$, then $a > b$ and so $(d_{ij}, m_{ij}) = (b, m_b) = (d_{i-1,\pi_i}+1, m_b)$. Now suppose $d_{i-1,j} = d_{i-1,\pi_i}+1$.

If $m_{i-1,j} \le m_b$ with $m_{i-1,j} > \pi_i$, then by the old updating rule, $a = d_{i-1,j} = b$ and so $(d_{ij}, m_{ij}) = (b, m_b) = (d_{i-1,\pi_i}+1, m_b)$. If $m_{i-1,j} \le m_b$ with $m_{i-1,j} \le \pi_i$, then by the old updating rule, $a = d_{i-1,j}+1 > b$ and so $(d_{ij}, m_{ij}) = (b, m_b) = (d_{i-1,\pi_i}+1, m_b)$.

If $m_{i-1,j} > m_b$, then $m_{i-1,j} > \pi_i$ and so $a = d_{i-1,j} = d_{i-1,\pi_i}+1 = b$. By the old updating rule, $(d_{ij}, m_{ij}) = (a, m_{i-1,j}) = (d_{i-1,j}, m_{i-1,j})$.

Case 2. $d_{i-1,j} = d_{i-1,\pi_i}$. In this case, if $j \ge \pi_i \ge m_{i-1,j}$, then by the old updating rules, $d_{ij} = \min\{a, b\}$ and $a = d_{i-1,j}+1 = d_{i-1,\pi_i}+1 = b$ and so $(d_{ij}, m_{ij}) = (d_{i-1,j}+1, \max(V_i))$.

Otherwise, if $j < \pi_i$, then by the old updating rules, $(d_{ij}, m_{ij}) = (d_{i-1,j}, m_{i-1,j})$. Or else $j \ge \pi_i$ but $\pi_i < m_{i-1,j}$, then by the old updating rules, $d_{ij} = \max\{a, b\}$ and $a = d_{i-1,j} < d_{i-1,\pi_i}+1 = b$ and so $(d_{ij}, m_{ij}) = (d_{i-1,j}, m_{i-1,j})$.

Case 3. $d_{i-1,j} < d_{i-1,\pi_i}$. In this case, $j < \pi_i$ by Lemma 8.3. By the old updating rule, $(d_{ij}, m_{ij}) = (d_{i-1,j}, m_{i-1,j})$. $\square$

column $i-1$

column i

$d_{i-1,\pi_i}+2$

$m_{ij}=m_b$

$d_{i-1,\pi_i}+1$

$m_{i-1,j}\leq m_b$

$m_{i-1,j}>m_b$

$d_{i-1,\pi_i}+1$

unchanged

d_{i-1,π_i}

$j\geq\pi_i\geq m_{i-1,j}$

$m_{ij}=\max(V_i)$

$j<\pi_i$
or
$\pi_i<m_{i-1,j}$

d_{i-1,π_i}

unchanged

$d_{i-1,\pi_i}-1$

$d_{i-1,\pi_i}-1$

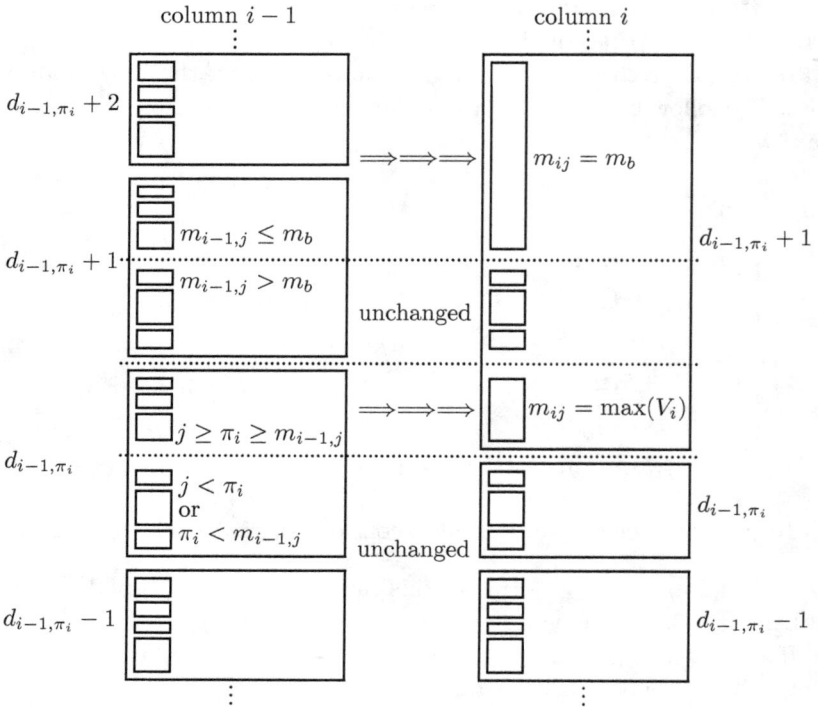

Fig. 8.6.   Pictorial description of the new updating rules.

In each iteration, the updating rules are used to update the markers of the ith column of M from the $(i-1)$th column. Some markers are deleted and some are inserted. To obtain an $O(n\log\log n)$-time algorithm, an efficient priority queue introduced by van Emde Boas (1977) is used to store the markers. In this data structure, with $O(n)$ preprocessing, each of the following operations can be executed in $O(\log\log n)$ time: insert(j), delete(j), predecessor(j) and successor(j). Furthermore, given a value j, an operation pseudoinsertion(j) can determine which (d,m)-block contains j. This operation is composed of insert(j), predecessor(j) and delete(j); and so also takes $O(\log\log n)$ time.

Each marker in the priority queue is associated with its corresponding m-value. Besides, a pointer array TOP[$1..n$] is used to store the information on d-blocks. For each value k, TOP[k] points to the largest marker in the d-block with the d-value k. A linear scan on all markers contained in a specified d-block can be performed by applying $\text{TOP}[k]$ to get the largest marker, and then applying predecessor$(\cdot)$ repeatedly to visit the subsequent markers before $\text{TOP}[k-1]$ is encountered. To split a (d, m)-block into two, a new marker is inserted into the queue. To merge two (d, m)-blocks, a marker is deleted from the queue.

Initially, there are at most two blocks. By the updating rules, also see Figure 8.6, at most three new blocks can be generated in each iteration. Hence, only $O(n)$ blocks are generated in the algorithm. Since the marker of each block can be inserted and deleted at most once, the running time is $O(n \log \log n)$.

Theorem 8.4 (Tsai and Hsu, 1993). *There is an $O(n \log \log n)$-time algorithm for the domination problem in permutation graphs.*

Achieving the best possible, Chao *et al.* (2000) established an amortized $O(n)$-time algorithm for the domination problem in permutation graphs.

On the other line, Liang *et al.* (1991) presented a new approach to give an $O(n(m+n))$-time algorithm for the weighted domination problem in permutation graphs. In the previous algorithms, two loops on i and j are used, resulting in time complexity $O(n^2)$ or $O(n^3)$. In their new approach, it only needs to deal with the case of $i = j$ or ij being an edge. This makes it possible to design a linear-time algorithm. By refining their method, Rhee *et al.* (1996) achieved an $O(n + m)$-time algorithm for the weighted domination problem in permutation graphs. In this book, we only present their approach in Section 8.3 for the weighted independent domination problem in permutation graphs.

8.2.3 *Total domination in permutation graphs*

Kratsch and Stewart (1997) established a transformation from a total domination problem to a domination problem. The *duplex graph* of a graph G is the graph G^* obtained from G by splitting every vertex into a pair of false twins. More precisely,

$$\text{vertex set } V(G^*) = \{v, v' : v \in V(G)\} \quad \text{and}$$
$$\text{edge set } E(G^*) = \{uv, u'v, uv', u'v' : uv \in E(G)\}.$$

Note that $vv' \notin E(G^*)$ and $N_{G^*}(v) = N_{G^*}(v') = \{u, u' : u \in V_G(v)\}$ for every vertex $v \in V(G)$.

Lemma 8.5. *For every graph G without isolated vertices, G has a total dominating set of size at most k if and only if G^* has a dominating set of size at most k.*

Proof.   Suppose D is a total dominating set of G. For every vertex $v \in V(G)$, there exists a vertex $u \in D$ adjacent to v. Note u is adjacent to v and v' in G^*. Hence, D is a dominating set of G^*.

Suppose D is a dominating set of G^*. If $v' \in D$ but $v \notin D$, then $(D \backslash \{v'\}) \cup \{v\}$ is also a dominating set of G^*. If $v' \in D$ and $v \in D$, then $(D \backslash \{v'\}) \cup \{u\}$ is also a dominating set of G^* for any neighbor u of v in G. Thus G^* has a dominating set $D^* \subseteq V(G)$ of size at most $|D|$. Then D^* is a total dominating set of G since for every $v \in V(G)$, v' is adjacent to some vertex $u \in D^*$ and so v is adjacent to some $u \in D$ in G. $\square$

If G is a permutation graph, then so is G^*.[2] To see this, suppose G is a permutation graph on $\pi = [\pi_1, \pi_2, \ldots, \pi_n]$. Consider the permutation $\pi' = [\pi'_1, \pi'_2, \ldots .\pi'_{2n}]$ defined by $\pi'_{2i-1} = 2\pi_i - 1$ and $\pi'_{2i} = 2\pi_i$ for $1 \le i \le n$. It is straightforward to check that G^* is the permutation graph on π'.

Therefore, any algorithm for domination in permutation graphs provides an algorithm for total domination in permutation graphs using the same time complexity.

8.3   Weighted independent domination in permutation graphs

The first algorithm for the weighted independent domination problem in permutation graphs was given by Farber and Keil (1985). They first presented an $O(n^3)$-time algorithm for weighted domination by using

---

[2]This is also true for cocomparability graphs, asteroidal triple-free graphs, k-polygon graphs, distance-hereditary graphs, dually chordal graphs, convex bipartite graphs, chordal bipartite graphs, homogeneously oracle graphs, bipartite graphs, comparability graphs and circle graphs.

a dynamic programming method. Then used a simple modification in one line to get an $O(n^3)$-time algorithm for weighted independent domination.

Brandstädt and Kratsch (1987) developed an $O(n^2)$-time algorithm for the weighted independent domination in permutation graphs. Their algorithm takes advantage of the observation that a set is an independent dominating set if and only if it is a maximal independent set. Since a maximal independent set in a permutation graph on π corresponds to a maximal increasing subsequence of π, it is only necessary to find such a subsequence of π of minimum weight.

Atallah *et al.* (1988) used the same idea to establish an $O(n \log^2 n)$-time algorithm for the independent domination problem in permutation graphs. Atallah and Kosaraju (1989) presented an $O(n \log n)$-time algorithm for the same problem through a maxdominance problem for points in the plane.

Liang and Rhee (1994) developed an $O(n + m)$-time algorithm for the weighted independent domination problem in permutation graphs (see also the presentation by Kratsch (1998)), where m is the number of edges of the graph.

8.3.1 An $O(n^2)$-time algorithm

This section presents the $O(n^2)$-time algorithm for weighted independent domination in permutation graphs by Brandstädt and Kratsch (1987) by means of finding a maximal increasing subsequence of the permutation π of minimum weight.

Suppose G is a permutation graph on $\pi = [\pi_1, \pi_2, \ldots, \pi_n]$ with vertex weights $w = [w_1, w_2, \ldots, w_n]$. For technical reasons, add to G an isolated vertex $n + 1$ to get a new permutation graph G' on $[\pi_1, \pi_2, \ldots, \pi_n, \pi_{n+1}]$, where $\pi_{n+1} = n + 1$ and $w_{n+1} = 0$. Note that $\gamma_i(G, w) = \gamma_i(G', w)$; and an independent dominating set always contains $\pi_{n+1} = n + 1$.

For i with $1 \leq i \leq n + 1$, let c_i be the minimum weight of a maximal increasing subsequence of $\pi_1, \pi_2, \ldots, \pi_i$ containing π_i. Note that $\gamma_i(G, w) = \gamma_i(G', w) = c_{n+1}$.

To calculate c_i, the following algorithm chooses all possible π_j, for j from $i - 1$ back to 1, which is the previous term right before π_i in the maximal subsequence.

Algorithm wIndDomPermutation: For a permutation graph G on $\pi = [\pi_1, \pi_2, \ldots, \pi_n]$ with vertex weights $[w_1, w_2, \ldots, w_n]$, find $\gamma_i(G, w)$ and a minimum w-dominating set D of G.

```
let π_{n+1} ← n + 1 and w_{n+1} ← 0;
for i = 1 to n + 1 do
{   c_i ← w_i;   p_i ← 0;   t ← 0;
    for j = i − 1 to 1 step −1 do
        if 0 = t < π_j < π_i then
        {     c_i ← c_j + w_i;   p_i ← j;
              t ← π_j;
        }
        else if 0 < t < π_j < π_i then
        {     if c_j + w_i < c_i then { c_i ← c_j + w_i;   p_i ← j; }
              t ← π_j;
        }
}
D ← ∅;
i ← p_{n+1};
while i > 0 do { D ← D ∪ {π_i};   i ← p_i; }
γ_i(G, w) ← c_{n+1};
```

Theorem 8.6 (Brandstädt and Kratsch, 1987). *Algorithm wIndDom-Permutation finds a minimum weighted independent dominating set of a permutation graph in $O(n^2)$ time.*

8.3.2 *An $O(n + m)$-time algorithm*

This section presents the $O(n+m)$-time algorithm for weighted independent domination in permutation graphs by Liang and Rhee. While most previous algorithms used two loops on i and j, their algorithm only considers those when $i = j$ or ij is an edge. This makes it possible to design a linear-time algorithm.

Suppose G is a permutation graph on $\pi = [\pi_1, \pi_2, \ldots, \pi_n]$ with vertex weights $w = [w_1, w_2, \ldots, w_n]$.

An *ordered cross pair* X_{ij} is an ordered pair (i, j) with $i \leq j$ and $\pi_j^{-1} \leq \pi_i^{-1}$, i.e., when $i \neq j$, line i crosses line j in the line diagram or ij is an edge in the graph G. For an ordered cross pair X_{ij}, let

$$V_{ij} = \{k : k \leq j, \ \pi_k^{-1} \leq \pi_i^{-1}\}.$$

Hence, V_{ij} contains all the vertices corresponding to the lines none of the upper (respectively, lower) points is on the right of line j (respectively, i). Note that $V_n = V_{\pi_n n}$.

It is clear that j is an isolated vertex in $G[V_{jj}]$. Let j_b be the largest number in $V_{jj} \backslash \{j\}$ and let i_b be the number among those $k \in V_{jj} \backslash \{j\}$ with the largest π_k^{-1}. Hence, $V_{jj} = V_{i_b j_b} \cup \{j\}$.

For X_{ij} with $i < j$, let j_c be the largest number among those k satisfying (i) $i \leq k < j$, (ii) $\pi_j^{-1} < \pi_k^{-1} \leq \pi_i^{-1}$ and (iii) there is no k' with $k < k' < j$ and $\pi_k^{-1} < \pi_{k'}^{-1} < \pi_i^{-1}$.

Lemma 8.7. *D is an independent dominating set of $G[V_{ij}]$ with $j \in D$ if and only if $D = D' \cup \{j\}$ where D' is an independent dominating set of $G[V_{i_b j_b}]$. D is an independent dominating set of $G[V_{ij}]$ with $j \notin D$ if and only if $i < j$ and D is an independent dominating set of $G[V_{ij_c}]$.*

Proof. The first statement follows from the fact that $N_{G[V_{ij}]}[j] = \{k : k \leq j, \pi_j^{-1} \leq \pi_k^{-1} \leq \pi_i^{-1}\}$ and $V_{ij} \backslash N_{G[V_{ij}]}[j] = \{k : k < j, \pi_k^{-1} < \pi_j^{-1}\} = V_{jj} \backslash \{j\} = V_{i_b j_b}$.

If D is an independent dominating set of $G[V_{ij}]$ with $j \notin D$, then $i < j$, otherwise $i = j$ would imply that j is an isolated vertex in $G[V_{jj}]$, contradicting $j \notin D$. Next prove that $D \subseteq V_{ij_c}$. Otherwise choose a largest $k \in D \backslash V_{ij_c}$. Then $j_c < k < j$ and so $i \leq k < j$. If $\pi_j^{-1} < \pi_k^{-1}$, then by the definition of j_c there is some k' with $k < k' < j$ and $\pi_k^{-1} < \pi_{k'}^{-1} < \pi_i^{-1}$. If $\pi_k^{-1} < \pi_j^{-1}$, then let $k' = j$. In either case, k' is not dominated by k. Choose $\ell \in D$ that dominates k'. Then $\ell < k$ by the maximality of k. Also, $\pi_{k'}^{-1} < \pi_\ell^{-1}$ and so $\pi_k^{-1} < \pi_\ell^{-1}$. Thus, D has two adjacent vertices k and ℓ, a contradiction. Hence D is an independent dominating set of V_{ij_c}.

On the other hand, suppose $i < j$ and D is an independent dominating set of $G[V_{ij_c}]$. If $k \in V_{ij} \backslash V_{ij_c}$, then $j_c < k \leq j$ and $\pi_k^{-1} < \pi_{j_c}^{-1}$. Choose $\ell \in D$ that dominates j_c. Then $\ell \leq j_c$ and $\pi_{j_c}^{-1} \leq \pi_\ell^{-1}$. And so $\ell < k$ and $\pi_k^{-1} \leq \pi_\ell^{-1}$, implying that k is dominated by $\ell \in D$. Hence, D is an independent dominating set of $G[V_{ij}]$. $\square$

By the lemma, the minimum weight of an independent dominating set of $G[V_{ij}]$ is equal to the minimum weight of an independent dominating set of either $G[V_{jj}]$ or $G[V_{ij_c}]$. These can be computed for all ordered cross pair X_{ij} in the following order: X_{ij} is scanned before $X_{x'y'}$ if $j < j'$ or $j = j'$ but $i > i'$.

To implement the lemma efficiently, it is desirable to compute i_b, j_b, j_c in $O(n + m)$ time. This can be done by using the so-called crossing lists ULCL_j, LLCL_j and LRCL_j for $j \in V_n$.

ULCL_j (the jth Upper Left Crossing List) denotes the ordered list $(k_0, k_1, \ldots, k_r)$ such that $k_0 < k_1 < \cdots < k_{r-1} < k_r = j$ and $\{k_0, k_1, \ldots, k_{r-1}\}$ is the set of all lines crossing line j, that have their upper points left to the upper point of j.

LLCL_j (the jth Lower Left Crossing List) denotes the ordered list $(k_0, k_1, \ldots, k_r)$ such that $\pi_{k_0}^{-1} < \pi_{k_1}^{-1} < \cdots < \pi_{k_{r-1}}^{-1} < \pi_{k_r}^{-1} = \pi_j^{-1}$ and $\{k_0, k_1, \ldots, k_{r-1}\}$ is the set of all lines crossing line j, that have their lower points left to the lower point of j.

LRCL_j (the jth Lower Right Crossing List) denotes the ordered list $(k_0, k_1, \ldots, k_r)$ such that $\pi_j^{-1} = \pi_{k_0}^{-1} < \pi_{k_1}^{-1} < \cdots < \pi_{k_{r-1}}^{-1} < \pi_{k_r}^{-1}$ and $\{k_1, k_2, \ldots, k_r\}$ is the set of all lines crossing line j, that have their lower points right to the lower point of j.

These lists can be established in $O(n+m)$ time. As an example, ULCL_j for $j \in V_n$ can be produced in the following way. The other two lists can be obtained in a similar way.

To compute ULCL_j for $j \in V_n$, we inspect the given permutation $\pi = (\pi_1, \pi_2, \ldots, \pi_n)$ in its reverse order from π_n to π_1. Let j be an element of the permutation in this order. We traverse a sorted list L, from the beginning, comparing each element of L with j until we insert j at the proper place in L, where L is maintained in the increasing order of its elements. Initially L is empty and $j = \pi_n$. When j is inserted into L, the sequence $k_0, k_1, \ldots, k_{r-1}$ of elements in L that are less than j, alone j itself, forms ULCL_j, since $k_0, k_1, \ldots, k_{r-1} < j$ and $\pi_{k_0}^{-1} > \pi_{k_1}^{-1} > \cdots > \pi_{k_{r-1}}^{-1} > \pi_j^{-1}$. For example, consider $\pi = [4, 1, 3, 2, 5]$. Initially L is empty. We first add $j = 5$ to L, making $L = (5)$. In this case, no comparison is required and we have $\text{ULCL}_5 = (5)$. Next, add $j = 2$ to the front of 5 since 2 is less than 5, making $L = (2, 5)$. Since there is no element less than 2 in L at this time, we have $\text{ULCL}_2 = (2)$. For $j = 3$, put 3 between 2 and 5, making $L = \{2, 3, 5\}$, which gives $\text{ULCL}_3 = (2, 3)$. For $j = 1$, add 1 to the beginning of L, making $L = (1, 2, 3, 5\}$, and get $\text{ULCL}_1 = (1)$. Finally for $j = 4$, we put 4 right after 3, making $L = (1, 2, 3, 4, 5)$, and get $\text{ULCL}_4 = (1, 2, 3, 4)$. Note that the number of comparisons made at the time of insertion of j into L is equal to the number $|\text{ULCL}_j|$ of elements in ULCL_j. Thus, the time complexity required in the above procedure is $\sum_{j=1}^{n} |\text{ULCL}_j| = O(n + m)$.

Finally, the lists $ULCL_j$, $LLCL_j$ and $LRCL_j$ can be used to compute i_b, j_b, j_c. For instance, for the i_b of V_{jj}, i_b is the largest ℓ for which $k_{r-s-1} < \ell < k_{r-s}$ for some $r - s$. That is, $k_r = j, k_{r-1} = j - 1, \ldots, k_{r-s} = j - s$ but $k_{r-s-1} < j - s - 1$ and $\ell = j - s - 1$. This step also takes $O(n + m)$ time. The other two indices j_b, j_c can be obtained similarly.

8.4 Weighted connected domination in permutation graphs

Colbourn and Stewart (1990) developed an $O(n^2)$-time algorithm for the connected domination problem and an $O(n^3)$-time algorithm for the weighted connected domination problem in permutation graphs. Arvind and Pandu Rangan (1992) later presented an $O(m + n)$-time algorithm for the connected domination problem and an $O(m + n \log n)$-time algorithm for the weighted connected domination problem in permutation graphs.

Suppose G is a connected permutation graph on $\pi = [\pi_1, \pi_2, \ldots, \pi_n]$ with vertex weights $w = [w_1, w_2, \ldots, w_n]$. By Lemma 1.3, we may assume that all $w_i \geq 0$. The algorithms for (weighted) connected domination in permutation graphs are based on the following theorem. The theorem is proved for minimum cardinality connected dominating sets D by Colbourn and Stewart. However, their proof is valid for the weighted case. The following is a revised proof.

Theorem 8.8. *A connected permutation graph G on π with nonnegative weights has a minimum weighted connected dominating set inducing a 4-cycle or a path.*

Proof. Choose a minimum weighted connected dominating set D of G. Assume that D is chosen such that $|D|$ is minimum.

First claim that D includes no 3-clique. Suppose to the contrary that D includes a 3-clique $\{i < j < k\}$. Then $\pi_i^{-1} > \pi_j^{-1} > \pi_k^{-1}$. Now consider any j' dominated by j. If $j' \leq j < k$, then $\pi_{j'}^{-1} \geq \pi_j^{-1} > \pi_k^{-1}$; and so j' is dominated by k. Similarly, if $j' > j > i$, then j' is dominated by i. Therefore, $D \backslash \{j\}$ is also a minimum weighted connected dominating set of G, a contradiction to the minimality of $|D|$. Hence, D includes no 3-clique.

If $|D| \leq 4$, then the theorem is true. Now suppose $|D| \geq 5$. Let $x = \min D$ and $y = \max D$. Choose a shortest x–y path $P = (x_1, x_2, \ldots, x_r)$ in

$G[D]$. It is the case that r is even with

$$x_1 < x_3 < x_2 < x_5 < x_4 < \cdots < x_{r-3} < x_{r-4} < x_{r-1} < x_{r-2} < x_r$$

$$\text{and} \quad \pi_{x_2}^{-1} < \pi_{x_1}^{-1} < \pi_{x_4}^{-1} < \pi_{x_3}^{-1} < \cdots < \pi_{x_{r-2}}^{-1} < \pi_{x_{r-3}}^{-1} < \pi_{x_r}^{-1} < \pi_{x_{r-1}}^{-1}.$$

Figure 8.7 shows an example with $r = 8$.

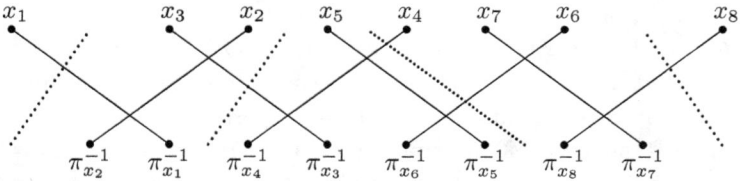

Fig. 8.7.   An example of a shortest x-y path of eight vertices.

If $|D| = r$, then the theorem holds. Assume $|D| > r$. Any $j \in D$ not in P has $x_1 < j < x_r$, say $x_i < j < x_{i+2}$ for some i. Since D includes no 3-cliques, either j is adjacent to x_1 (as the left most dot line in Figure 8.7 shows) or x_r (as the right most dot line in Figure 8.7 shows) but no other vertices in P otherwise it would form a $K_{1,3}$ with three vertices in P (as shown by the middle dot lines in Figure 8.7).

If $G[D]$ contains a $K_{1,3}$, then by symmetry, we may assume the left case of Figure 8.8 holds. Consider any j' dominated by j. If $j' \leq j < \ell$, then $\pi_{j'}^{-1} \geq \pi_j^{-1} > \pi_\ell^{-1}$; and so j' is dominated by ℓ. Similarly, if $j' > j$, then j' is dominated by i or k or ℓ. Therefore, $D\backslash\{j\}$ is also a minimum weighted connected dominating set of G, a contradiction to the minimality of $|D|$. Hence, $G[D]$ contains no $K_{1,3}$ and so is an induced path as desired. Note that if $r = 2$ and $|D| = 4$, it is possible that $G[D]$ is a 4-cycle. $\square$

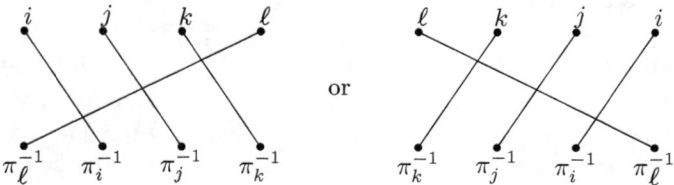

Fig. 8.8. $K_{1,3}$.

Using this theorem (for the cardinality case) and the concept of an "initial line," Colbourn and Stewart gave an $O(n^2)$-time algorithm for the connected domination problem in permutation graphs. They then employed dynamic programming methods to establish $O(n^3)$-time algorithms for the Steiner tree problem and the weighted connected domination problems in permutation graphs.

Using this theorem and the shortest path techniques, Arvind and Pandu Rangan (1992) gave an $O(m + n \log n)$-time (respectively, $O(m + n)$-time) algorithm for the weighted (respectively, cardinality) connected domination problem in permutation graphs. First, some useful lemmas.

Lemma 8.9. *For any $S \subseteq V(G)$ in a permutation graph G on $\pi = [\pi_1, \pi_2, \ldots, \pi_n]$ such that $G[S]$ is connected, if $\min S \leq i \leq \max S$, then i is dominated by S.*

Proof. If $i \in S$, then i is certainly dominated by S. Now suppose $i \notin S$ and so $\min S < i < \max S$. Choose a $\min S$–$\max S$ path $(j_1, j_2, \ldots, j_r)$ in $G[S]$. Then $j_s < i < j_{s+1}$ for some s. As $\pi_{j_s} > \pi_{j_{s+1}}$, either $\pi_i < \pi_{j_s}$ and so i is dominated by j_s or else $\pi_{j_{s+1}} < \pi_i$ and so i is dominated by j_{s+1}. $\square$

Lemma 8.10. *If $\{i < j < k < \ell\}$ is a connected dominating set of a permutation graph on $\pi = [\pi_1, \pi_2, \ldots, \pi_n]$ that induces a 4-cycle (i, k, j, ℓ, i), then k dominates all vertices $x < i$ and j dominated all vertices $y > \ell$.*

Proof. Note that $\pi_k^{-1} < \pi_\ell^{-1} < \pi_i^{-1} < \pi_j^{-1}$. For $x < i$ dominated by $\{i, j, k, \ell\}$, it is the case that $\pi_x^{-1} > \pi_k^{-1}$ and so x is dominated by k. For $y > \ell$ dominated by $\{i, j, k, \ell\}$, it is the case that $\pi_y^{-1} < \pi_j^{-1}$ and so y is dominated by j. $\square$

The proof of Theorem 8.8 also gives the following property.

Lemma 8.11. *If D is a connected dominating set of a connected permutation graph on $\pi = [\pi_1, \pi_2, \ldots, \pi_n]$ that induces a path $(x_1, x_2, \ldots, x_r)$ with $x_1 \leq x_r$, then $\min D \in \{x_1, x_2\}$ and $\max D \in \{x_{r-1}, x_r\}$. Furthermore, every $x < \min D$ is dominated by $\{x_1, x_2\}$ and every $y > \max D$ is dominated by $\{x_{r-1}, x\}$.*

Based on the above lemmas, an edge dominates all $x < \min D$ or all $y > \max D$ is important. An *initial edge* is an edge ij such that $\{i, j\}$ dominates all vertices $x < \min\{i, j\}$; a *terminal edge* is an edge ij such that $\{i, j\}$ dominates all vertices $y > \max\{i, j\}$.

Note that an edge ij with $i < j$ is an initial edge if and only if $\pi_j^{-1} < \pi_x^{-1}$ for all $x < i$. So for each vertex i, compute

$$L(i) = \min\{\pi_x^{-1} : 1 \leq x < i\}.$$

All values $L(i)$ can be computed in the order $L(1), L(2), L(3), \ldots$ in $O(n)$-time. As an edge ij with $i < j$ is an initial edge if and only if $\pi_j^{-1} < L(i)$, all initial edges can be identified in $O(m)$ time.

Similarly, an edge ij with $i < j$ is a terminal edge if and only if $\pi_i^{-1} > \pi_y^{-1}$ for all $y > j$. So for each vertex j, compute

$$R(j) = \max\{\pi_y^{-1} : j > y \geq n\}.$$

All values $R(j)$ can be computed in the order $R(n), R(n-1), R(n-2), \ldots$ in $O(n)$ time. As an edge ij with $i < j$ is a terminal edge if and only if $\pi_i^{-1} > R(j)$, all terminal edges can be identified in $O(m)$ time.

Next, construct an auxiliary graph G' with $V(G') = V(G) \cup \{s, t\}$, where s and t are two new vertices, and

$$E(G') = E(G) \cup \{(s, i), (s, j) : ij \text{ is an initial edge}\}$$
$$\cup \{(i, t), (j, t) : ij \text{ is a terminal edge}\}.$$

Assign nonnegative weight to the edges of G' by:

$$w(i, j) = (w(i) + w(j))/2 \quad \text{if } ij \in E(G);$$
$$w(s, i) = \min\{w(i)/2 + w(j) : ij \text{ is an initial edge}\};$$
$$w(i, t) = \min\{w(i)/2 + w(j) : ij \text{ is a terminal edge}\}.$$

Note that a minimum weighted connected dominating set is one of the following.

(1) A vertex of degree $n - 1$. This can be computed in $O(n)$ time.
(2) Two vertices that form both an initial edge and a terminal edge. This can be computed in $O(m)$ time.
(3) A path $P = (x_1, x_2, \ldots x_r)$ with $r \geq 3$ and $x_1 x_2$ an initial edge and $x_{r-1} x_r$ a terminal edge. This corresponds to an s–t path

$(s, x_2, \ldots x_{r-1}, t)$ in G' with total edge weights

$$(w(x_1) + w(x_2)/2) + (w(x_2) + w(x_3))/2 + \cdots$$
$$+ (w(x_{r-2}) + w(x_{r-1}))/2 + (w(x_{r-1})/2 + w(x_r)),$$

which is the total vertex weights of P. This can be computed by a shortest path algorithm in $O(m + n \log n)$ time, see the algorithm by Fredman and Tarjan (1987), in fact $O(m + n)$ time when all weights $w(i) = 1$.

8.5 Exercises

(8.1) Show that G_π is isomorphic to $G_{\pi^{-1}}$ for any permutation π.

(8.2) (a) Find a permutation π for which the complete graph K_n is the permutation graph of π.

   (b) Find a permutation π for which the complement $\overline{K}_n$ of K_n is the permutation graph of π.

   (c) Find a permutation π for which the n-path P_n is the permutation graph of π.

(8.3) Provide examples of comparability graphs which are not permutation graphs.

(8.4) (a) Provide examples of disconnected permutation graphs.

   (b) Characterize connected permutation graphs in terms of the representing permutations.

   (c) Find the number of components of a permutation graph in term of its representing permutation.

(8.5) (a) Show that for every permutation π, set $S = \{\pi_{i_1}, \pi_{i_2}, \ldots, \pi_{i_r}\}$ is a clique (respectively, stable set) if and only if $\pi_{i_1} < \pi_{i_2} < \cdots < \pi_{i_r}$ (respectively, $\pi_{i_1} > \pi_{i_2} > \cdots > \pi_{i_r}$).

   (b) Suppose G is a permutation graph of n vertices. Show that either G or $\overline{G}$ has a clique of size $\lceil \sqrt{n} \rceil$ (Erdős and Szekeres, 1935).

(8.6) Show that in a permutation graph any cycle of length at least five has a chord.

(8.7) Show that a C_4-free permutation graph is an interval graph.

(8.8) Find a minimum dominating set of the permutation graph G_π with $\pi = [3, 4, 1, 5, 9, 6, 2, 8, 7]$ by using Farber and Keil's algorithm.

(8.9) Find a minimum weighted dominating set of the permutation graph G_π with $\pi = [3, 4, 1, 5, 9, 6, 2, 8, 7]$, where $w_i = i$ for all vertices i.

(8.10) Prove that the duplex graph of a cocomparability graph (respectively, AT-free graph, bipartite graph, comparability graph) is also a cocomparability graph (respectively, AT-free graph, bipartite graph, comparability graph).

(8.11) Let $\mathcal{F}$ be the family of graphs starting from K_1 by repeatedly applying one of the following two graph operations: add a new leaf, split one vertex into a pair of false twin. Characterize graphs in $\mathcal{F}$.

(8.12) Prove that in a permutation graph, an independent dominating set corresponds to a maximal increasing subsequence of the permutation for the graph.

(8.13) Prove Theorem 8.6.

(8.14) Describe a linear-time algorithm to compute LLCL_j and LRCL_j for $j \in V_n$, as defined in Section 8.3.

(8.15) Describe a linear-time algorithm to compute j_b and j_c for $j \in V_n$, as defined in Section 8.3.

Chapter 9

Distance-Hereditary Graphs

9.1 Connection to trees and cographs

A graph G is *distance-hereditary* if every connected induced subgraph H of G has the property that $d_H(u,v) = d_G(u,v)$ for all $u, v \in V(H)$. It is not hard to see that an induced u–v path is a shortest u–v path in a distance-hereditary graph. Note that distance-hereditary graphs are parity graphs, which are graphs satisfying that if u and v are in the same component of the graph, then the lengths of every two induced u–v paths have the same parity, see Burlet and Uhry (1982).

Distance-hereditary graphs were introduced by Howorka (1977), who gave the following characterization.

Theorem 9.1 (Howorka, 1977). *A graph G is distance-hereditary if and only if every cycle of length at least five in G has two crossing chords.*

Proof. ($\Rightarrow$) Suppose G is distance-hereditary, but it has a cycle $C = (v_0, v_1, \ldots, v_r)$, where $v_r = v_0$, of length $r \geq 5$ without two crossing chords.

If C has no chord, then $(v_0, v_1, \ldots, v_{r-2})$ and (v_0, v_{r-1}, v_{r-2}) are two induced v_0-v_{r-1} paths of lengths $r - 2 \neq 2$, a contradiction.

Now, C has at least one chord, say $v_0 v_i$. By symmetry, assume that $3 \leq i \leq r - 2$. Since C has no crossing chords, $v_a v_b \notin E(G)$ for $0 < a < i$ and $i < b < r$. There are two induced v_1-v_{r-1} paths. One is (v_1, v_r, v_{r-1}) of length 2; and the other is $(v_1, \ldots, v_i, \ldots, v_{r-1})$ which must also be of length 2, and then in fact it is (v_1, v_i, v_{r-1}). Hence, $v_1 v_i \in E(G)$. Similarly, the two induced v_{i+1}-v_{i-1} paths, (v_{i+1}, v_i, v_{i-1}) and $(v_{i+1}, \ldots, v_0, \ldots, v_{i-1})$, imply that $v_0 v_{i-1} \in E(G)$, which crosses $v_1 v_i \in E(G)$, a contradiction.

($\Leftarrow$) Suppose every cycle of length at least five in G has two crossing chords, but there are two vertices u and v and two induced u-v paths $(u = x_0, x_1, \ldots, x_r = v)$ and $(u = y_0, y_1, \ldots, y_s = v)$ with $2 \le r < s$. Assume that the two paths are chosen so that $r + s$ is minimum. Then they only have common vertices at two end vertices, and so together form a cycle of length $r + s \ge 5$. Choose two crossing chords $x_i y_{j'}$ and $y_{i'} x_j$, where $0 < i < j < r$ and $0 < i' < j' < s$. Assume that these two chords are chosen such that i is minimum and j is maximum.

Then $(x_0, x_1, \ldots, x_i, y_{j'})$ and $(y_0, y_1, \ldots, y_{j'})$ are two induced u-$y_{j'}$ paths whose length sum is less than $r + s$. By the minimality of $r + s$, we have $i + 1 = j'$. Similarly, $(y_{i'}, x_j, x_{j+1}, \ldots, x_r)$ and $(y_{i'}, y_{i'+1}, \ldots, y_s)$ are two induced $y_{i'}$-v paths whose length sum is less than $r + s$. By the minimality of $r+s$, we have $1 + r - j = s - i'$. Summing up the two equalities gives

$$r + 1 \ge (i + 1) + (1 + r - j) = j' + (s - i') \ge s + 1,$$

contradicting $r < s$. $\square$

Small graphs can be checked whether they are distance-hereditary or not just by definition. Examples of graphs that are not distance-hereditary include n-cycles C_n of length $n \ge 5$ together with chords passing a same vertex of the n-cycle, see Figure 9.1 for examples.

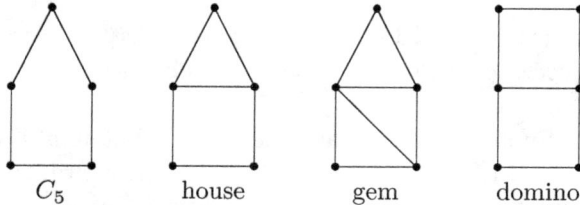

C_5 house   gem   domino

Fig. 9.1. Examples of graphs that are not distance-hereditary.

Bandelt and Mulder (1986) considered *one-vertex extensions* by which all distance-hereditary graphs can be constructed. For a graph G, extend it to a graph G' in one of the following operations (see Figure 9.2).[1]

---

[1]Bandelt and Mulder did not consider operation (α), since the distance-hereditary they considered are connected.

(α) Add a new isolated vertex x'.
(β) Add a new leaf x' adjacent to some $x \in V(G)$.
(γ) Add a new vertex x' adjacent to all $y \in N[x]$ for some $x \in V(G)$.
(δ) Add a new vertex x' adjacent to all $y \in N(x)$ for some $x \in V(G)$.

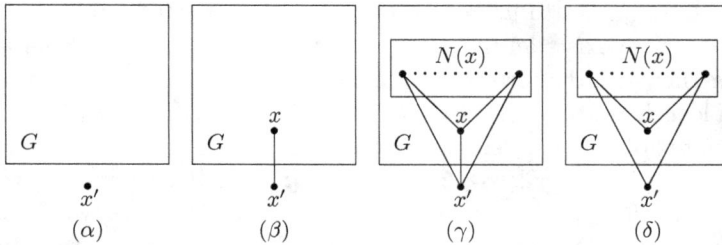

Fig. 9.2. Operations (α), (β), (γ) and (δ).

We say that in case (β) G' is obtained from G by *attaching a leaf*, and in case (γ) or (δ) G' is obtained from G by *splitting a vertex*. Operations of this kind are quite popular and occur in the literature. For instance, Cornell *et al.* (1981) called vertices x and x' of a split pair *strong siblings* (if x and x' are adjacent) or *weak siblings* (if not adjacent), Burlet and Uhry (1982) called them *true twins* (if adjacent) or *false twins* (if not adjacent).

We may also consider the inverses of operations (α), (β), (γ) and (δ). We say that in case of the inverse of (α) (respectively, (β)) G is obtained from G' by deleting an isolated vertex (respectively, a leaf). In case of the inverse of (γ) or (δ) G is obtained from G' by *identifying* twins.

It is not hard to see that any one-vertex extension of a distance-hereditary graph is again distance-hereditary. The converse that every distance-hereditary graph can be obtained from K_1 by a sequence of one-vertex extensions was proved by Bandelt and Mulder. The proof here uses some techniques from the papers by D'Atri and Moscarini (1988) and Hammer and Maffray (1990), in which distance-hereditary graphs were called completely separable graphs.

First, a useful lemma for P_4-free graphs. Note that a P_4-free graph, which is also called a *cograph*, is distance-hereditary since, in such a graph, every induced path has length at most two.

Lemma 9.2 (Corneil *et al.*, 1981). *Every P_4-free graph G of at least two vertices has at least two (true or false) twins.*

Proof.   The lemma is proved by induction on $n := |V(G)|$. The case of $n = 2$ is clear. Suppose $n \geq 3$ and the lemma holds for P_4-free graphs of less than n vertices.

We may assume that G is connected. The lemma is clear when G is complete. Otherwise, let S be a minimal cutset of G. Then for every vertex $u \in S$ and every component C of $G - S$, there exists at least one vertex $u_C \in V(C)$ adjacent to u. In fact, every vertex $v \in V(C)$ is adjacent to u, otherwise there exists an induced path of length at least two from v to u, this together with the vertex $u_{C'}$ of another component C' form an induced path of length at least three, contradicting the fact that G is P_4-free. Hence, every vertex of S is adjacent to every vertex in $V(G) - S$.

Since $n \geq 3$, either $|S| \geq 2$ or $|V(G) - S| \geq 2$. By symmetry, we may assume that $|S| \geq 2$. By the induction hypothesis, $G[S]$ has twins u and v. Since u and v are adjacent to all vertices in $V(G) - S$, they are also twins in G. $\square$

Theorem 9.3 (Bandelt and Mulder, 1986). *A graph G is distance-hereditary if and only if every induced subgraph H of G has twins or a vertex of degree at most one.*

Consequently, a graph is distance-hereditary if and only if it is obtained from K_1 by a sequence of one-vertex extensions: adding isolated vertices, attaching leaves and splitting vertices.

In other words, a graph is distance-hereditary if and only if by a sequence of one-vertex reductions it becomes K_1: deleting isolated vertices, deleting leaves and identifying twins.

Proof.   ($\Rightarrow$) Note that H is also distance-hereditary. We prove the assertion by induction on $n := |V(H)|$. The case of $n \leq 2$ is easy. Now suppose $n \geq 3$. We may also assume that H is connected and is not complete. Choose a minimal cutset Q of H.

We first prove that $N(u)\backslash Q = N(v)\backslash Q$ for $u, v \in Q$. Consider any two components C and C' of $H - Q$. Suppose $x \in V(C) \cap N(u)$. Choose a vertex $y \in V(C)$ nearest to x that is adjacent to v. Then a shortest x–y path P together with v form a chordless x–v path (P, v). Choose a vertex $x' \in V(C')$ adjacent to u and a vertex $y' \in V(C')$ nearest to x' that is adjacent to v. Then v together with a shortest y'–x' path P' form a chordless v–x' path (v, P'). Then (x, u, x') is an x–x' path of length two, and (P, v, P') is also an x–x' path, which lead to $x = y$ and so $x \in N(v)\backslash Q$. Considering all components of $H - Q$ gives that $N(u)\backslash Q = N(v)\backslash Q$ for $u, v \in Q$.

Next suppose Q is chosen such that $H - Q$ has a component C of smallest size. We prove that every vertex in $V(C)$ is adjacent to all vertices in Q. Suppose to the contrary there is some vertex u in $V(C)$ which is not adjacent to all vertices of Q and hence is adjacent to no vertices of Q. Then $Q' := \{v : v \in V(C)$ is adjacent to all vertices of $Q\}$ is a minimal cutset of H for which the component C' of $H - Q'$ containing u has size smaller than $|V(C)|$, a contradiction. Hence, every vertex of $V(C)$ is adjacent to all vertices of Q.

If $C = \{x\}$, then $N(x) = Q$. If $|Q| = 1$, then x is a leaf in H as desired. Otherwise $|Q| \geq 2$. Then $H[Q]$ is P_4-free, for otherwise there exists a u–v– path of length 3, contradicting that (u, x, v) is a u–v path of length 2. By Lemma 9.2, $H[Q]$ has twins p and q. Since $N(p)\backslash Q = N(q)\backslash Q$, p and q are also twins of H.

Now, $|V(C)| \geq 2$. Since every vertex of C is adjacent to all vertices of Q, again C is P_4-free and hence has twins, which are also twins of H.

($\Leftarrow$) Suppose every induced subgraph H of G has twins or a vertex of degree at most one, but G is not distance-hereditary. By Theorem 9.1, G has a cycle $(v_0, v_1, \ldots, v_r)$ without crossing chords, where $r \geq 5$ and $v_0 = v_r$. In $H = G[\{v_1, v_2, \ldots, v_r\}]$, there exist no vertices of degree at most one, and so there exist twins, say v_0 and v_i with $i \geq 3$. Then v_i is adjacent to v_1 as v_0 is adjacent to v_1; and v_0 is adjacent to v_{i-1} as v_i is adjacent to v_{i-1}. Hence, the cycle has two crossing chords $v_0 v_{i-1}$ and $v_1 v_i$, a contradiction. This proves that G is distance-hereditary. $\square$

The cutset Q and the component C of $H - Q$ in the proof of Theorem 9.3 can be viewed as follows. Choose a vertex u. Consider a component C of the subgraph induced by all vertices farthest to u, and Q is the neighbor of $V(C)$. More precisely, consider the BFS of a connected graph starting from a root vertex. The *hanging* of a connected graph G at a vertex $u \in V(G)$ is $h_u = [L_0(u), L_1(u), \ldots, L_t(u)]$ (or $h_u = [L_0, L_1, \ldots, L_t]$ if there is no ambiguity), where $t = \max_{v \in V(G)} d(u, v)$ and

$$L_i(u) = \{v \in V(G) : d(u, v) = i\} \text{ for } 0 \leq i \leq t.$$

For convenience, let $L_{-1}(u) = L_{t+1}(u) = \emptyset$. The set $L_i(u)$ is called the ith level of the hanging, and a vertex in $L_i(u)$ is said to have *level* i.

Note that a vertex of level $i \geq 1$ is adjacent to at least one vertex of level $i - 1$, and is only adjacent to vertices of level $i - 1$, i or $i + 1$.

A shortest path from the root vertex u to a vertex v of level i is of the form $(u = v_0, v_1, \ldots, v_i = v)$, where $v_j \in L_j$ for $0 \le j \le i$.

For every vertex v of level $i \ge 1$, let $N'(v) = N(v) \cap L_{i-1}$. Here is a characterization of connected distance-hereditary graphs in term of hanging.

Theorem 9.4 (D'Atri and Moscarini, 1988). *A connected graph G is distance-hereditary if and only if for every hanging $h_u = [L_0, L_1, \ldots, L_t]$ of G and every pair of vertices $x, y \in L_i$, $1 \le i \le t$, that are in the same component of $G - L_{i-1}$, we have $N'(x) = N'(y)$.*

Proof.   ($\Rightarrow$) Suppose G is distance-hereditary. For $x, y \in L_i$ that are in a same component of $G - L_{i-1}$ with $1 \le i \le t$ and every $x' \in N'(x)$ and every $y' \in N'(y)$, choose

> a shortest u–x path $(x_0, x_1, \ldots, x_i)$, where $x_0 = u$, $x_{i-1} = x'$ and $x_i = x$,
> a shortest u–y path $(y_0, y_1, \ldots, y_i)$, where $y_0 = u$, $y_{i-1} = y'$ and $y_i = y$,
> a shortest x–y path $(z_0, z_1, \ldots z_r)$ in $G - L_{i-1}$, where $z_0 = x$ and $z_r = y$.

Let m be the maximum index with $0 \le m \le r$ for which x' is adjacent to z_m. Then $(x_0, x_1, \ldots, x_{i-1}, z_m, z_{m+1} \ldots, z_r)$ and $(y_0, y_1, \ldots, y_i)$ are induced u–y paths of lengths $i + r - m$ and i, respectively. Therefore, $i + r - m = i$ and so $m = r$, implying $x' \in N'(y)$. Hence, $N'(x) \subseteq N'(y)$. Similarly, $N'(y) \subseteq N'(x)$. These give that $N'(x) = N'(y)$.

($\Leftarrow$) Suppose the property for hangings holds. Consider two induced u–v paths $P = (u = x_0, x_1, \ldots, x_r = v)$ and $(u = y_0, y_1, \ldots, y_s = v)$. We shall prove that $x_i, y_i \in L_i$ for all i. The assertion is clearly true for $i \le 1$. Suppose $i \ge 2$ and the assertion is true for all $i' < i$. Then $x_i \in L_{i-2} \cup L_{i-1} \cup L_i$. If $x_i \in L_{i-2}$, then $N'(x_i) = N'(x_{i-2})$ by the property for hangings. Hence, $x_i = x_0$ (when $i = 2$) or $x_i x_{i-3} \in E(G)$ (when $i \ge 3$), contradicting the assumption that P is an induced path. If $x_i \in L_{i-1}$, then $N'(x_i) = N'(x_{i-1})$ by the property for hangings. Hence, $x_i x_{i-2} \in E(G)$, contradicts that P is an induced path. In conclusion, $x_i \in L_i$. Similarly, $y_i \in L_i$. Therefore, $x_i, y_i \in L_i$ for all i and so $r = s$.   □

9.2   Recognition algorithms for distance-hereditary graphs

By implementing Theorems 9.3 and 9.4, D'Atri and Moscarini (1988) described two polynomial algorithms for recognizing distance-hereditary graphs. Their implementations are naive and so are not efficient.

For instance, to implement Theorem 9.4 it is necessary to check the condition in the theorem for *every* hanging L_u. Note that the theorem is not necessarily true if the condition holds for only one particular hanging L_u. An easy example is the graph G obtained from C_5 by adding two chords meet at a vertex u. Although the condition holds for L_u, the graph G is not distance-hereditary.

Bandelt and Mulder (1986) in fact proved that a connected graph is distance-hereditary if some five conditions hold for *any fixed* hanging L_u. Although only one hanging L_u is needed to be considered, the five conditions are slightly complicated and not easy to be checked.

Additional properties of hangings derived from the proof of Theorem 9.3 are as follows. Using these necessary conditions together with Theorem 9.3, Hammer and Maffray (1990) gave a linear-time algorithm for recognizing distance-hereditary graphs.

A vertex $x \in L_i$ is said to have a *minimal neighborhood* in L_{i-1} if $N'(y)$ is not a proper subset of $N'(x)$ for any other $y \in L_i$. Note that such a vertex x certainly exists. A vertex set S is *homogeneous* if $N(x) \setminus S = N(y) \setminus S$ for $x, y \in S$.

Theorem 9.5. *Suppose* $h_u = (L_0, L_1, \ldots, L_t)$ *is a hanging of a connected distance-hereditary graph at u.*

(1) **(Bandelt and Mulder, 1986).** *For any two vertices $x, y \in L_i$ with $i \geq 1$, we have $N'(x) \cap N'(y) = \emptyset$ or $N'(x) \subseteq N'(y)$ or $N'(x) \supseteq N'(y)$.*

(2) **(Hammer and Maffray, 1990).** *If $v \in L_i$ with $i \geq 1$ has a minimal neighborhood in L_{i-1}, then $N'(v)$ is homogeneous.*

(3) *$G[L_i]$ is P_4-free for $0 \leq i \leq t$.*

Proof.   (1) Suppose to the contrary that there exist $z \in N'(x) \cap N'(y)$, $x' \in N'(x) \setminus N'(y)$ and $y' \in N'(y) \setminus N'(x)$. It must be $i \geq 2$. By Theorem 9.4, x is not adjacent to y, for otherwise $N'(x) = N'(y)$ as desired. Then x' is not adjacent to y', for otherwise (x', y', y, z, x, x') is a 5-cycle with no crossing chords, contradicting Theorem 9.1. Since there exists an x'–y' path (x', x, z, y, y') in $H - L_{i-2}$, by Theorem 9.4, $N'(x') = N'(y')$ and there exists a vertex w in this set. Hence, (x, z, y) and (x, x', w, y', y) are two induced x–y paths of different lengths, a contradiction.

(2) Suppose to the contrary that there exists $z \in N(x) \backslash N'(v)$ but $z \notin N(y) \backslash N'(v)$ for some $x, y \in N'(v)$. It must be $i \geq 2$.

If $z \in L_{i-2}$, then $z \in N'(x)\backslash N'(y)$, contradicting $N'(x) = N'(y)$ by Theorem 9.4 as (x, v, y) is a path in $H - L_{i-2}$.

If $z \in L_{i-1}$, then there exists some $w \in N'(z) = N'(y)$ by Theorem 9.4, since (z, x, v, y) is a path in $H - L_{i-2}$. Then there exist two induced z-v paths (z, x, v) and (z, w, y, v) of different lengths, a contradiction.

If $z \in L_i$, then $N'(z) \cap N'(v) = \emptyset$ or $N'(z) \subseteq N'(v)$ or $N'(z) \supseteq N'(v)$ by (1). The facts that $x \in N'(z) \cap N'(v)$ and $y \in N'(v)\backslash N'(z)$ imply that $N'(z)$ is a proper subset of $N'(v)$, contradicting the minimality of v.

(3) $G[L_0]$ is certainly P_4-free. Suppose $1 \leq i \leq t$ and $G[L_i]$ has a P_4 (a, b, c, d). By Theorem 9.4, $N'(a) = N'(d)$. Choose a vertex $e \in N'(a)$. Then there are two induced a–d paths (a, b, c, d) and (a, e, d) of different lengths, a contradiction. Hence, G is P_4-free. $\square$

The class of P_4-free graphs play an important role in the study of distance-hereditary graphs. Note that it is easy to see that P_4-free graphs are distance-hereditary, since every two induced paths between two non-adjacent vertices are of length 2. As shown below, P_4-free graphs have several equivalent descriptions.

A *complement reducible graph* (called a *cograph* for short) is defined recursively as follows.

(C1) A graph of a single vertex is a cograph.
(Cu) If $G_1, G_2, \ldots, G_k$ are cographs, then so is their union $G_1 \cup G_2 \cup \cdots \cup G_k$.
(Cc) If G is a cograph, then so is its complement $\overline{G}$.

As $G_1 + G_2 + \cdots + G_k = \overline{\overline{G_1} \cup \overline{G_2} \cup \cdots \cup \overline{G_k}}$, condition (Cc) can be replaced by (Cj) without changing the definition.

(Cj) If $G_1, G_2, \ldots, G_k$ are cographs, then so is their join $G_1 + G_2 + \cdots + G_k$.

Note that the definition remains equivalent if conditions (Cu) and (Cj) are replaced by taking $k = 2$.

In fact, cographs are exactly the P_4-free graphs as shown in the following.

Theorem 9.6. *A graph is a cograph if and only if it is P_4-free.*

Proof.   ($\Rightarrow$) This follows from that K_1 is P_4-free and that operations union and join do not create new induced P_4.

($\Leftarrow$) It suffices to prove that a connected P_4-free graph G of at least two vertices is the join of two graphs. Choose a vertex u. Since G is connected and P_4-free, $V(G) = \{u\} \cup N(u) \cup N_2(u)$, where $N_2(u)$ is the set of all vertices at distance two from u.

We first claim that if $x, y \in N(u)$ and $xy \notin E(G)$, then $N(x) \cap N_2(u) = N(y) \cap N_2(u)$. If $z \in N(x) \cap N_2(u)$, then the path (z, x, u, y) has a chord, implying that $z \in N(y)$. Thus, $N(x) \cap N_2(u) \subseteq N(y) \cap N_2(u)$. Similarly, $N(y) \cap N_2(u) \subseteq N(x) \cap N_2(u)$. These give that $N(x) \cap N_2(u) = N(y) \cap N_2(u)$.

Choose a vertex $v \in N(u)$ such that $|N(v) \cap N_2(u)|$ is maximum. Let $A = \{x \in N(u) : N(x) \cap N_2(u) = N(v) \cap N_2(u)\}$ and $B = N(u) \backslash A$. By the claim above, if $y \in B$, then y is adjacent to all vertices $x \in A$. We claim that $N_2(u) \subseteq N(v)$. Otherwise, there exists some vertex $r \in N_2(u) \backslash N(v)$, which is adjacent to some vertex $z \in B$. By the maximality of $|N(v) \cap N_2(u)|$, $N(v) \cap N_2(u)$ contains some vertex s not adjacent to z. Then the path (r, z, v, s) has a chord, which is rs and so (u, z, r, s) is an induced path, a contradiction. This proves that $N_2(u) \subseteq N(v)$ and so G is the join of $G[A]$ and $G - A$. □

Using conditions (C1), (Cu) and (Cj), a cograph G with at least two vertices can be represented by a cotree T rooted at some node r. Each leaf node v of T is a vertex in $V(G)$. Each node a of T corresponds to the subgraph $G[L_a]$ of G, where

$$L_a = \{v : v \text{ is a descendant of } a \text{ that is a leaf node of } T\}.$$

An internal node a of T is called a $\cup$-node (respectively, $+$-node) if a has s children $c_1, c_2, \ldots, c_s$ such that $G[L_a] = G[L_{c_1}] \cup G[L_{c_2}] \cup \cdots \cup G[L_{c_s}]$ (respectively, $G[L_a] = G[L_{c_1}] + G[L_{c_2}] + \cdots + G[L_{c_s}]$).

Since consecutive union (respectively, join) operations can be combined into just one set of union (respectively, join) operations, every child of a $\cup$-node (respectively, $+$-node) is either a leaf node or a $+$-node (respectively, $\cup$-node).

Since a union or a join operation on just one graph can be ignored, every internal node of T has at least two children if the graph G has at least two vertices.

Figure 9.3 shows a cograph G of eight vertices and its cotree T.

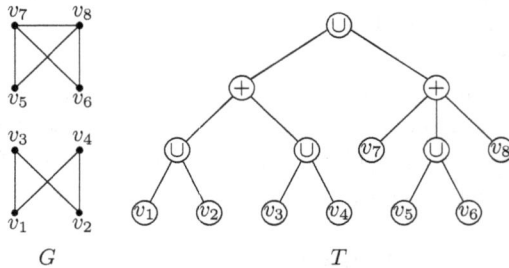

Fig. 9.3.　A cograph G of eight vertices and its cotree T.

Corneil *et al.*'s (1985) recognition algorithm for cographs is incremental in the sense that the vertices are processed one at a time until the entire graph has been handled. Since cographs are hereditary (i.e., every induced subgraph is a cograph), assume G is a cograph with a cotree T. Then design an algorithm which determines if $G+x$ is also a cograph. In case of a positive answer, modify T to get a cotree of $G + x$. Given the list of the vertices adjacent to x we percolate this adjacency information up the tree from the leaves to the root. For every node a in T, let

$$\widehat{L}_a = \{v \in L_a : vx \in E(G + x)\}.$$

It is clear that the following facts hold.

If $\widehat{L}_a = \emptyset$, then $\widehat{L}_d = \emptyset$ for every descendant d of a.
If $\widehat{L}_a = L_a$, then $\widehat{L}_d = L_d$ for every descendant d of a.
If $\emptyset \neq \widehat{L}_a \subsetneq L_a$, then $\emptyset \neq \widehat{L}_b \subsetneq L_b$ for every ancestor b of a.

Theorem 9.7 (Corneil *et al.*, 1985). *If G is a cograph with cotree T rooted at r, then $G + x$ is a cograph if and only if*

(1) *$\widehat{L}_r = \emptyset$ or $\widehat{L}_r = L_r$, or*
(2) *there exists a path $(a_0, a_1, \ldots, a_m = r)$ such that for $0 \leq i \leq m$:*
　　(2.1) *if $i = 0$, then $\widehat{L}_a = \emptyset$ or $\widehat{L}_a = L_a$ for every child a of a_0, and a_0 has some child b with $\widehat{L}_b = \emptyset$ and some child c with $\widehat{L}_c = L_c$;*
　　(2.2) *if $i \geq 1$ and a_i is a $\cup$-node, then $\widehat{L}_b = \emptyset$ for every child $b \neq a_{i-1}$ of a_i;*
　　(2.3) *if $i \geq 1$ and a_i is a $+$-node, then $\widehat{L}_c = L_c$ for every child $c \neq a_{i-1}$ of a_i.*

Proof. ($\Rightarrow$) Suppose $G + x$ is a cograph but condition (1) does not hold, that is, $\emptyset \neq \widehat{L}_r \subsetneqq L_r$. Choose a node a_0 farthest from r such that $\emptyset \neq \widehat{L}_{a_0} \subsetneqq L_{a_0}$. Consider the a_0-r path $(a_0, a_1, \ldots, a_m = r)$ formed by the ancestors of a_0. Then $\emptyset \neq \widehat{L}_{a_i} \subsetneqq L_{a_i}$ for $0 \leq i \leq m$.

We shall prove condition (2) by induction on i.

For $i = 0$, (2.1) holds by the fact that a_0 is a node farthest to r such that $\emptyset \neq \widehat{L}_{a_0} \subsetneqq L_{a_0}$.

Suppose $i \geq 1$ and the claim holds for $i' < i$. Consider two cases.

Case 1. a_i is a $\cup$-node.

We shall prove (2.2). Suppose to the contrary that (2.2) does not hold. Then $\widehat{L}_b \neq \emptyset$ for some child $b \neq a_{i-1}$ of a_i. Choose $y \in \widehat{L}_b$. Since a_{i-1} is a $+$-node, by the induction hypothesis, $\widehat{L}_c = L_c$ for some child c of a_{i-1}. Choose $z \in \widehat{L}_c$. By (2.1), $\widehat{L}_{b'} = \emptyset$ for some child b' of a_0, which is also a descendant of a_{i-1}. Choose $w \in L_{b'}$. Then (w, z, x, y) is an induced P_4, a contradiction.

Case 2. a_i is a $+$-node.

We shall prove (2.3). Suppose to the contrary that (2.3) does not hold. Then $\widehat{L}_c \subsetneqq L_c$ for some child $c \neq a_{i-1}$ of a_i. Choose $y \in L_c \backslash \widehat{L}_c$. Since a_{i-1} is a $\cup$-node, by the induction hypothesis, $\widehat{L}_b = \emptyset$ for some child b of a_{i-1}. Choose $z \in L_b$. By (2.1), $\widehat{L}_{c'} = L_{c'}$ for some child c' of a_0, which is also a descendant of a_{i-1}. Choose $w \in \widehat{L}_{c'}$. Then (x, w, y, z) is an induced P_4, a contradiction.

($\Leftarrow$) First consider the case when condition (1) holds.

If $\widehat{L}_r = \emptyset$, then $G + x$ is $G \cup K_1$ and so is a cograph. The cotree T' for $G + x$ can be obtained from T as follows. If r is a $\cup$-node, then add x as a new child of r to get T'. If r is a $+$-node, then add a new $\cup$-node r' as the root of T' and make x and r as children of r'.

If $\widehat{L}_r = L_r$, then $G + x$ is $G + K_1$ is also a cograph. The cotree T' for $G + x$ can be obtained from T as follows. If r is a $+$-node, then add x as a new child of r to get T'. If r is a $\cup$-node, then add a new $+$-node r' as the root of T' and make x and r as children of r'.

Suppose condition (2) holds. By (2.1), assume that a_0 has $s \geq 1$ children $b_1, b_2, \ldots, b_s$ with each $\widehat{L}_{b_i} = \emptyset$ and $t \geq 1$ children $c_1, c_2, \ldots, c_t$ with each $\widehat{L}_{c_j} = L_{c_j}$. Updating T according to the following two cases gives a cotree T' for $G + x$.

(a) a_0 **is a $\cup$-node:** In this case, consider several subcases. If $t = 1$ and c_1 is a $+$-node, then add x as a new child of c_1. Otherwise either $t = 1$ with c_1 being a leaf node or else $t \geq 2$. Now create a new $+$-node a as a child of a_0 and make x as a child of a. If $t = 1$ with c_1 being a leaf node, then move c_1 to be a child of a. If $t \geq 2$, then create a new $\cup$-node b as a child of a and move $c_1, c_2, \ldots, c_t$ to be children of b. These give T'.

(b) a_0 **is a $+$-node**: In this case, consider several subcases. If $s = 1$ and b_1 is a $\cup$-node, then add x as a new child of b_1. Otherwise either $s = 1$ with b_1 being a leaf node or else $s \geq 2$. Now create a new $\cup$-node a as a child of a_0 and make x as a child of a. If $s = 1$ with b_1 being a leaf node, then move b_1 to be a child of a. If $s \geq 2$, then create a new $+$-node c as a child of a and move $b_1, b_2, \ldots, b_s$ to be children of c. These give T'.

The tree T' is a cotree of $G + x$ according to (2.2) and (2.3). □

A linear-time algorithm for recognizing cographs then follows. The correctness of the algorithm can be seen from the proof of the above theorem. At iteration i, vertex v_i is added into the current G. The number of vertices adjacent to $x := v_i$ is $f = O(\deg_G(v_i))$ in the algorithm. The total number of nodes managed in M and F is $O(f)$. Hence, the running time is linear.

Algorithm RecogCographs: Determine if a graph G of at least two vertices is a cograph. In case of a positive answer, a cotree T is given.

Give an arbitrary ordering $v_1, v_2, \ldots, v_n$ of $V(G)$;
create a tree T rooted at r with two children v_1 and v_2; $\mathrm{md}(r) \leftarrow 0$;
if $v_1 v_2 \notin E(G)$ **then** label r as a $\cup$-node **else** label r as a $+$-node;
for $i = 3$ to n **do** //* Extend to the next vertex v_i. *//
{ $\quad x \leftarrow v_i$; $\quad F \leftarrow \{v_j : 1 \leq j < i, v_j x \in E(G)\}$ and $f \leftarrow |F|$;
$\quad$ **if** $f = 0$ or $f = i - 1$ **then**
$\quad$ { **if** $(f = 0$ and r is a $\cup$-node) or $(f = i - 1$ and r is a $+$-node)
$\quad$ **then** add x as a child of r;
$\quad$ **else** { create a root r' with two children r and x; $\quad \mathrm{md}(r') \leftarrow 0$;
$\quad\quad\quad\quad$ **if** $f = 0$ **then** label r' by $\cup$ **else** label r' by $+$; }
$\quad$ } **else** //* Now $0 < f < i - 1$. *//
$\quad$ { $M \leftarrow \emptyset$;
$\quad\quad$ **while** F is not empty **do**
$\quad\quad$ { $\quad$ remove a node a out of F and let p be the parent of a;
$\quad\quad\quad\quad$ $\mathrm{md}(a) \leftarrow 0$; $\quad \mathrm{md}(p) \leftarrow \mathrm{md}(p) + 1$; $\quad M \leftarrow M \cup \{p\}$;
$\quad\quad\quad\quad$ **if** $\mathrm{md}(p) = d(p)$ **then** move p from M to F;
$\quad\quad$ } //* Now $M \neq \emptyset$. *//
$\quad\quad$ **for** all $a \in M$ **do** let b be the parent (resp., grandparent) of a if it
$\quad\quad$ is a $\cup$-node (resp., $+$-node), mark b if $b \in M$ and $\mathrm{md}(b) = d(b) - 1$;
$\quad\quad$ **if** M has two unmarked nodes **then** { G is not a cograph; **stop**; }
$\quad\quad$ choose the only unmarked nodes a_0 in M;

if a_0 has a proper ancestor b of $+$-node with $\mathrm{md}(b) \neq d(b) - 1$
then { G is not a cograph; **stop**; }
let a_0 has $s \geq 1$ children $b_1, b_2, \ldots, b_s$ with each $\widehat{L}_{b_i} = \emptyset$
    and $t \geq 1$ children $c_1, c_2, \ldots, c_t$ with each $\widehat{L}_{c_i} = L_{c_i}$;
(a) a_0 is a $\cup$-node: create a new $+$-node a and a new $\cup$-node b;
  if $t = 1$, c_1 is a $\cup$-node **then** add x as a new child of c_1;
  if $t = 1$, c_1 is a leaf node **then** make x, c_1 as children of a;
  if $t \geq 2$ **then** move $c_1, c_2, \ldots, c_t$ as children of b
        and make b as a child of a;
(b) a_0 is a $+$-node: create a new $\cup$-node a and a new $+$-node c;
  if $s = 1$, b_1 is a $+$-node **then** add x as a new child of b_1;
  if $s = 1$, b_1 is a leaf node **then** make x, b_1 as children of a;
  if $s \geq 2$ **then** move $b_1, b_2, \ldots, b_s$ as children of c
        and make c as a child of a;
for all $a \in M$ **do** $\mathrm{md}(a) \leftarrow 0$;
}
}

More recognition algorithms for cographs were given by Damiand *et al.*
(2001), Habib and Paul (2005) and Bretscher *et al.* (2008). Having this
cograph recognition in mind, now we are back to distance-hereditary graph
recognition by Hammer and Maffray.

Algorithm RecogDistance: Determine if a connected graph G is
distance-hereditary.

Choose a vertex u and build the hanging $h_u = [L_1, L_2, \ldots, L_t]$;
for $j = t$ **to** 2 **step by** -1 **do**
  (1) find the components of L_j;
  (2) for each component A of L_j with $|A| \geq 2$ call REDUCE(A);
  (3) (L_j now is a stable set.)
     order the vertices x of L_j by increasing degree $\deg'(x)$;
  (4) delete those with $\deg'(x) = 1$;
     (they are leaves of the remaining graph.)
  (5) for each vertex x of L_j taken by increasing degree $\deg'(x)$,
     call REDUCE($N'(x)$) (now $\deg'(x) = 1$), and delete x;

REDUCE(A): This subroutine checks whether A is homogeneous
and P_4-free or not (for instance, using the algorithm by Corneil,
Perl and Stewart). If the answer is negative, then stop and report
that G is not distance-hereditary, else replace A by a single vertex.

Distance-hereditary graph recognition algorithms were also offered by Damiand *et al.* (2001) and Gioan and Paul (2012). In fact, Damiand *et al.* (2001) presented two counterexamples to the recognition algorithm above. But it is not clear how exactly their examples break the algorithm's steps.

9.3   Weighted connected domination in distance-hereditary graphs

D'Atri and Moscarini (1988) developed $O(nm)$-time algorithms for the connected domination and the Steiner tree problems in distance-hereditary graphs. Brandstädt and Dragan (1998) presented a linear-time algorithm for the connected r-domination and Steiner tree problems in distance-hereditary graphs.

This section presents a linear-time algorithm by Yeh and Chang (1998) for finding a minimum weighted connected dominating set of a connected distance-hereditary graph G, where each vertex v has a weight $w(v)$, which may be assumed to be nonnegative by Lemma 1.3.

Theorem 9.8 (Yeh and Chang, 1998). *Suppose G is a connected distance-hereditary graph in which every vertex v has a non-negative weight $w(v)$. Let $h_u = [L_0, L_1, \ldots, L_t]$ be a hanging at a vertex u of minimum weight. Consider the set $\mathcal{A} = \{N'(v) : v \in L_i \text{ with } 2 \le i \le t \text{ and } N'(v) \text{ is the minimal neighborhood of } v \text{ in } L_{i-1}\}$. For each $N'(v) \in \mathcal{A}$, choose one vertex $v' \in N'(v)$ of minimum weight, and let D be the set of all such v'. Then D or $D \cup \{u\}$ or some $\{v\}$ with $v \in V(G)$ is a minimum weighted connected dominating set of G.*

Proof.   Choose a minimum weighted connected dominating set M of G. If $|M| = 1$, then some $\{v\}$ with $v \in V(G)$ is a minimum weighted connected dominating set of G as desired. Now suppose $|M| \ge 2$.

Suppose $N'(v) \in \mathcal{A}$, where $v \in L_i$ and $2 \le i \le t$. Choose a vertex $y \in M$ dominating v. Then $y \in L_{i-1} \cup L_i \cup L_{i+1}$. If $y \in L_{i-1}$, then $y \in M \cap N'(v)$. If $y \in L_i \cup L_{i+1}$, then choose a vertex $x \in M \cap (L_0 \cup L_1)$ and an x-y path

$$P = (x = v_1, v_2, \ldots, v_r = y)$$

in $G[M]$. Let j be the smallest index such that $\{v_j, v_{j+1}, \ldots, v_r\} \subseteq L_i \cup L_{i+1} \cup \ldots \cup L_t$. Then $v_j \in L_i$, $v_{j-1} \in N'(v_j)$, and v and v_j are in the same component of $G - L_{i-1}$. By Theorem 9.4, $N'(v) = N'(v_j)$ and so $v_{j-1} \in M \cap N'(v)$. In any case, there is some vertex $v'' \in M \cap N'(v)$ for every $N'(v) \in \mathcal{A}$. Since sets in $\mathcal{A}$ are pairwise disjoint, v, v', v'' are one-to-one correspondence. Hence, $|M| \geq |\mathcal{A}| = |D|$. Consider two cases.

Case 1. $|M| > |D|$. In this case, there is at least one vertex x in M that is not a v''. Then

$$w(M) \geq \sum_{v''} w(v'') + w(x) \geq \sum_{v'} w(v') + w(u) = w(D \cup \{u\}).$$

Next to show that $D \cup \{u\}$ is a dominating set of G. If $x \in L_0 \cup L_1$, then it is dominated by u. If $x \in L_i$ with $i \geq 2$, then $N'(x) \supseteq N'(v)$ for some $N'(v) \in \mathcal{A}$, and so x is dominated by $v' \in D$.

Finally, suppose $x \in D$. If $x \in L_1$, then it is adjacent to u. If $x \in L_i$ with $i \geq 2$, then $N'(x) \supseteq N'(v)$ for some $N'(v) \in \mathcal{A}$, and so x is adjacent to some $v' \in D$. In any case, there is an x–u path using vertices only in $D \cup \{u\}$. Therefore, $G[D \cup \{u\}]$ is connected.

Hence, $D \cup \{u\}$ is a minimum weighted connected dominating set of G.

Case 2. $|M| = |D|$. In this case, $M = \{v'' : v'' \in N'(v) \in \mathcal{A}\}$. Then

$$w(M) = \sum_{v''} w(v'') \geq \sum_{v'} w(v') = w(D).$$

Next to show that $G[D]$ is connected. Suppose x'' is adjacent to y'' in $G[M]$, where $x'' \in N'(x) \in \mathcal{A}$ and $y'' \in N'(y) \in \mathcal{A}$. By Theorem 9.5 (2), $N'(x)$ and $N'(y)$ are homogeneous. Hence, x' is adjacent to y'. The connectivity of $G[M]$ then implies that $G[D]$ is connected.

For every $x \in V(G)$, since M is a dominating set of G, x is dominated by some $v'' \in N'(v) \in \mathcal{A}$. If $x \notin N'(v)$, then x is also dominated by $v' \in D$, since $N'(v)$ is homogeneous. Now suppose $x \in N'(v)$. Since $|D| \geq 2$ and $G[D]$ is connected, v' is adjacent to some $y' \in D \backslash N'(v)$. Then x is also adjacent to y', since $N'(v)$ is homogeneous. In any case, D is a dominating set of G.

Therefore, D is a minimum weighted connected dominating set of G. □

Theorem 9.8 provides an efficient algorithm for solving the weighted connected domination problem in distance-hereditary graphs. To implement the algorithm efficiently, the set $\mathcal{A}$ is not actually found. Instead, the following step is performed for each $2 \leq i \leq t$. Sort the vertices in L_i such that

$$|N'(x_1)| \le |N'(x_2)| \le \ldots \le |N'(x_j)|.$$

Then process $N'(x_k)$ for k from 1 to j. At iteration k, if $N'(x_k) \cap D = \emptyset$, then $N'(x_k)$ is in $\mathcal{A}$ and choose a vertex of minimum weight to put it into D; otherwise $N'(x_k) \notin \mathcal{A}$ and do nothing.

Algorithm wConDomDH: Find a minimum weighted connected domination set D of a connected distance-hereditary graph G in which every vertex v has a nonnegative weight $w(v)$.

$D \leftarrow \emptyset$;
let u be a vertex of minimum weight;
determine the hanging $h_u = [L_0, L_1, \ldots, L_t]$ of G at u;
for $i = t$ **to** 2 **step by** -1 **do**
{   let $L_i = \{x_1, x_2, \ldots, x_j\}$;
     sort L_i and assume that $|N'(x_{i_1})| \le |N'(x_{i_2})| \le \ldots \le |N'(x_{i_j})|$;
     for $k = 1$ **to** j **do**
          if $N'(x_{i_k}) \cap D = \emptyset$
          then $D \leftarrow D \cup \{y\}$ where y is of minimum weight in $N'(x_{i_k})$;
}
if not $(L_1 \subseteq N[D]$ and $G[D]$ is connected) **then** $D \leftarrow D \cup \{u\}$;
if there is some $v \in V(G)$ dominates G and $w(v) < w(D)$ **then** $D \leftarrow \{v\}$;

Theorem 9.9. *Algorithm wConDomDH gives a minimum weighted connected dominating set of a connected distance-hereditary graph in linear time.*

Proof.   The correctness of the algorithm follows from Theorem 9.8. For each i, sort L_i by using a bucket sort. Then the algorithm runs in $O(|V(G)| + |E(G)|)$ time. $\qquad\square$

Later, Yeh and Chang (2001) gave $O(nm)$-time algorithms for the weighted connected k-domination problem and the weighted k-dominating clique problems in distance-hereditary graphs.

9.4   Domination in distance-hereditary graphs

Chang *et al.* (2002) established a linear-time algorithm for the domination problem in distance-hereditary graphs. They used a labeling method in which a one-vertex-extension ordering is used. At each iteration, the algorithm decides if a vertex v_i is to be included in a minimum solution according to its label. During the iterations, some information of deleted vertices are still kept in the remaining graph. For this purpose, they introduced the following concept called L-domination.

Suppose G is a graph whose vertex set is a subset of a *ground set* V'. Furthermore, each vertex v in G is associated with a *regular label* $L(v) = S(v)$ or a *conditional label* $L(v) = S(v) * p(v)$, where $S(v) \subseteq \{\{0\}, \{1\}, \{0,1\}\}$ and $p(v) \in V' \backslash V(G)$ with the property that $p(x) \neq p(y)$ for two distinct vertices x and y. An L-*dominating set* of G is a subset $D \subseteq V'$ such that (LD) hold when v has a regular label $L(v) = S(v)$, and (LD) holds or $p(v) \in D$ when v has a conditional label $L(v) = S(v) * p(v)$.

(LD) For each $A \in S(v)$, there exists some $x \in D$ such that $d_G(v, x) \in A$.

In other words, if $L(v) = \{\{0\}\}$, then $v \in D$; if $L(v) = \{\{1\}\}$, then v has a neighbor in D; if $L(v) = \{\{0,1\}\}$, then v has a closed neighbor in D; if $L(v) = \{\{0,1\}\} * p(v)$, then $p(v) \in D$ or v has a closed neighbor in D; ... etc. When $V' = V(G)$ and each vertex has a regular label $\{\{0,1\}\}$ (respectively, $\{\{1\}\}$), L-domination is just the usual domination (respectively, total domination).

As distance-hereditary graphs are combinations of trees and cographs, the labeling method shares similarities with the method described in Section 3.2. The significant difference comes from the conditional label, which is new. This new label increases the complexity of the argument, as discussed in their paper.

9.5 Exercises

(9.1) Show that a graph is distance-hereditary if and only if for every two vertices u and v with $d(u, v) = 2$, there is no induced u–v path of length greater than two.

(9.2) Show that a graph is distance-hereditary if and only if every cycle of length at least five has at least two chords and every 5-cycle has two crossing chords.

(9.3) Show that a graph is distance-hereditary if and only if it is C_n-free for all $n \geq 5$, house-free, gem-free and domino-free, see Figure 9.1 for these forbidden subgraphs.

(9.4) For any two vertices u and v, define

$$I(u, v) = \{x : x \text{ is a vertex in a shortest } u\text{–}v \text{ path}\}.$$

Show that a graph is distance-hereditary if and only if it is house-free, gem-free and domino-free, see Figure 9.1 for these forbidden subgraphs, and for every three vertices u, v and w,

$$I(u, v) \cap I(v, w) = \{v\}$$

implies

$$d(u, w) \geq d(u, v) + d(v, w) - 1.$$

(9.5) Show that a graph is distance-hereditary if and only if for every four vertices u, v, w and x, at least two of the following distance sums are equal:

$$d(u, v) + d(w, x), \qquad d(u, w) + d(v, x), \qquad d(u, x) + d(v, w).$$

(9.6) Show that a graph is bipartite and distance-hereditary if and only if it is C_n-free for all $n \geq 5$, K_3-free and domino-free, see Figure 9.1 for the domino graph.

(9.7) Show that a graph is bipartite and distance-hereditary if and only if for every three vertices u, v and w,

$$I(u, v) \cap I(u, w) \cap I(v, w) \neq \emptyset,$$

and at least two of the following unions are equal:

$$I(u, v) \cup I(u, w), \qquad I(u, v) \cup I(v, w), \qquad I(u, w) \cup I(v, w).$$

(9.8) Provide examples of distance-hereditary graphs that are not cocomparability graphs.

(9.9) Provide examples of permutation graphs that are not distance-hereditary graphs.

(9.10) Show that a cograph is a graph all of whose induced subgraphs have the property that every maximal clique intersects every maximal stable set at exactly one vertex.

(9.11) Design an efficient algorithm for the weighted domination problem in cographs.

(9.12) Design an efficient algorithm for the weighted independent domination problem in cographs.

(9.13) Design an efficient algorithm for the weighted total domination problem in cographs.

(9.14) Design an efficient algorithm for the weighted domination clique problem in cographs.

(9.15) While Yeh and Chang (2001) gave an $O(nm)$-time algorithm for the weighted connected k-domination problem in distance-hereditary graphs, discuss the possibility of designing a linear-time algorithm for the cardinality connected k-domination problem in distance-hereditary graphs. If not possible, explain why.

(9.16) While Yeh and Chang (2001) gave an $O(nm)$-time algorithm for the weighted k-domination clique problem in distance-hereditary graphs, discuss the possibility of designing a linear-time algorithm for the cardinality k-domination clique problem in distance-hereditary graphs. If not possible, explain why.

(9.17) Design a linear-time algorithm for the total domination problem in distance-hereditary graphs.

(9.18) Design an efficient algorithm for the k-domination problem in distance-hereditary graphs.

(9.19) Design an efficient algorithm for the total k-domination problem in distance-hereditary graphs.

Bibliography

Note: At the end of each reference, the section/chapter number is listed to indicate the section/chapter whose content is related to this reference, and a * means that the reference is precisely cited.

Abramson, B. and Yung, M. (1989). Divide and conquer under global constraints: a solution to the N-queens problem, *J. Parallel Distrib. Comput.*, **6**, pp. 649–662. (1.1 chess)

Aho, A. V., Hopcroft, J. E. and Ullman, J. D. (1974). *The Design and Analysis of Computer Algorithms*, Addison-Wesley, Reading, Massachusetts. (2.1*)

Ahrens, W. (1910). *Mathematische Unterhaltungen und Spiele*, Leipzig-Berlin. (1.1 chess)

Alman, J. and Williams, V. V. (2024). A refined laser method and faster matrix multiplication, *Theor. Comput. Sci.*, **3**, Article 21, pp. 1–32. (7.1)

Alon, N., Fellows, M. R. and Hare, D. R. (1996). Vertex transversals that dominate, *J. Graph Theory*, **21**, pp. 21–31. (4.4*)

Anstee, R. P. and Farber, M. (1984). Characterizations of totally balanced matrices, *J. Algorithms*, **5**, pp. 215–230. (6.2)

Ao, S. (1994). *Independent Domination Critical Graphs*, Master's Thesis, Univ. Victoria, Victoria, BC, Canada. (1.1 thesis)

Araki, T. and Yamanaka, R. (2019). Secure domination in cographs, *Discrete Appl. Math.*, **262**, pp. 179–184. (9)

Arnborg, S. and Proskurowski, A. (1989). Linear time algorithms for NP-hard problems restricted to partial k-trees, *Discrete Appl. Math.*, **23**, pp. 11–24. (3.7)

Arvind, K., Breu, H., Chang, M.-S., Kirkpatrick, D. G., Lee, F. Y., Liang, Y. D., Madhukar, K., Pandu Rangan, C. and Srinivasan, A. (1996). Efficient algorithms in cocomparability and trapezoid graphs, manuscript. (7.2* 7.3* 7.4*)

Arvind, K. and Pandu Rangan, C. (1992). Connected domination and Steiner set on weighted permutation graphs, *Inform. Process. Lett.*, **41**, pp. 215–220. (8.4*)

Atallah, M. J. and Kosaraju, S. R. (1989). An efficient algorithm for maxdominance, with applications, *Algorithmica*, **4**, pp. 221–236.   (8.3)

Atallah, M. J., Manacher, G. K. and Urrutia J. (1988). Finding a minimum independent dominating set in a permutation graph, *Discrete Appl. Math.*, **21**, pp. 177–183.   (8.3*)

Balakrishnan, R. and Paulraja, P. (1983). Powers of chordal graphs, *J. Austr. Math. Soc.*, **35**, pp. 211–217.   (6.1*)

Balakrishnan, H., Rajaraman, A. and Pandu Rangan, C. (1993). Connected domination and Steiner set on asteroidal triple-free graphs, in Dehne, F., Sack, J. R., Santoro, N. and Whitesides, S. eds., *Proc. Workshop on Algorithms and Data Structures (WADS'93), Montreal, Canada, Lecture Notes in Comput. Sci.*, **709**, Springer-Verlag, Berlin, pp. 131–141.   (7.5* 7.6*)

Bandelt, H.-J. and Mulder, H. M. (1986). Distance-hereditary graphs, *J. Combin. Theory, Ser. B*, **41**, pp. 182–208.   (9.1* 9.2*)

Banerjee, S., Chaudhary, J. and Pradhan, D. (2023). Unique response Roman domination: complexity and algorithms, *Algorithmica*, **85**, pp. 3889–3927.
                                                                          (5, 9)

Banerjee, S., Henning, M. A. and Pradhan, D. (2020). Algorithmic results on double Roman domination in graphs, *J. Combin. Optim.*, **39**, pp. 90–114.
                                                                          (3.7, 5)

Banerjee, S., Henning, M. A. and Pradhan, D. (2021). Perfect Italian domination in cographs, *Appl. Math. Comput.*, **391**, 125703.   (9)

Banerjee, S., Keil, J. M. and Pradhan, D. (2019). Perfect Roman domination in graphs, *Theor. Comput. Sci.*, **796**, pp. 1–21.   (3.7, 5 9)

Bange, D. W., Barkauskas, A. E., Host, L. H. and Slater, P. J. (1996). Generalized domination and efficient domination in graphs. *Discrete Math.*, **159**, pp. 1–11.   (4.4)

Bange, D. W., Barkauskas, A. E. and Slater, P. J. (1988). Efficient dominating sets in graphs, in Ringeisen, R. D. and Roberts, F. S. eds., *Applications of Discrete Math.*, SIAM, Philadelphia, PA, pp. 189–199.   (3.3, 4.4)

Bar-Yehuda, R. and Vishkin, U. (1982). Complexity of finding k-path-free dominating sets in graphs, *Inform. Process. Lett.*, **14**, pp. 228–232.   (4.4)

Bean, T. J., Henning, M. A. and Swart, H. C. (1994). On the integrity of distance domination in graphs, *Austral. J. Combin.*, **10**, pp. 29–43.   (4.4)

Bellman, R. E. and Dreyfus, S. E. (1962). *Applied Dynamic Programming*, Princeton University Press.   (3.3*)

Benzer, S. (1959). On the topology of the genetic fine structure, *Proc. Nat. Acad. Sci. U.S.A.*, **45**, pp. 1607–1620.   (5.1*)

Berge, C. (1960). Les problémes de coloration en théorie des graphes, *Publ. Inst. Stat. Univ. Paris*, **9**, pp. 123–160.   (1.5, 4.1*)

Berge, C. (1961). Färbung von Graphen dern sämtliche bzw. dern ungerade Kreise starr sind (Zusammenfassung), *Wiss. Z. Martin-Luther Univ. Halle-Wittenberg, Math.-Natur. Reihe*, **10**, pp. 114–115.   (1.5)

Berge, C. (1963). Some classes of perfect graphs, *Six Papers on Graph Theory*, Indian Stat. Inst., Calcutta, pp. 1–21.   (1.5*)

Berge, C. (1973). *Graphs and Hypergraphs*, North-Holland, Amsterdam; translated by Minieka, E. from the French version. (1.2 book)

Berge, C. (1997). Motivations and history of some of my conjectures, *Discrete Math.*, **165/166**, pp. 61–70. (1.5*)

Berge, C. and Ramírez Alfonsín, J. L. (2001). Origins and genesis, in Ramírez Alfonsín, J. L. and Reed, B. A., eds., *Perfect Graphs*, John Wiley & Sons, Chichestrer, pp. 1–12. (1.5*)

Bernhardsson, B. (1991). Explicit solutions to the N-queens problem for all N, *SIGART Bull.*, **2**, pp. 7. (1.1 chess)

Bertossi, A. A. (1984). Dominating sets for split and bipartite graphs, *Inform. Process. Lett.*, **19**, pp. 37–40. (4.4)

Bertossi, A. A. (1986). Total domination in interval graphs, *Inform. Process. Lett.*, **23**, pp. 131–134. (5.2*)

Bertossi, A. A. (1988). On the domatic number of interval graphs, *Inform. Process. Lett.*, **28**, pp. 275–280. (5.5*)

Bertossi, A. A. and Gori, A. (1988). Total domination and irredundance in weighted interval graphs, *SIAM J. Discrete Math.*, **1**, pp. 317–327. (5.3*)

Beyer, T., Proskurowski, A., Hedetniemi, S. T. and Mitchell, S. (1977). Independent domination in trees, *Proc. 8th S. E. Conf. Combin., Graph Theory and Comput., Congr. Numer.*, **19**, pp. 321–328. (3.3*)

Bezzel, M. (1848). Schachfreund, *Berliner Schachzeitung*, **3**, pp. 363. (1.1*)

Biggs, N. L., Lloyd, E. K. and Wilson, R. J. (1986). *Graph Theory 1736–1936*, Clarendon Press, Oxford. (1.1*)

Bird, W. H. (2017). *Computational Methods for Domination Problems*, Ph.D. Thesis, Dept. Comput. Sci., Univ. Victoria. (1.1 thesis chess)

Blitch, P. (1983). *Domination in Graphs*, Ph.D. Thesis, Univ. South Carolina. (1.1 thesis)

Bondy, J. A. and Murty, U. S. R. (1976). *Graph Theory with Applications*, North Holland, New York. (1.2 book)

Bondy, J. A. and Murty, U. S. R. (2008). *Graph Theory*, Graduate Texts in Mathematics, **244**, Springer. (1.2 book 3.1*)

Bonuccelli, M. A. (1985). Dominating sets and domatic number of circular arc graphs, *Discrete Appl. Math.*, **12**, pp. 203–213. (5.6*)

Booth, K. S. (1980). *Dominating Sets in Chordal Graphs*, Ph.D. Thesis, Univ. Waterloo. (1.1 thesis)

Booth, K. S. and Johnson, J. H. (1982). Dominating sets in chordal graphs, *SIAM J. Comput.*, **11**, pp. 191–199. (4.4*)

Booth, K. S. and Lueker, G. S. (1976). Testing for the consecutive ones property, interval graphs, and graph planarity using PQ-tree algorithms, *J. Comput. Sys. Sci.*, **13**, pp. 335–379, (5.1*)

Borie, R. B. (1988). *Recursively Constructed Graph Families: Membership and Linear Algorithms*, Ph.D. Thesis, Georgia Institute of Technology. (1.1 thesis)

Bozeman, C., Brimkov, B., Erickson, C., Ferrero, D., Flagg, M. and Hogben, L. (2019). Restricted power domination and zero forcing problems, *J. Combin. Optim.*, **37**, pp. 935–956. (3.5)

Brandstädt, A. (1990). On the domination problem for bipartite graphs, in Bodendieck, R. and Henn, R. eds., *Topics in Combinatorics and Graph Theory*, pp. 145–152, Physica-Verlag, Heidelberg. (4.4)

Brandstädt, A. (1993). On improved bounds for permutation graph problems, in Mayr, E. W. ed., *Proc. WG'92, Lecture Notes in Comput. Sci.*, **657**, Berlin, Springer Verlag, pp. 1–10. (8.3)

Brandstädt, A., Chepoi V. D. and Dragan, F. F. (1995). The algorithmic use of hypertree structure and maximum neighbourhood orderings, in Mayr, E. W., Schmidt, G. and Tinhofer, G. eds., *20th Internat. Workshop Graph-Theoretic Concepts in Computer Science (WG'94), Lecture Notes in Comput. Sci.*, **903**, Berlin, Springer-Verlag, pp. 65–80. (6)

Brandstädt, A., Chepoi V. D. and Dragan, F. F. (1997). Clique r-domination and clique r-packing problems on dually chordal graphs, *SIAM J. Discrete Math.*, **10**, pp. 109–127. (6)

Brandstädt, A., Chepoi V. D. and Dragan, F. F. (1998). The algorithmic use of hypertree structure and maximum neighbourhood orderings, *Discrete Appl. Math.*, **82**, pp. 43–77. (6)

Brandstädt, A. and Dragan, F. F. (1998). A linear-time algorithm for connected r-domination and Steiner tree on distance-hereditary graphs, Networks, **31**, pp. 177–182. (9.3*)

Brandstädt, A., Dragan, F. F., Chepoi, V. D. and Voloshin, V. I. (1993). Dually chordal graphs, in *19th Internat. Workshop Graph-Theoretic Concepts in Computer Science (WG'93), Lecture Notes in Comput. Sci.*, **790**, Berlin, Springer-Verlag, pp. 237–251. (6)

Brandstädt, A., Dragan, F. F., Chepoi, V. D. and Voloshin, V. I. (1998). Dually chordal graphs. *SIAM J. Discrete Math.*, **11**, pp. 437–455. (6)

Brandstädt, A., Fičur, P., Leitert, A. and Milanič, M. (2015). Polynomial-time algorithms for weighted efficient domination problems in AT-free graphs and dually chordal graphs, *Inform. Process. Lett.*, **115**, pp. 256–262. (6, 7.7)

Brandstädt, A. and Kratsch, D. (1985). On the restriction of some NP-complete graph problems to permutation graphs, in Budach, L. ed., *Proc. FCT'85, Lecture Notes in Comput. Sci.*, **199**, Berlin, Springer-Verlag, pp. 53–62. (8.2, 8.3, 8.4)

Brandstädt, A. and Kratsch, D. (1987). On domination problems on permutation and other graphs, *Theor. Comput. Sci.*, **54**, pp. 181–198. (8.3*)

Brešar, B., Henning, M. A. and Rall, D. F. (2008). Rainbow domination in graphs, *Taiwanese J. Math.*, **12**, pp. 213–225. (3.2)

Bretscher, A., Corneil, D., Habib, H. and Paul, C. (2008). A simple linear time LexBFS cograph recognition algorithm, *SIAM J. Discrete Math.*, **22**, pp. 1277–1296. (9.2*)

Breu, H. and Kirkpatrick, D. G. (1993). Algorithms for dominating and Steiner set problems in cocomparability, manuscript. (7.2∗ 7.3∗)

Brimkov, B., Mikesell, D. and Smith, L. (2019). Connected power domination in graphs, *J. Combin. Optim.*, **38**, pp. 292–315. (3.5)

Broere, I., Hattingh, J. H., Henning, M. A. and McRae, A. A. (1995). Majority domination in graphs, *Discrete Math.*, **138**, pp. 125–135. (4.4)

Buneman, P. (1974). A characterization of rigid circuit graphs, *Discrete Math.*, **9**, pp. 205–212. (4.1∗)

Burger, A. P. (1993). *Domination in the Queen's Graph*, Master's Thesis, Univ. South Africa. (1.1 thesis chess)

Burger, A. P., Cockayne, E. J. and Mynhardt, C. M. (1997). Domination and irredundance in the queen's graph, *Discrete Math.*, **163**, pp. 47–66. (1.1 chess)

Burger, A. P., de Villiers, A. P. and van Vuurenb, J. H. (2014). A linear algorithm for secure domination in trees, *Discrete Appl. Math.*, **171**, pp. 15–27. (3.2)

Burger, A. P., Mynhardt, C. M. and Cockayne, E. J. (1994). Domination numbers for the queen's graph, *Bull. Inst. Combin. Appl.*, **10**, pp. 73–82. (1.1 chess)

Burlet, M. and Uhry, J. P. (1982). Parity graphs, in Bachem, A., Grötschel, M. and Korte, B. eds., Bonn Workshop on Combinatorial Optimization, *Annals of Discrete* **16**, North-Holland, Amsterdam, pp. 1–26. (9.1∗)

Campbell, P. J. (1977). Gauss and the eight queens problem: a study in miniature of the propagation of historical error, *Historia Math.*, **4**, pp. 397–404. (1.1 chess)

Carrington, J. R. (1992). *Global Domination of Factors of a Graph*, Ph.D. Thesis, Univ. Central Florida, Orlando. (1.1 thesis)

Carson, D. I. (1995). *Computational Aspects of Some Generalized Domination Parameters*, Ph.D. Thesis, Univ. Natal. (1.1 thesis)

Chang, G. J. (1982). *k-Domination and Graph Covering Problems*, Ph.D. Thesis, Cornell Univ., Ithaca, NY. (1.1 thesis 6.1∗)

Chang, G. J. (1988/89). Labeling algorithms for domination problems in sun-free chordal graphs, *Discrete Appl. Math.*, **22**, pp. 21–34. (6.3∗ 6.5∗)

Chang, G. J. (1989). Total domination in block graphs, *Oper. Res. Lett.*, **8**, pp. 53–57. (3.7)

Chang, G. J. (1990). Knuth numbers, *Mathmedia*, **14 (3)**, pp. 11–14. (in Chinese) (1.7∗)

Chang, G. J. (1991). Centers of chordal graphs, *Graphs and Combin.*, **7**, pp. 305–313. (6.5∗)

Chang, G. J. (1998). Algorithmic aspects of domination in graphs, in Du, D. Z. and Pardalos, P. eds., *Handbook of Combinatorial Optimization*, **3**, Kluwer Academic Pub., Dordrecht, the Netherlands, pp. 339–405. (1.1∗)

Chang, G. J. (2004). The weighted independent domination problem is NP-complete for chordal graphs, *Discrete Appl. Math.*, **143**, pp. 351–352. (4.5∗)

Chang, G. J. (2013). Algorithmic aspects of domination in graphs, in Pardalos, P., Du, D. Z. and Graham, R. eds., *Handbook of Combinatorial Optimization, Second Edition*, **1**, Springer, New York, NY, pp. 221–282. (1.1*)

Chang, G. J. (2017). *Graph Theory, With An Algorithmic Perspective* (in Chinese), National Taiwan Univ. Press. (1.2 book)

Chang, G. J., Dorbec, P., Montassier, M. and Raspaud, A. (2012). Generalized power domination of graphs, *Discrete Appl. Math.*, **160**, pp. 1691–1698. (3.5, 4.4)

Chang, G. J., Farber, M. and Tuza, Z. (1993). Algorithmic aspects of neighborhood numbers, *SIAM J. Discrete Math.*, **6**, pp. 24–29. (6)

Chang, G. J., Li, B.-J. and Wu, J.-J. (2013). Rainbow domination and related problems on strongly chordal graphs, *Discrete Appl. Math.*, **161**, pp. 1395–1401. (3.7)

Chang, G. J. and Nemhauser, G. L. (1982). R-domination of block graphs, *Oper. Res. Lett.*, **1**, pp. 214–218. (3.7)

Chang, G. J. and Nemhauser, G. L. (1984). The k-domination and k-stability on sun-free chordal graphs, *SIAM J. Algeb. Discrete Methods*, **5**, pp. 332–345. (6.1* 6.3* 6.5*)

Chang, G. J. and Nemhauser, G. L. (1985). Covering, packing and generalized perfection, *SIAM J. Algeb. Discrete Methods*, **6**, pp. 109–132. (6.1* 6.3*)

Chang, G. J., Panda, B. S. and Pradhan, D. (2012). Complexity of distance paired-domination problem in graphs, *Theor. Comput. Sci.*, **459**, pp. 89–99. (6)

Chang, G. J., Pandu Rangan, C. and Coorg, S. R. (1995). Weighted independent perfect domination on cocomparability graphs, *Discrete Appl. Math.*, **63**, pp. 215–222. (7)

Chang, G. J. and Roussel, N. (2010). On the k-power domination of hypergraphs, *J. Combin. Optim.*, **30**, pp. 1095–1106. (3.5)

Chang, G. J. and Tsai, M.-T. (2020). *Graph Theory, With An Algorithmic Perspective*, Revised Edition (in Chinese), National Taiwan Univ. Press. (1.2 book)

Chang, G. J., Wu, J. and Zhu, X. (2010). Rainbow domination on trees, *Discrete Appl. Math.*, **158**, pp. 8–12. (3.2*)

Chang, M.-S. (1995). Weighted domination on cocomparability graphs, in *Algorithms and Computation, Proc. ISAAC'95, Lecture Notes in Comput. Sci.*, **1004**, Berlin, Springer-Verlag, pp. 122–131. (7.3*)

Chang, M.-S. (1997). Weighted domination of cocomparability graphs, *Discrete Appl. Math.*, **80**, pp. 135–148. (7.3* 7.8*)

Chang, M.-S. (1998). Efficient algorithms for the domination problems on interval graphs and circular-arc graphs, *SIAM J. Comput.*, **27**, pp. 1671–1694. (5.2*)

Chang, M.-S. and Hsu, C.-C. (1997). On minimum intersection of two minimum dominating sets of interval graphs, *Discrete Appl. Math.*, **78**, pp. 41–50. (5)

Chang, M.-S. and Liu, Y.-C. (1993). Polynomial algorithms for the weighted perfect domination problems on chordal and split graphs, *Inform. Process. Lett.*, **48**, pp. 205–210. (4.6)

Chang, M.-S. and Liu, Y.-C. (1994). Polynomial algorithms for weighted perfect domination problems on interval and circular-arc graphs, *J. Inform. Sci. Engin.*, **10**, pp. 549–568. (5.6*)

Chang, M.-S., Peng, S.-L. and Liaw, J.-L. (1993). Deferred-query—an efficient approach for problems on interval and circular-arc graphs, in Dehne, F., Sack, J.-R., Santoro, N. and Whitesides, S. eds., *Proc. Third Workshop, Algorithms and Data Structures, Lecture Notes in Comput. Sci.*, **709**, pp. 222–233. (5.5*)

Chang, M.-S., Peng, S.-L. and Liaw, J.-L. (1999). Deferred-query—an efficient approach for problems on interval and circular-arc graphs, *Networks*, **34**, pp. 1–10. (5.5*)

Chang, M.-S., Wu, S.-C., Chang, G. J. and Yeh, H.-G. (2002). Domination in distance-hereditary graphs, *Discrete Appl. Math.*, **116**, pp. 103–113. (9.4*)

Chang, T. Y. (1992). *Domination Numbers of Grid Graphs*, Ph.D. Thesis, Univ. South Florida. (1.1 thesis chess)

Chang, T. Y. and Clark, W. E. (1993). The domination numbers of the $5 \times n$ and $6 \times n$ grid graphs, *J. Graph Theory*, **17**, pp. 81–107. (1.1 chess)

Chang, T. Y., Clark, W. E. and Hare, E. O. (1994). Domination numbers of complete grid graphs, I, *Ars Combin.*, **38**, pp. 97–111. (1.1 chess)

Chao, H. S., Hsu, F. R. and Lee, R. C. T. (2000). An optimal algorithm for finding the minimum cardinality dominating set on permutation graphs, *Discrete Appl. Math.*, **102**, pp. 159–173. (8.2*)

Chapelle, M., Cochefert, M., Couturier, J.-F., Kratsch, D., Letourneur, R., Liedloff, M. and Perez, A. (2018). Exact algorithms for weak Roman domination, *Discrete Appl. Math.*, **248**, pp. 79–92. (5)

Charles Pierre Trémaux (1859–1882). École polytechnique of Paris (X: 1876), French engineer of the telegraph in Public conference, December 2, 2010 – by professor Jean Pelletier-Thibert in Académie de Mâcon (Burgundy – France) – (Abstract published in the Annals academic, March 2011.) (3.1*)

Chartrand, G. and Zhang, P. (2005). *Introduction to Graph Theory*, McGraw-Hill, Boston. (1.2 book)

Chartrand, G., Haynes, T. W., Henning, M. A. and Zhang, P. (2019). *From Domination to Coloring, Stephen Hedetniemi's Graph Theory and Beyond*, Springer Nature Switzerland AG, Cham, Switzerland. (1.1 book)

Chaudhary, J. and Pradhan, D. (2024). Roman 3-domination in graphs: complexity and algorithms, *Discrete Appl. Math.*, **354**, pp. 301–325. (9)

Chen, H. and Lu, C. (2022). Roman 2-domination problem in graphs, *Discuss. Math. Graph Theory*, **42**, pp. 641–660. (3.3)

Chen, L., Lu, C. and Zeng Z. (2009). Hardness results and approximation algorithms for (weighted) paired-domination in graphs, *Theor. Comput. Sci.*, **410**, pp. 5063–5071. (3.3)

Chen, L., Lu, C. and Zeng Z. (2009). Distance paired-domination problems on subclasses of chordal graphs, *Theor. Comput. Sci.*, **410**, pp. 5072–5081. (3.7, 5.4)

Chen, L., Lu, C. and Zeng Z. (2009). A linear-time algorithm for paired-domination problem in strongly chordal graphs, *Inform. Process. Lett.*, **110**, pp. 20–23.   (6.5)

Chen, L., Lu, C. and Zeng Z. (2010). Labelling algorithms for paired-domination problems in block and interval graphs, *J. Combin. Optim.*, **19**, pp. 457–470.   (3.7, 5.4)

Chen, L., Lu, C. and Zeng Z. (2012). Vertices in all minimum paired-dominating sets of block graphs, *J. Combin. Optim.*, **24**, pp. 176–191.   (3.7)

Cheng, C., Lu, C. and Zhou, Y. (2020). The k-power domination problem in weighted trees, *Theor. Comput. Sci.*, **809**, pp. 231–238.   (3.3, 3.5)

Cheng, T. C. E., Kang, L. Y. and Ng, C. T. (2007). Paired domination on interval and circular-arc graphs, *Discrete Appl. Math.*, **155**, pp. 2077–2086.   (5.4*)

Cheng, T. C. E., Kang, L. Y. and Shan, E. (2009). A polynomial-time algorithm for the paired-domination problem on permutation graphs, *Discrete Appl. Math.*, **157**, pp. 262–271.   (8)

Cheston, G. A., Fricke, G. H., Hedetniemi, S. T. and Jacobs, D. P. (1990). On the computational complexity of upper fractional domination, *Discrete Appl. Math.*, **27**, pp. 195–207.   (3.4*)

Chiarelli, N., Hartinger, T. R., Leoni, V. A., Pujato, M. I. L. and Milanič, M. (2019). New algorithms for weighted k-domination and total k-domination problems in proper interval graphs, *Theor. Comput. Sci.*, **795**, pp. 128–141.   (5.3)

Chin, F. Y. L., Ting, H.-F., Tsin, Y. H. and Zhang, Y. (2023). A linear-time certifying algorithm for recognizing generalized series-parallel graphs, *Discrete Appl. Math.*, **325**, pp. 152–171.   (4.6)

Chudnovsky, M., Robertson, N., Seymour, P. and Thomas, R. (2006). The strong perfect graph theorem, *Annals Math., Ser.* 2, **164**, pp. 51–229.   (1.5*)

Cockayne, E. J. (1990). Chessboard domination problems, *Discrete Math.*, **86**, pp. 13–20.   (1.1 chess)

Cockayne, E. J., Gamble, B. and Shepherd, B. (1986). Domination parameters for the bishops graph. *Discrete Math.*, **58**, pp. 221–227.   (1.1 chess)

Cockayne, E., Goodman, S. and Hedetniemi, S. (1975). A linear algorithm for the domination number of a tree, *Inform. Process. Lett.* **4**, pp. 41–44.
                                                                  (1.1* 3.1* 3.2*)

Cockayne, E. J., Hare, E. O., Hedetniemi, S. T. and Wimer, T. V. (1985). Bounds for the domination number of grid graphs, *Congr. Numer.*, **47**, 217–228.
                                                                       (1.1 chess)

Cockayne, E. J. and Hedetniemi, S. T. (1986). On the diagonal queens domination problem, *J. Combin. Theory, Ser. A*, **42**, pp. 137–139.   (1.1 chess)

Cockayne, E. J., MacGillivray G. and Mynhardt, C. M. (1991). A linear algorithm for 0-1 universal minimal dominating functions in trees, *J. Combin. Math. Combin. Comput.*, **10**, pp. 23–31.   (3.2)

Cockayne, E. J. and Spencer, P. H. (1988). On the independent queens covering problem, *Graphs Combin.*, **4**, pp. 101–110.   (1.1 chess)

Colbourn, C. J., Keil, J. M. and Stewart, L. K. (1985). Finding minimum dominating cycles in permutation graphs. *Oper. Res. Lett.*, **4**, pp. 13–17.   (8)

Colbourn, C. J. and Stewart, L. K. (1985). Dominating cycles in series-parallel graphs, *Ars Combin.*, **19A**, pp. 107–112.   (4.6)

Colbourn, C. J. and Stewart, L. K. (1990). Permutation graphs: connected domination and Steiner trees, *Discrete Math.*, **86**, pp. 179–189.   (8.4*)

Colbourn, C. J., Slater, P. J. and L. K. Stewart, L. K. (1987). Locating-dominating sets in series-parallel networks, *Congr. Numer.*, **56**, pp. 135–162.   (4.4)

Cook, S. A. (1971). The complexity of theorem proving procedures, *Proc. 3rd Annual ACM Symposium on Theory of Computing*, pp. 151–158.   (2.3*)

Cormen, T. H, Leiserson, C. E., Rivest, R. L. and Stein, C. (2022). *Introduction to Algorithms*, Fourth Edition, The MIT Press, Cambridge, Mass.   (2.1*)

Corneil, D. G. (2004). Lexicographic breadth first search–a survey. *Lecture Notes Comput. Sci.*, **3353**, pp. 1–19.   (4.3)

Corneil, D. G. (2004). A simple 3-sweep LBFS algorithm for the recognition of unit interval graphs, *Discrete Appl. Math.*, **138**, pp. 371–379.   (5.1)

Corneil, D. G. and Keil, J. M. (1987). A dynamic programming approach to the dominating set problem on k-trees, *SIAM J. Algeb. Discrete Methods*, **8**, pp. 535–543.   (3.7, 4.4)

Corneil, D. G., Lerchs, H. and Stewart Burlingham, L. (1981). Complement reducible graphs, *Discrete Appl. Math.*, **3**, pp. 163–174.   (9.1*)

Corneil, D. G., Olariu, S. and Stewart, L. (1992). Asteroidal triple-free graphs, Technical Report 262/92, Univ. of Toronto, June 1992.   (7.5*)

Corneil, D. G., Olariu, S. and Stewart, L. (1994). Asteroidal triple-free graphs, in *Proc. 19th Internat. Workshop on Graph-Theoretic Concepts in Computer Science (WG'93)*, *Lecture Notes in Comput. Sci.*, **790**, Springer-Verlag, Berlin, pp. 211–224.   (7.5)

Corneil, D. G., Olariu, S. and Stewart, L. (1995). A linear time algorithm to compute dominating pairs in asteroidal triple-free graphs (Extended Abstract), in *Proc. 22nd Internat. Colloq. on Automata, Languages and Programming (ICALP'95)*, *Lecture Notes in Comput. Sci.*, **944**, Springer-Verlag, Berlin, pp. 292–302.   (7.6)

Corneil, D. G., Olariu, S. and Stewart, L. (1995). Computing a dominating pair in an asteroidal triple-free graph in linear time, in *Proc. 4th Algorithms and Data Structures Workshop*, *Lecture Notes in Comput. Sci.*, **955**, Springer, pp. 358–368.   (7.6)

Corneil, D. G., Olariu, S. and Stewart, L. (1995). A linear time algorithm to compute a dominating path in an AT-free graph, *Inform. Process. Lett.*, **54**, pp. 253–258.   (7.6)

Corneil, D. G., Olariu, S. and Stewart, L. (1997). Asteroidal triple-free graphs, *SIAM J. Discrete Math.*, **10**, pp. 399–430.   (7.5*)

Corneil, D. G., Olariu, S. and Stewart, L. (1998). The ultimate interval graph recognition algorithm? in *Proc. Ninth Annual ACM-SIAM Symp. Discrete Alg.*, ACM, New York, SIAM, Philadelphia, pp. 175–180.   (5.1*)

Corneil, D. G., Olariu, S. and Stewart, L. (1999). Linear time algorithms for dominating pairs in asteroidal triple-free graphs, *SIAM J. Comput.*, **28**, pp. 1284–1297.   (7.5∗ 7.6∗)

Corneil, D. G., Olariu, S. and Stewart, L. (2009). The LBFS structure and recognition of interval graphs, *SIAM J. Discrete Math.*, **23**, pp. 1905–1953. (5.1∗)

Corneil, D. G. and Perl, Y. (1984). Clustering and domination in perfect graphs, *Discrete Appl. Math.*, **9**, pp. 27–39.   (4.4)

Corneil, D. G., Perl, Y. and Stewart, L. (1984). Cographs: recognition, application and algorithms, *Congr. Numer.*, **43**, pp. 249–258.   (9.2)

Corneil, D. G., Perl, Y. and Stewart, L. K. (1985). A linear recognition algorithm for cographs, *SIAM J. Comput.*, **14**, pp. 926–934.   (9.2∗)

Corneil, D. G. and Stewart, L. K. (1990). Dominating sets in perfect graphs, *Discrete Math.*, **86**, pp. 145–164.   (9.4)

Dabney, J. R. (2007). *A Linear-Time Algorithm for Broadcast Domination in a Tree*, Master thesis, Computer Sci., Cleson Univ.   (1.1 thesis)

Dabney, J., Dean, B. C. and Hedetniemi, S. T. (2009). A linear-time algorithm for broadcast domination in a tree, *Networks*, **53**, pp. 160–169.   (3)

Damaschke, P., Müller, H. and Kratsch, D. (1990). Domination in convex and chordal bipartite graphs, *Inform. Process. Lett.*, **36**, pp. 231–236.   (4.4)

Damiand, G., Habib, M. and Paul, C. (2001). A simple paradigm for graph recognition: application to cographs and distance hereditary graphs, *Theor. Computer Sci.*, **263**, pp. 99–111.   (9.2∗)

D'Atri, A. and Moscarini, M. (1988). Distance-hereditary graphs, Steiner trees, and connected domination, *SIAM J. Comput.*, **17**, pp. 521–538.
                                                                    (9.1∗ 9.2∗ 9.3∗)

Day, D. P. Oellermann, O. R. and Swart, H. C. (1994). Steiner distance-hereditary graphs, *SIAM J. Discrete Math.*, **7**, pp. 437–442.   (9.1)

de Jaenisch, C. F. (1862). *Traité Des Applications De L'analyse Mathématique Au Jeu Des Échecs: Précédé D'une Introduction À L'usage Des Lecteurs Soit Étrangers Aux Échecs, Soit Peu Versés Dans L'analyse*, I, Dufour et cie.   (1.1 chess)

Demirors, O., Rafraf, N. and Tanik, M. (1992). Obtaining N-queens solutions from magic squares and constructing magic squares from N-queens solutions, *J. Recreational Math.*, **24**, pp. 272–280.   (1.1 chess)

Dettlaff, M., Gözüpek, D. and Raczek, J. (2022). Paired domination versus domination and packing number in graphs, *J. Combin. Optim.*, **44**, pp. 921–933.   (4.3)

Diaz, G. C. (2013). Mathematical Induction, Archived 2 May 2013 at the Wayback Machine, Harvard University. http://www.math.harvard.edu/ archive/23a_fall_05/Handouts/induction.pdf   (1.4∗)

Dilworth, R. P. (1950). A decomposition theorem for partially ordered sets, *Annals Math.*, **51**, pp. 161–166.   (7.1∗)

Dirac, G. A. (1961). On rigid circuit graphs, *Abh. Math. Sem. Univ. Hamburg*, **25**, pp. 71–76.   (4.1∗)

Diestel, R. (2000). *Graph Theory*, Second Edition, Springer-Verlag, New York.
(1.2 book)

Domke, G. S., Hattingh, J. H., Hedetniemi, S. T., Laskar, R. C. and Markus, L. R. (1999). Restrained domination in graphs, *Discrete Math.*, **203**, pp. 61–69.
(3.3, 4.4)

Dragan, F. F. (1991). Dominating and packing in triangulated graphs, *Metody Diskret. Analiz.*, **1 (51)**, pp. 17–36. (4)

Dragan, F. F. (1993). Domination in Helly graphs without quadrangles, *Cybernet. Systems Anal.*, **6**, pp. 47–57. (6)

Dragan, F. F. (1993). HT-graphs: centers, connected *r*-domination and Steiner trees, *Comput. Sci. J. Moldova (Kishinev)*, **1**, pp. 64–83. (6)

Dragan, F. F. (1994). Dominating cliques in distance-hereditary graphs, in *Proc. Algorithm Theory - SWAT'94: 4th Scandinavian Workshop on Algorithm Theory, Lecture Notes in Comput. Sci.*, **824**, Berlin. Springer-Verlag, pp. 370–381. (9)

Dragan, F. F. and Brandstädt, A. (1994). Dominating cliques in graphs with hypertree structure, in *Proc. 11th Annual Symp. on Theoretical Aspects of Computer Science (STACS'94), Lecture Notes in Comput. Sci.*, **775**, Berlin. Springer-Verlag, pp. 735–746. (6)

Dragan, F. F. and Brandstädt, A. (1996). *r*-Dominating cliques in graphs with hypertree structure, *Discrete Math.*, **162**, pp. 93–108. (6.5)

Dragan, F. F. and Nicolai, F. (1995). *r*-domination problems on homogeneously orderable graphs, in Reichel, H. ed., *Proc. Fundamentals of Computation Theory, FCT'95, Lecture Notes in Comput. Sci.*, **965**, pp. 201–210, Berlin, Springer-Verlag. (6 9)

Dreyfus, S. E. and Law, A. M. (1977). *The Art and Theory of Dynamic Programming*, Academic Press, New York. (3.3*)

Duan, Z., Jiang, H., Liu, X., Wu, P. and Shao, Z. (2022). Independent Roman domination: the complexity and linear-time algorithm for trees, *Symmetry*, **14**, 404. (3.3, 4.4)

Duan, R., Wu, H. and Zhou, R. (2023). Faster matrix multiplication via asymmetric hashing, *2023 IEEE 64th Annual Sym. Found. Comput. Sci. (FOCS)*. Santa Cruz, CA, USA, 2023, pp. 2129–2138. (7.1*)

Duchet, P. (1984). Classical perfect graphs, an introduction with emphasis on triangulated and interval graphs, *Topics on Perfect Graphs*, Berge, C. and Chvátal, V. eds., *North-Holland Math. Studies* **88**/*Annals of Discrete Math.*, **21**, pp. 67–96. (6.1*)

Duffin, R. J. (1965). Topology of series-paralle networks, *J. Math. Analy. Appl.*, **10**, pp. 303–318. (4.6*)

Dunbar, J., Goddard, W., Hedetniemi, S., McRae, A. and Henning, M. A. (1996) The algorithmic complexity of minus domination in graphs, *Discrete Appl. Math.*, **68**, pp. 73–84. (3.3, 4.4)

Dunbar, J. E., Grossman, J. W., Hattingh, J. H., Hedetniemi, S. T. and McRae, A. A. (1997). On weakly connected domination in graphs, *Discrete Math.*, **167/168**, pp. 261–269. (3, 4.4)

Eisenstein, M., Grinstead, C. M., Hahne, B. and Van Stone, D. (1992). The queen domination problem, *Congr. Numer.*, **91**, pp. 189–193.   (1.1 chess)

El-Mallah, E. S. (1987). *Decomposition and Embedding Problems for Restricted Networks*, Ph.D. Thesis, Univ. Waterloo.   (5.1 thesis)

Elmallah, E. S. and Stewart, L. K. (1990). Domination in polygon graphs, *Congr. Numer.*, **77**, pp. 63–76.   (5.6)

Elmallah, E. S. and Stewart, L. K. (1993). Independence and domination in polygon graphs, *Discrete Appl. Math.*, **44**, pp. 65–77.   (5.6)

Erbas, C. and Tanik, M. M. (1992). Storage schemes for parallel memory systems and the N-queens problem, in *Proc. 15th Ann. ASME ETCE Conf., Computer Applications Symp., Houston, TX*.   (1.1 chess)

Erbas, C., Tanik, M. M. and Aliyazicioglu, Z. (1992). Linear congruence equations for the solutions of the N-queens problem, *Inform. Process. Lett.*, **41**, pp. 301–306.   (1.1 chess)

Erdős, P. (1964). Extremal problems in graph theory, in *Theory of Graphs and its Applications (Proc. Sympos. Smolenice, 1963)*, Publ. House Czechoslovak Acad. Sci., Prague, pp. 29–36.   (3.1*)

Erdős, P. and Szekeres, G. (1935) A combinatorial problem in geometry, *Compositio. Math.*, **2**, pp. 463–470.   (8.2*)

Euler, L. (1736). Solutio problematis ad geometriam situs pertinentis, *Comment. Acad. Sci. Imp. Petropol.*, **8**, pp. 128–140.   (1.1*)

Even, S., Pnueli A. and Lempel A. (1972). Permutation graphs and transitive graphs. *J. Assoc. Comput. Mach.*, **19**, pp. 400–410.   (8.1*)

Falkowski, B. J. and Schmitz, L. (1986). A note on the queens problem, *Inform. Process. Lett.*, **23**, pp. 39–46.   (1.1 chess)

Farber, M. (1981). *Applications of Linear Programming Duality to Problems Involving Independence and Domination*, Ph.D. Thesis, Rutgers Univ.   (1.1 thesis)

Farber, M. (1981). Domination and duality in weighted trees, *Congr. Numer.*, **33**, pp. 3–13.   (3.4*)

Farber, M. (1982). Independent domination in chordal graphs, *Oper. Res. Lett.*, **1**, pp. 134–138.   (4.5*)

Farber, M. (1983). Characterizations of strongly chordal graphs, *Discrete Math.*, **43**, pp. 173–189.   (6.1* 6.2*)

Farber, M. (1984). Domination, independent domination and duality in strongly chordal graphs, *Discrete Appl. Math.*, **7**, pp. 115–130.   (6.2* 6.3*)

Farber, M. and Jamison, R. E. (1986). Convexity in graphs and hypergraphs, *SIAM J. Algeb. Disc. Meth.*, **7**, pp. 433–444.   (4.1*)

Farber, M. and Keil, J. M. (1985). Domination in permutation graphs, *J. Algorithms*, **6**, pp. 309–321.   (8.2* 8.3*)

Farley, A., Hedetniemi, S. and Proskurowski, A. (1981). Partitioning trees: matching, domination and maximum diameter, *Internat. J. Comput. Inform. Sci.*, **10**, pp. 55–61.   (3.3)

Fomin, F. V. and Kratsch, D. (2010). *Exact Exponential Algorithms*, Springer-Verlag Berlin Heidelberg. (1.1 book)

Fellows, M. R. and Hoover, M. N. (1991). Perfect domination, *Australas. J. Combin.*, **3**, pp. 141–150. (4.4)

de Figueiredo, C. M. H., Meidanis, J. and de Mello, C. P. (1995). A linear-time algorithm for proper interval graph recognition, *Inform. Process. Lett.*, **56**, pp. 179–184. (5.1)

Fredman, M. L. and Tarjan, R. E. (1987). Fibonacci heaps and their uses in improved networks optimization algorithms, *J. ACM*, **34**, pp. 596–615. (8.4∗)

Fricke, G. H., Hedetniemi, S. M., Hedetniemi, S. T., McRae, A. A., Wallis, C. K., Jacobson, M. S., Martin, W. W. and Weakley, W. D. (1995). Combinatorial problems on chessboards: a brief survey, in Alavi, Y. and Schwenk, A. J. eds., *Graph Theory, Combinatorics, Algorithms and Applications*, (Kalamazoo, MI 1992), Wiley, **1**, pp. 507–528. (1.1 chess)

Fulkerson, D. R. and Gross, O. A. (1965). Incidence matrices and interval graphs, *Pacific J. Math.*, **15**, pp. 835–855. (4.1∗ 4.2∗ 5.1∗ 5.7∗)

Gamble, B., Shepherd, B. and Cockayne, E. J. (1985). Domination of chessboards by queens on a column, *Ars Combin.*, **19**, pp. 105–118. (1.1 chess)

Garey, M. R. and Johnson, D. S. (1979). *Computers and Intractability: A Guide to the Theory of NP-Completeness*, Freeman, New York. (1.1∗ 2.3∗)

Garnick, D. K. and Nieuwejaar, N. A. (1995). Total domination of the $m \times n$ chessboard by kings, crosses, and knights, *Ars. Combin.*, **41**, pp. 65–75. (1.1 chess)

Gavril, F. (1974). The intersection graphs of subtrees of trees are exactly the chordal graphs, *J. Combin. Theory, Ser. B*, **16**, pp. 47–56. (4.1∗)

Gera, R., Hedetniemi, S. and Larson, C. eds. (2016) *Graph Theory: Favorite Conjectures and Open Problems–1*, Springer International Publishing, Cham, Switzerland. (1.2 book)

Gera, R., Hedetniemi, S. and Larson, C. eds. (2019) *Graph Theory: Favorite Conjectures and Open Problems–2*, Springer International Publishing, Cham, Switzerland. (1.2 book)

Gibbons, P. B. and Webb, J. A. (1997). Some new results for the queens domination problem, *Australian J. Combin.*, **15**, pp. 145–160. (1.1 chess)

Gilmore, P. C. and Hoffman, A. J. (1964). A characterization of comparability graphs and of interval graphs, *Canad. J. Math.*, **16**, pp. 539–548. (4.1∗ 5.1∗ 7.1∗)

Ginsburg, J. (1938). Gauss's arithmetization of the problem of 8 queens, *Scripta Math.*, **5**, pp. 63–66. (1.1 chess)

Gioan, E. and Paul, C. (2012). Split decomposition and graph-labelled trees: characterizations and fully dynamic algorithms for totally decomposable graphs, *Discrete Appl. Math.*, **160**, pp. 708–733. (9.2∗)

Goddard, W., Henning, M. A. and McPillan, C. A. (2014). The disjunctive domination number of a graph, *Quaestiones Math.*, **37**, pp. 547–561. (5.4∗)

Golumbic, M. C. (1980). *Algorithmic Graph Theory and Perfect Graphs*, Academic Press, New York.   (1.5* 1.6* 4.1* 4.3* 5.1* 5.6*)

Goncalves, D., Pinlou, A., Rao, M. and Thomassé, S. (2011). The domination number of grids, *SIAM J. Discrete Math.*, **25**, pp 1443–1453.   (1.1 chess)

Gould, R. (1988). *Graph Theory*, the Benjamin/Cummings, Menlo Park, CA.
(1.2 book)

Grinstead, C. M., Hahne, B. and Van Stone, D. (1990). On the queen domination problem, *Discrete Math.*, **86**, pp. 21–26.   (1.1 chess)

Grinstead, D. L. (1989). *Algorithmic Templates and Multiset Problems in Graphs*, Ph.D. Theses, Univ. Alabama in Huntsville.   (1.1 thesis)

Grinstead, D. L. and Slater, P. J. (1990). On minimum dominating sets with minimum intersection, *Discrete Math.*, **86**. pp. 239–254.   (3.3)

Grinstead, D. L., Slater, P. J., Sherwani, N. A. and Holmes, N. D. (1993). Efficient edge domination problems in graphs, *Inform. Process. Lett.*, **48**, pp. 221–228.   (4.6)

Gröbler, P. J. P. (1994). *Functional Generalisations of Dominating Sets of Graphs*, Master's Thesis, Univ. South Africa.   (1.1 thesis)

Guo, J., Niedermeier, R. and Raible, D. (2008). Improved algorithms and complexity results for power domination in graphs, *Algorithmica*, **52**, pp. 177–202.   (3.5* 4.4)

Habib, M., McConnell, R., Paul, C. and Viennot, L. (2000). Lex-BFS and partition refinement, with applications to transitive orientation, interval graph recognition and consecutive ones testing, *Theor. Comput. Sci.*, **234**, pp. 59–84.   (5.1)

Habib, M. and Paul, C. (2005). A simple linear time algorithm for cograph recognition, *Discrete Appl. Math.*, **145**, pp. 183–197.   (9.2*)

Hajnal, A. and Surányi, J. (1958). Über die Auflösung von Graphen in vollständige Teilgraphen, *Ann. Univ. Sci. Budapest Eötvös Sect. Math.*, **1**, pp. 113–121.
(4.1*)

Hajós, G. (1957). Über eine Art von Graphen, *Nachrichten der Österreichischen Mathematischen Gesellschaft (bericht uber den IV Osterreichischen Mathematikerkongres, Wien, 17–22, Sept. 1956)*, **11**, April 1957, Nr. 47/48, pp. 65.   (5.1*)

Hammer, P. L. and Maffray, F. (1990). Completely separable graphs, *Discrete Appl. Math.*, **27**, pp. 85–99.   (9.1* 9.2*)

Harary, F. (1969). *Graph Theory*, Addison-Wesley, Reading, Mass.   (1.2 book)

Hare, E. O. (1989). *Algorithms for Grids and Grid-like Graphs*, Ph.D. Thesis, Clemson Univ.   (1.1 thesis chess)

Hare, E. O. (1994). Fibonacci numbers and fractional domination of $P_m \times P_n$, *Fibonacci Quart.*, **32**, pp. 69–73.   (1.1 chess)

Hare, E. O. and Hedetniemi, S. T. (1987). A linear algorithm for computing the knight's domination number of a $k \times n$ chessboard, *Congr. Numer.*, **59**, pp. 115–130.   (1.1 chess)

Hare, E. O., Hedetniemi, S. T. and Hare, W. R. (1986). Algorithms for computing the domination number of $k \times n$ complete grid graphs, *Congr. Numer.*, **55**, pp. 81–92. (1.1 chess)

Hare, E. O., Hedetniemi, S. T., Laskar, R. C., Peters, K. and Wimer, T. V. (1987). Linear-time computability of combinatorial problems on generalized-series-parallel graphs, in Johnson, D., Nishizeki, T., Nozaki, A. and Wilf, H. eds., *Discrete Algorithms and Complexity*, pp. 437–457, Academic Press. (4.6)

Haskell, D., Hrushovski, E. and Macpherson, D. (2007). *Stable Domination and Independence in Algebraically Closed Valued Field*, Cambridge Univ. Press, New York. (1.1 book)

Hattingh, J. H. (1997). Majority domination and its generalizations, in Haynes, T. W., Hedetniemi, S. T. and Slater, P. J. eds., *Domination in Graphs: Advanced Topics*, Chapter 4, Marcel Dekker, Inc., New York. (4.4)

Hattingh, J. H., Henning, M. A. and Slater, P. J. (1995). On the algorithmic complexity of signed domination in graphs, *Australas. J. Combin.*, **12**, pp. 101–112. (3.3)

Hattingh, J. H., Henning, M. A. and Walters, J. L. (1993). On the computational complexity of upper distance fractional domination. *Australas. J. Combin.*, **7**, pp. 133–144. (4.4)

Hayes, P. (1992). A problem of chess queens, *J. Recreational Math.*, **24**, pp. 264–271. (1.1 chess)

Haynes Rice, T. (1988). *On k-γ-Insensitive Domination*, Ph.D. Thesis, Univ. Central Florida. (1.1 thesis)

Haynes, T. W., Hedetniemi, S. M., Hedetniemi, S. T. and Henning, M. A. (2002). Domination in graphs applied to electric power networks, *SIAM J. Disc. Math.*, **15**, pp. 519–529. (3.5* 4.4)

Haynes, T. W., Hedetniemi, S. T. and Henning, M. A. eds. (2020). *Topics in Domination in Graphs*, Springer Nature Switzerland AG, Cham, Switzerland. (1.1 book)

Haynes, T. W., Hedetniemi, S. T. and Henning, M. A. eds. (2021). *Structures of Domination in Graphs*, Springer Nature Switzerland AG, Cham, Switzerland. (1.1*)

Haynes, T. W., Hedetniemi, S. T. and Henning, M. A. (2023). *Domination in Graphs: Core Concepts*, Springer Nature Switzerland AG, Cham Switzerland. (1.1*)

Haynes, T. W., Hedetniemi, S. T. and Slater,P. J. eds. (1998). *Domination in Graphs: Advanced Topics*, Marcel Dekker, Inc., New York. (1.1 book)

Haynes, T. W., Hedetniemi, S. T. and Slater, P. J. (1998a). *Fundamentals of Domination in Graphs*, Marcel Dekker, Inc., New York. (1.1*)

Haynes, T. W. and Slater, P. J. (1998). Paired-domination in graphs, *Networks*, **32**, pp. 199–206. (3.6* 7.3*)

Haynes, T. W. and Yeo, A. (2013). *Total Domination in Graphs*, Springer, New York. (1.1 book)

Hedetniemi, S. T. (2016). My top 10 graph theory conjectures and open problems, *Graph Theory, Favorite Conjectures and Open Problems*, Volume 1, Gera, R., Hedetniemi, S. T. and Larson, C. eds., *Graph Theory, Problem Books in Mathematics*, Springer, Cham, 2016, pp. 109–134.   (1.1 chess)

Hedetniemi, S. M. Hedetniemi, S. T. and Jacobs, D. P. (1990). Private domination: theory and algorithms, *Congr. Numer.*, **79**, pp. 147–157.   (3.3, 4.4)

Hedetniemi, S. M., Hedetniemi, S. T. and Laskar, R. C. (1985). Domination in trees: models and algorithms, in Alavi, A., Chartrand, G., Lesniak, L., Lick, D. R. and Wall, C. E. eds., *Graph Theory with Applications to Algorithms and Computer Science*, pp. 423–442. Wiley, New York.   (3.2)

Hedetniemi, S. M., Hedetniemi, S. T. and Reynolds, R. (1997). Combinatorial problems on chessboards: II, in Haynes, T. W., Hedetniemi, S. T. and Slater, P. J. eds., *Domination in Graphs: Advanced Topics*, Chapter 6, Marcel Dekker, Inc., New York.   (1.1 chess)

Hedetniemi, S. T. and Laskar, R. C. (1990). Bibliography on domination in graphs and some basic definitions of domination parameters. *Discrete Math.*, **86**, pp. 257–277.   (1.3)

Hedetniemi, S. T. and Laskar, R. C. eds (1990). *Topics on Domination, Annals of Discrete Math.*, **48**, North Holland, New York.   (1.1*)

Hedetniemi, S. T. Laskar, R. C. and Pfaff, J. (1986). A linear algorithm for finding a minimum dominating set in a cactus, *Discrete Appl. Math.*, **13**, pp. 287–292.   (3.7)

Hedetniemi, S. T., McRae, A. A. and Parks, D. A. (1997). Complexity results. in Haynes, T. W., Hedetniemi, S. T. and Slater, P. J. eds., *Domination in Graphs: Advanced Topics*, Chapter 9. Marcel Dekker, Inc., 1997.   (4.4)

Hell, P. and Huang, J. (2005). Certifying LexBFS recognition algorithms for proper interval graphs and proper interval bigraphs, *SIAM J. Discrete Math.* **18**, pp. 554–570.   (5.1)

Hendry, G. R. T. (1990). Extending cycles in graphs, *Discrete Math.* **85**, pp. 59–72.   (4.7*)

Henning, M. A., Kusum, Pandey, A. and Paul, K. (2023). Complexity of total dominator coloring in graphs, *Graphs Combin.*, **39**, 128.   (4.4)

Henning, M. A., Pal, S. and Pradhan, D. (2020). Algorithm and hardness results on hop domination in graphs, *Inform. Process. Lett.*, **153**, 105872.   (8)

Henning, M. A., Pal, S. and Pradhan, D. (2022). The semitaotal domination problem in block graphs, *Discuss. Math. Graph Theory*, **42**, pp. 231–248. (3.7)

Henning, M. A. and Pandey, A. (2019). Algorithmic aspects of semitotal domination in graphs, *Theor. Comput. Sci.*, **766**, pp. 46–57.   (5)

Henning, M. A., Pandey, A., Sharma, G. and Tripathi, V. (2024). Algorithms and hardness results for edge total domination problem in graphs, *Theor. Comput. Sc.*, **982**, Article 114270.   (5.2)

Henning, M. A., Pandey, A. and Tripathi, V. (2020). Complexity and algorithms for semitotal domination in graphs, *Theory Comput. Sys.*, **64**, pp. 1225–1241.   (5.4)

Henning, M. A., Pandey, A. and Tripathi, V. (2023). Algorithmic aspects of paired disjunctive domination in graphs, *Theor. Comput. Sci.*, **966–967**, 113990.
(5.4*)

Henning, M. A., Pandey, A. and Tripathi, V. (2025). More on the complexity of defensive domination in graphs, *Discrete Appl. Math.*, **362**, pp. 167–179.
(4.4)

Henning, M. A. and Pradhan, D. (2020). Algorithmic aspects of upper paired-domination in graphs, *Theor. Comput. Sci.*, **804**, pp. 98–114.   (5.4)

Henning, M. A. and Yeo, A. (2013). *Total Domination in Graphs*, Springer Science+Business Media, New York.   (1.1*)

Hoffman, A. J., Kolen, A. W. J. and Sakarovitch, M. (1985). Totally-balanced and greedy matrices, *SIAM J. Algeb. Discrete Methods*, **6**, pp. 721–730.
(6.2* 6.3*)

Hoffman, E. J., Loessi, J. C. and Moore, R. C. (1969). Constructions for the solution of the *m* queens problem, *Math. Mag.*, **42**, pp. 66–72.   (1.1 chess)

Hollander, D. H. (1973). An unexpected two-dimensional space-group containing seven of the twelve basic solutions to the eight queens problem, *J. Recreational Math.*, **6**, pp. 287–291.   (1.1 chess)

Horton, J. D. and Kilakos, K. (1993). Minimum edge dominating sets, *SIAM J. Discrete Math.*, **6**, pp. 375–387.   (4.6)

Howorka, E. (1977). A characterization of distance-hereditary graphs, *Quart. J. Math. Oxford Ser. 2*, **28**, pp. 417–420.   (9.1*)

Hsu, W.-L. (1980). On the domination numbers of trees, Technical report, *Dept. Indus. Eng. and Manag. Sci.*, Northwestern Univ.   (3.3)

Hsu, W.-L. (1982). The distance-domination numbers of trees, *Oper. Res. Lett.*, **1**, pp. 96–100.   (3.3)

Hsu, W.-L. (1993). A simple test for interval graphs, *Lecture Notes Comput. Sci.*, **657**, pp. 11–16.   (5.1*)

Hsu, W.-L. (2002). A simple test for the consecutive ones property, *J. Algorithms*, **42**, pp. 1–16.   (5.1*)

Hsu, W.-L. and Ma, T.-H. (1999). Fast and simple algorithms for recognizing chordal comparability graphs and interval graphs, *SIAM J. Comput.*, **28**, pp. 1004–1020.   (5.1*)

Hsu, W.-L. and McConnell, R. M. (2003). PC trees and circular-ones arrangements, *Theor. Comput. Sci.*, **296**, pp. 99–116.   (5.1*)

Hsu, W.-L. and Tsai, K.-H. (1988). Linear time algorithms on circular-arc graphs, in *Proc. 26th Ann. Ailerton Conf*, pp. 842–851.   (5.6*)

Hsu, W.-L. and Tsai, K.-H. (1991). Linear time algorithms on circular-arc graphs, *Inform. Process. Lett.*, **40**, pp. 123–129.   (5.6*)

Hung, R.-W. (2010). A linear-time algorithm for the restricted paired-domination problem in cographs, *Inter. J. Innov. Comput. Inform. Control*, **6**, pp. 4957–4978.   (9)

Hung, R.-W. (2012). Linear-time algorithm for the paired-domination problem in convex bipartite graphs, *Theory Comput. Syst.*, **50**, pp. 721–738.   (6)

Hung, R.-W. and Yao, C.C. (2011). Linear-time algorithm for the matched-domination problem in cographs, *Inter. J. Comput. Math.*, **88**, pp. 2042–2056.   (9)

Hwang, S.-F. (1991) *Domination and Related Topics*, Ph.D. Thesis, National Chiao-Tung Univ., Hsinchu, Taiwan.   (1.1 thesis)

Hwang, S.-F. and Chang, G.J. (1991). The k-neighbor domination problem in block graphs, *European J. Oper. Res.*, **52**, pp. 373–377.   (3.7, 4.4)

Hwang, S.-F. and Chang, G.J. (1995). The edge domination problem, *Discuss. Math.–Graph Theory*, **15**, pp. 51–57, 1995.   (3.7)

Ibarra, O.H. and Zheng, Q. (1994). Some efficient algorithms for permutation graphs, *J. Algorithms*, **16**, pp. 453–469.   (7.3∗)

Jackson, A.H. and Pargas, R.P. (1991). Solutions to the $N \times N$ knights cover problem, *Rec. Math.*, **23**, pp. 255–267.   (1.1 chess)

Jacobson, M.S. and Peters, K. (1989). Complexity questions for n-domination and related parameters, *Congr. Numer.*, **68**, pp. 7–22.   (4.4)

Jayaram, T. S., Sri Karishna, G. and Pandu Rangan, C. (1990). A unified approach to solving domination problems on block graphs, Report TR-TCS-90-09, Dept. of Computer Science and Eng., Indian Inst. of Technology.   (3.7)

Jha, A., Pradhan, D. and Banerjee, S. (2019). The secure domination problem in cographs, *Inform. Process. Lett.*, **145**, pp. 30–38.   (9)

Johnson, S. M. (1972). A new lower bound for coverings by rook domains, *Utilitas Math.*, **1**, pp. 121–140.   (1.1 chess)

Johnson, D.S. (1981). The NP-completeness column: An ongoing guide, *J. Algorithms*, **2** (**4**), pp. 393–405.   (2.3)

Johnson, D.S. (1982). The NP-completeness column: An ongoing guide, *J. Algorithms*, **3** (**1**), pp. 89–99.   (2.3)

Johnson, D.S. (1982). The NP-completeness column: An ongoing guide, *J. Algorithms*, **3** (**2**), pp. 182–195.   (2.3)

Johnson, D.S. (1982). The NP-completeness column: An ongoing guide, *J. Algorithms*, **3** (**3**), pp. 288–300.   (2.3)

Johnson, D.S. (1982). The NP-completeness column: An ongoing guide, *J. Algorithms*, **3** (**4**), pp. 381–395.   (2.3)

Johnson, D.S. (1983). The NP-completeness column: An ongoing guide, *J. Algorithms*, **4** (**1**), pp. 87–100.   (2.3)

Johnson, D.S. (1983). The NP-completeness column: An ongoing guide, *J. Algorithms*, **4** (**2**), pp. 189–203.   (2.3)

Johnson, D.S. (1983). The NP-completeness column: An ongoing guide, *J. Algorithms*, **4** (**3**), pp. 286–300.   (2.3)

Johnson, D.S. (1983). The NP-completeness column: An ongoing guide, *J. Algorithms*, **4** (**4**), pp. 397–411.   (2.3)

Johnson, D.S. (1984). The NP-completeness column: An ongoing guide, *J. Algorithms*, **5** (**1**), pp. 147–160.   (2.3)

Johnson, D. S. (1984). The NP-completeness column: An ongoing guide, *J. Algorithms*, **5 (2)**, pp. 284–299. (2.3)

Johnson, D. S. (1984). The NP-completeness column: An ongoing guide, *J. Algorithms*, **5 (3)**, pp. 433–447. (2.3)

Johnson, D. S. (1984). The NP-completeness column: An ongoing guide, *J. Algorithms*, **5 (4)**, pp. 595–609. (2.3)

Johnson, D. S. (1985). The NP-completeness column: An ongoing guide, *J. Algorithms*, **6 (1)**, pp. 145–159. (2.3)

Johnson, D. S. (1985). The NP-completeness column: An ongoing guide, *J. Algorithms*, **6 (2)**, pp. 291–305. (2.3)

Johnson, D. S. (1985). The NP-completeness column: An ongoing guide, *J. Algorithms*, **6 (3)**, pp. 434–451. (2.3)

Johnson, D. S. (1986). The NP-completeness column: An ongoing guide, *J. Algorithms*, **7 (2)**, pp. 289–305. (2.3)

Johnson, D. S. (1986). The NP-completeness column: An ongoing guide, *J. Algorithms*, **7 (4)**, pp. 584–601. (2.3)

Johnson, D. S. (1987). The NP-completeness column: An ongoing guide, *J. Algorithms*, **8 (2)**, pp. 285–303. (2.3)

Johnson, D. S. (1987). The NP-completeness column: An ongoing guide, *J. Algorithms*, **8 (3)**, pp. 438–448. (2.3)

Johnson, D. S. (1988). The NP-completeness column: An ongoing guide, *J. Algorithms*, **9 (3)**, pp. 426–444. (2.3)

Johnson, D. S. (1990). The NP-completeness column: An ongoing guide, *J. Algorithms*, **11 (1)**, pp. 144–151. (2.3)

Johnson, D. S. (1992). The NP-completeness column: An ongoing guide, *J. Algorithms*, **13 (3)**, pp. 502–524. (2.3)

Johnson, D. S. (2005). The NP-completeness column, *ACM Trans. Algorithms*, **1**, pp. 160–176. (2.3)

Jordan, C. (1869). Sur les assemblages de lignes, *J. Reine Angew. Math.*, **70**, pp. 185–190. (6.5*)

Kang, L. (2013). Algorithmic aspects of domination in graphs, in Pardalos, P., Du, D. Z. and Graham, R. eds., *Handbook of Combinatorial Optimization*, Second Edition, **5**, pp. 3363–3394 Springer, New York, NY. (1.1*)

Kang, L., Sohn, M. Y. and Cheng, T. C. E. (2004). Paired-domination in inflated graphs, *Theor. Comput. Sci.*, **320**, pp. 485–494. (3.6)

Kao, M. J., Liao, C. S. and Lee, D. T. (2011). Capacitated domination problem, *Algorithmica*, **60**, pp. 274–300. (3.3)

Karp, R. M. (1972). Reducibility among combinatorial problems, in Miller, R. E. and Thatcher, J. W. eds., *Complexity of Computer Computations*, pp. 85–104, Plenum Press, New York. (2.3*)

Keil, J. M. (1986). Total domination in interval graphs, *Inform. Process. Lett.*, **22**, pp. 171–174. (5.2*)

Keil, J. M. (1993). The complexity of domination problems in circle graphs, *Discrete Appl. Math.*, **42**, pp. 51–63. (5.6)

Keil, J. M. and Schaefer, D. (1982). An optimal algorithm for finding dominating cycles in circular-arc graphs, *Discrete Appl. Math.*, **36**, pp. 25–34. (5.6)

Kelleher, L. L. (1985). *Domination in Graphs and its Application to Social Network Theory*, Ph.D. Thesis, Northeastern Univ. (1.1 thesis)

Kikuno, T., Yoshida, N. and Kakuda, Y. (1980). The NP-completeness of the dominating set problem in cubic planar graphs. *IEICE Trans.*, **63**, pp. 443–444. (4.4)

Kikuno, T., Yoshida, N. and Kakuda, Y. (1983). A linear algorithm for the domination number of a series-parallel graph, *Discrete Appl. Math.*, **5**, pp. 299–311. (4.6)

Kilakos, K. (1988). *On the Complexity of Edge Domination*, Master's Thesis, Univ. New Brunswick. (1.1 thesis)

Klee, V. (1969). What are the intersection graphs of arcs in a circle? *Amer. Math. Monthly*, **76**, pp. 810–813. (5.6*)

Kloks, T., Kratsch, D. and Müller, H. (1995). Approximating the bandwidth for asteroidal triple-free graphs, *Proc. ESA'95, Lecture Notes in Comput. Sci.*, **979**, Springer, Berlin, pp. 434–447.

Kloks, T., Kratsch, D. and Müller, H. (1999) Approximating the bandwidth for asteroidal triple-free graphs, *J. Algorithm*, **32**, pp. 41–57. (7.7*)

Kneis, J., Mölle, D., Richter, S. and Rossmanith, P. (2006). Parameterized power domination complexity, *Inform. Process. Lett.*, **98**, pp. 145–149. (3.5*)

Köhler, E. G. (1994). *Domination Problems in Three-dimensional Chessboard Graphs*, Master's Thesis, Clemson Univ. (1.1 thesis chess)

Köhler, E. (1996). Connected domination on trapezoid graphs in $O(n)$ time, Manuscript. (7.3*)

Köhler, E. (2000). Connected domination and dominating clique in trapezoid graphs, *Discrete Appl. Math.*, **99**, pp. 91–100. (7.3*)

Kolen, A. (1982). *Location Problems on Trees and the Rectilinear Plane*, Ph.D. Thesis, Univ. Amsterdam. (1.1 thesis)

Kőnig, D. (1936). *Theory of Finite and Infinite Graphs*, translated by Mcloart, R. with commentary by Tutte, W. T., Birkhäuser, Boston, 1990. Originally published as *Theorie der endlichen und unendlichen Graphen*, Akademische Verlagsgesellschaft Leipzig 1936. German edition 1986. (1.1*)

Korneyenko, N. M. (1994). Combinatorial algorithms on a class of graphs, *Discrete Appl. Math.*, **54**, pp. 215–217. (4.6)

Korte, N. and Mohring, R. H. (1989). An incremental linear-time algorithm for recognizing interval graphs, *SIAM J. Comput.*, **18**, pp. 68–81. (5.1*)

Kratochvil, J., Manuel, P. and Miller, M. (1995). Generalized domination in chordal graphs, *Nordic J. Comput.*, **2**, pp. 41–50. (4)

Kratsch, D. (1990). Finding dominating cliques efficiently, in strongly chordal graphs and undirected path graphs, *Discrete Math.*, **86**, pp. 225–238. (6.3* 6.5*)

Kratsch, D. (1996). Domination and total domination in asteroidal triple-free graphs, Technical Report Math/Inf/96/25, F.-Schiller-Univ. Jena. (7.7)

Kratsch, D. (1998). Algorithms, in Haynes, T. W., Hedetniemi, S. T. and Slater, P. J. eds., *Domination in Graphs: Advanced Topics*, Chapter 8, Marcel Dekker, Inc., New York. (7.2∗ 7.3∗ 7.4∗ 8.3∗)

Kratsch, D. (2000). Domination and total domination in asteroidal triple-free graphs, *Discrete Appl. Math.*, **99**, pp. 111–123. (7.7∗)

Kratsch, D., Damaschke, P. and Lubiw, A. (1994). Dominating cliques in chordal graphs, *Discrete Math.*, **128**, pp. 269–275. (6.5∗)

Kratsch, D., McConnel, R. M., Mehlhorn, K. and Spinrad, J. P. (2006). Certifying algorithms for recognizing interval graphs and permutation graphs. *SIAM J. Comput.*, **36**, pp. 326–353. (5.1, 8.1)

Kratsch, D. and Stewart, L. (1993). Domination on cocomparability graphs, *SIAM J. Discrete Math.*, **6**, pp. 400–417. (7.2∗ 7.3∗ 7.4∗)

Kratsch, D. and Stewart, L. (1997). Total domination and transformation, *Inform. Proc. Lett.*, **63**, pp. 167–170. (8.2∗)

Kumar, H. N., Pradhan, D. and Venkatakrishnan, Y. B. (2021). Double vertex-edge domination in graphs: complexity and algorithms, *J. Appl. Math. Comput.*, **66**, pp. 245–262. (5)

Kumar, M. and Reddy, P. V. S. (2023). Total 2-rainbow domination in graphs: complexity and algorithms, *Inter. J. Found. Comput. Sci.*, **35**, pp. 887–906. (9)

Lan, J. K. and Chang, G. J. (2013). On the mixed domination problem in graphs, *Theor. Comput. Sci.*, **476**, pp. 84–93. (3.4)

Lappas, E., Nikolopoulos, S. D. and Palios, L. (2013). An $O(n)$-time algorithm for the paired domination problem on permutation graphs, *Europ. J. Combin.*, **34**, pp. 593–608. (8)

Larson, L. C. (1977). A theorem about primes proved on a chessboard, *Math. Mag.*, **50**, pp. 69–74. (1.1 chess)

Laskar, R., Pfaff, J., Hedetniemi, S. M. and Hedetniemi, S. T. (1984) On the algorithmic complexity of total domination, *SIAM J. Algeb. Discrete Methods*, **5**, pp. 420–425. (3.2∗ 4.4∗)

Laskar, R. and Shier, D. (1983). On powers and centers of chordal graphs, *Discrete Appl. Math.*, **6**, pp. 139–147. (6.1∗)

Laskar, R. C. and Wallis, C. K. (1994). Domination parameters of graphs of three-class association schemes and variations of chessboard graphs, *Congr. Numer.*, **100**, pp. 199–213. (1.1 chess)

Lee, C. Y. (1961). An algorithm for path connections and its applications, *IRE Trans. Electr. Computers*, **EC-10**, pp. 346–365. (3.1∗)

Lee, C. (1994). *On the Domination Number of a Digraph*, Ph.D. Thesis, Michigan State Univ. (1.1 thesis)

Lekkerkerker, C. G. and Boland, J. Ch. (1962). Representation of a finite graph by a set of intervals on the real line, *Fund. Math.*, **51**, pp. 45–64. (4.1∗ 5.1∗ 7.5∗)

Li, P. and Wu., Y. (2014). A four-sweep LBFS recognition algorithm for interval graphs, *Discrete Math. & Theor. Comput. Sci.*, **16**, pp. 23–50. (5.1∗)

Liang, Y. D. (1994). Domination in trapezoid graphs, *Inform. Process. Lett.*, **52**, pp. 309–315. (8.2)

Liang, Y. D. (1995). Steiner set and connected domination in trapezoid graphs, *Inform. Process. Lett.*, **56**, pp. 101–108. (8.4)

Liang, Y. D. and Rhee, C. (1994). Linear algorithms for two independent set problems in permutation graphs, *Proc. 22nd Computer Science Conference*, pp. 90–93. (8.3*)

Liang, Y. D., Rhee, C., Dall, S. K. and Lakshmivarahan, S. (1991). A new approach for the domination problem on permutation graphs, *Inform. Process. Lett.*, **37**, pp. 219–224. (8.2*)

Liao, C.-S. (2016). Power domination with bounded time constraints, *J. Combin. Optim.*, **31**, pp. 725–742. (3.7)

Liao, C.-S. and Chang, G. J. (2002). Algorithmic aspect of k-tuple domination in graphs, *Taiwanese J. Math.*, **6**, pp. 415–420. (3.2)

Liao, C.-S. and Chang, G. J. (2003). k-tuple domination in graphs, *Inform. Process. Lett.*, **87**, pp. 45–50. (6)

Liao, C.-S. and Lee, D. T. (2013). Power domination in circular-arc graphs, *Algorithmica*, **65**, pp. 443–466. (5.6)

Liedloff, M., Kloks, T., Liu, J. and Peng, S.-L. (2008). Efficient algorithms for Roman domination on some classes of graphs, *Discrete Appl. Math.*, **156**, pp. 3400–3415. (7.7)

Lin, C.-C., Hsieh, C. Y. and Mu, T. Y. (2022). A linear-time algorithm for weighted paired-domination on block graphs, *J. Combin. Optim.*, **44**, pp. 269–286. (3.7)

Lin, C.-C., Ku, K.-C. and Hsu, C.-H. (2020). Paired-domination problem on distance-hereditary graphs, *Algorithmica*, **82**, pp. 2809–2840. (9)

Lin, C.-C. and Tu, H.-L. (2015). A linear-time algorithm for paired-domination on circular-arc graphs, *Theor. Comput. Sci.*, **591**, pp. 99–105. (5.6)

Liu, C. L. (1968). *Introduction to Combinatorial Mathematics*, McGraw-Hill, New York. (8.1*)

Liu, C.-S., Peng, S.-L. and Tang, C.-Y. (2010). Weak Roman domination on block graphs, *The 27th Workshop on Combinatorial Math. and Comput. Theory*, pp. 86–89. (3.7)

Lovász, L. (1972). Normal hypergraphs and the perfect graph conjecture, *Discrete Math.*, **2**, pp. 253–257. (1.5*)

Lovász, L. (1972a). A characterization of perfect graphs, *J. Combin. Theory, Ser. B*, **13**, pp. 95–98. (1.5*)

Lovász, L. (1979). On the Shannon capacity of a noisy channel, *IRE Trans. Inform. Theory*, **IT-25**, pp. 1–7. (1.5*)

Lu, C., Wang, B. and Wang, K. (2018). Algorithm complexity of neighborhood total domination and $(\rho, \gamma nt$-graphs, *J. Combin. Optim.*, **35**, pp. 424–435. (3.3)

Lu, C. L., Ko, M.-T. and Tang, C. Y. (2002). Perfect edge domination and efficient edge domination in graphs, *Discrete Appl. Math.*, **119**, pp. 227–250. (4.4, 4.6*)

Lu, T.-L. Ho, P.-H. and Chang, G. J. (1990). The domatic number problem in interval graphs, *SIAM J. Discrete Math.*, **3**, pp. 531–536. (5.5*)

Lubiw, A. (1982). *Γ-free Matrices*, Master's thesis, Dept. Combinatorics and Optimization, Univ. Waterloo. (6.1*)

Lubiw, A. (1987). Doubly lexical orderings of matrices, *SIAM J. Comput.*, **16**, pp. 854–879. (6.1* 6.2*)

Lyu, Y. (2023). Labeling algorithm for power domination problem of trees, *Appl. Math. Comput.*, **457**, 128163. (3.5*)

Ma, K. L. (1990). *Partition Algorithm for the Dominating Set Problem*, Master's Thesis, Concordia Univ. (1.1 thesis)

Majumdar, A. (1992). *Neighborhood Hypergraphs: a Framework for Covering and Packing Parameters in Graphs*, Ph.D. Thesis, Clemson Univ. (1.1 thesis)

Manacher, G. K. and Mankus, T. A. (1996). Finding a domatic partition of an interval graph in time $O(n)$, *SIAM J. Discrete Math.*, **9**, pp. 167–172. (5.5*)

McConnell, R. M. and Spinrad, J. P. (1999). Modular decomposition and transitive orientation, *Discrete Math.*, **201**, pp. 189–241. (7.1*)

McRaee, A. A. (1994). *Generalizing NP-Completeness Proofs for Bipartite and Chordal Graphs*, Ph.D. Thesis, Clemson Univ. (1.1 thesis)

McRae, A. A. and Hedetniemi, S. T. (1991). Finding n-independent dominating sets, *Congr. Numer.*, **85**, pp. 235–244. (3.3, 4.4)

Meir, M. and Moon, J. W. (1975). Relations between packing and covering numbers of a tree, *Pacific J. Math.*, **61**, pp. 225–233. (3.4)

Messick, B. C. (1995). *Finding Irredundant Sets in Queen's Graphs*, Master's Thesis, Clemson Univ. (1.1 thesis chess)

Mirsky, L. (1971). A dual of Dilworth's decomposition theorem, *Amer. Math. Monthly*, **78**, pp. 876–877. (7.1*)

Mitchell, S. L. (1977). *Linear Algorithms on Trees and Maximal Outerplanar Graphs: Design, Complexity Analysis and Data Structures Study*, Ph.D. thesis, Univ. Virginia. (1.1 thesis)

Mitchell, S. L., Cockayne, E. J. and Hedetniemi, S. T. (1977). Linear algorithms on recursive representaions of trees, *J. Comput. Syst. Sci.*, **18**, pp. 76–85. (3.2)

Mitchell, S. and Hedetniemi, S. (1977). Edge domination in trees, *Proc. 8th S. E. Conf. Combin., Graph Theory and Computing, Congr. Numer.*, **19**, pp. 489–509. (3.2)

Mitchell, S., Hedetniemi, S. and Goodman, S. (1975). Some linear algorithms on trees, *Congr. Numer.*, **14**, pp. 467–483. (3.2)

Moore, E. F. (1959). The shortest path through a maze, *Proc. of the International Symposium on the Theory of Switching*, Harvard University Press. pp. 285–292. (3.1*)

Moscarini, M. (1993). Doubly chordal graphs, Steiner trees and connected domination. *Networks*, **23**, pp. 59–69. (6.5)

Müller, H. and Brandstädt, A. (1987). The NP-completeness of Steiner tree and dominating set for chordal bipartite graphs, *Theor. Comput. Sci.*, **53**, pp. 257–265. (4.4)

Nadel, B. A. (1990). Representation selection for constraint satisfaction: a case study using the N-queens, *IEEE Expert*, **5**, pp. 16–23.   (1.1 chess)

Natarajan, K. S. and White, L. J. (1978). Optimum domination in weighted trees, *Inform. Process. Lett.*, **7**, pp. 261–265.   (3.3*)

Nauck, F. (1950). Schach: eine in das gebiet der mathematik fallende aufgabe von herrn dr. Nauck in Schleusingen, *Illustrirte Zeitung* **14**, pp. 416.   (1.1 chess)

Nauck, F. (1850). Briefwechseln mit allen für alle, *Illustrirte Zeitung*, **377**, pp. 182.   (1.1 chess)

Nemhauser, G. L. (1966). *Introduction to Dynamic Programming*, John Wiley & Sons.   (3.3*)

Östergård, P. R. D. and Weakley, W. D. (2001). Values of domination numbers of the queen's graph, *Electron. J. Combin. Research*, **8**, #R29.   (1.1 chess)

Padamutham, P. and Palagiri, V. S. R. (2020). Algorithmic aspects of Roman domination in graphs, *J. Appl. Math. Comput.*, **64**, pp. 89–102.   (4.4)

Paige, R. and Tarjan, R. E. (1987). Three partition refinement algorithms, *SIAM J. Comput.*, **16**, pp. 973–989.   (6.2*)

Panda, B. S. and Paul, S. (2013). Liar's domination in graphs: complexity and algorithm, *Discrete Appl. Math.*, **161**, pp. 1085–1092.   (3.2)

Panda, B. S. and Paul, S. (2013). A linear time algorithm for liar's domination problem in proper interval graphs, *Inform. Process. Lett.*, **113**, pp. 815–822.   (5)

Panda, B. S. and Pradhan, D. (2013). A linear time algorithm for computing a minimum paired-dominating set of a convex bipartite graph, *Discrete Appl. Math.*, **161**, pp. 1776–1783.   (6)

Pauls, E. (1874). Das maximal problem der Damen auf dem Schachbrett, *Deutsche Schachzeitung*, **29**, pp. 129–134.   (1.1 chess)

Peng, S.-L. and Chang, M.-S. (1991). A new approach for domatic number problem on interval graphs, *Proc. National Comp. Symp. R.O.C.*, pp. 236–241.   (5.5*)

Peng, S.-L. and Chang, M.-S. (1992). A simple linear time algorithm for the domatic partition problem on strongly chordal graphs, *Inform. Process. Lett.*, **43**, pp. 297–300.   (6.4*)

Peters, K. W. (1986). *Theoretical and Algorithmic Results on Domination and Connectivity*, Ph.D. Thesis, Clemson Univ.   (1.1 thesis)

Pfaff, J. (1984). *Algorithmic Complexities of Domination-related Graph Parameters*, Ph.D. Thesis, Clemson Univ.   (1.1 thesis)

Pfaff, J., Laskar, R. and Hedetniemi, S. T. (1984). Linear algorithms for independent domination and total domination in series-parallel graphs, *Congr. Numer.*, **45**, pp. 71–82.   (4.6)

Piotrowski, W. (1992). *Combinatorial Optimization: Scheduling, Facility Location and Domination*, Ph.D. Thesis, North Dakota State Univ.   (1.1 thesis)

Pnueli, A., Lempel, A. and Even, S. (1971). Transitive orientation of graphs and identification of permutation graphs, *Canad. J. Math.*, **23**, pp. 160–175.   (7.1* 8.1*)

Poureidi, A. (2022). Algorithmic results in secure total dominating sets on graphs, *Theor. Comput. Sci.*, **918**, pp. 1–17. (3.3)

Pradhan, D., Banerjee, S. and Liu, J. B. (2022). Perfect Italian domination in graphs: complexity and algorithms, *Discrete Appl. Math.*, **319**, pp. 271–295. (3.7)

Pradhan, D. and Jha, A. (2018). On computing a minimum secure dominating set in block graphs, *J. Combin. Optim.*, **35**, pp. 613–631. (3.7)

Pradhan, D. and Panda, B. S. (2019). Computing a minimum paired-dominating set in strongly orderable graphs, *Discrete Appl. Math.*, **253**, pp. 37–50. (6)

Proskurowski, A. (1979). Minimum dominating cycles in 2-trees, *Internat. J. Comput. Inform. Sci.*, **8**, pp. 405–417. (3.7)

Proskurowski, A. and Syslo, M. M. (1981). Minimum dominating cycles in outerplanar graphs, *Internat. J. Comput. Inform. Sci.*, **10**, pp. 127–139. (3.7)

Qiao, H., Kang, L., Cardei, M. and Du, D.-Z. (2003). Paired-domination of trees, *J. Global Optim.*, **25**, pp. 43–54. (3.6*)

Raczek, J. (2024). Complexity issues on of secondary domination number, *Algorithmica*, **86**, pp. 1163–1172. (3.3)

Rajbaum, S. (2002). Improved tree decomposition based algorithms for domination-like problems, *Lecture Notes Comput. Sci.*, **2286**, pp. 613–627. (3.7)

Ramalingam, G. and Pandu Rangan, C. (1988a). Total domination in interval graphs revisited, *Inform. Process. Lett.*, **27**, pp. 17–21. (5.2*)

Ramalingam, G. and Pandu Rangan, C. (1988b). A unified approach to domination problems in interval graphs, *Inform. Process. Lett.*, **27**, pp. 271–274. (5.1* 5.3*)

Raychaudhuri, A. (1987). On powers of interval and unit interval graphs, *Cong. Numer.*, **59**, pp. 235–242. (6.1)

Reichling, M. (1987). A simplified solution for the N queens problem, *Inform. Process. Lett.*, **25**, pp. 253–255. (1.1 chess)

Rhee, C. (1990). Parallel Algorithms for Special Graphs, Ph.D. Thesis, Univ. Oklahoma, Norman, OK. (1.1 thesis 5.6*)

Rhee, C., Liang, Y. D., Dhall, S. K. and Lakshmivarahan, S. (1996). An $O(n+m)$ algorithm for finding a minimum-weight dominating set in a permutation graph, *SIAM J. Comput.*, **25**, pp. 404–419. (8.2*)

Richey, M. B. (1985). *Combinatorial Optimization on Series-Parallel Graphs: Algorithms and Complexity*, Ph.D. Thesis, Georgia Institute of Technology. (1.1 thesis)

Rivin, I. and Zabih, R. (1992). A dynamic programming solution to the N-queens problem, *Inform. Process. Lett.*, **41**, pp. 253–256. (1.1 chess)

Roberts, F. S. (1969). Indifference graphs, *Proof Techniques in Graph Theory* (*Proc. Second Ann Arbor Graph Theory Conf., Ann Arbor, Mich., 1968*), Academic Press, New York, pp. 139–146. (5.1* 5.7*)

Roberts, F. S. (1971). On the compatibility between a graph and a simple order, *J. Combin. Theory, Ser. B*, **11**, pp. 28–38.   (5.1)

Rodemich, E. R. (1970). Coverings by rook domains, *J. Combin. Theory*, **9**, pp. 117–128.   (1.1 chess)

Rohl, J. S. (1983). A faster lexicographic N queens algorithm, *Inform. Process. Lett.*, **17**, pp. 231–233.   (1.1 chess)

Rose, D. J., Tarjan, R. E. and Lueker, G. S. (1976). Algorithmic aspects of vertex elimination on graphs, *SIAM J. Comput.*, **5**, pp. 266–283.   (4.3*)

Rouse Ball, W. W. (1892). *Mathematical Recreations and Problems of Past and Present Times*, MacMillan, London.   (1.1 chess)

Rouse Ball, W. W. (1939). *Mathematical Recreations & Essays*, revision by Coxeter, H. S. M. of the original 1892 Edition, Chapter 6, Minimum Pieces Problem, Macmillan, London.   (1.1 chess)

Rozhoň, V. (2019). A local approach to the Erdős–Sós conjecture, *SIAM J. Discrete Math.*, **33**, pp. 643–664.   (3.1*)

Ryser, H. J. (1969). Combinatorial configurations, *SIAM J. Appl. Math.*, **17**, pp. 593–602.   (5.7*)

Scheffler, P. (1987). Linear-time algorithms for NP-complete problems restricted to partial k-trees, Technical Report 03/87, IMATH, Berlin.   (3.7)

Scheffler, P. (1989). *Die Baumweite von Graphen als ein Mass für die Kompliziertheit Algorithmischer Probleme*, Ph.D. Thesis, Akademie der Wissenschaften der DDR, Berlin.   (1.1 thesis)

Shannon, C. E. (1956). The zero error capacity of a noisy channel, *IRE Trans. Inform. Theory*, **IT-2**, pp. 8–19. Reprinted in Slepian, D. ed., *Key Papers in the Development of Information Theory*, New York, IEEE Press, 1974.
                                                                          (1.5*)

Simon, K. (1991). A new simple linear algorithm to recognize interval graphs, *Lecture Notes Comput. Sci.*, **553**, pp. 289–308.   (5.1)

Slater, P. J. (1976). R-domination in graphs, *J. Assoc. Comput. Mach.*, **23**, pp. 446–450.   (3.2*)

Slater, P. J. (1987). Domination and location in acyclic graphs, *Networks*, **17**, pp. 55–64.   (3.2)

Smart, C. B. (1996). *Studies of Graph Based IP/LP Parameters*, Ph.D. Thesis, Univ. Alabama in Huntsville.   (1.1 thesis)

Sosic, R. and Gu, J. (1990). A polynomial time algorithm for the N-queens problem, *SIGART Bull.*, **2**, pp. 7–11.   (1.1 chess)

Spinrad, J. (1983). Transitive orientation in $O(n^2)$ time, *Proc. 15th Ann. ACM Symp. on Theory of Computing*, pp. 457–466.   (7.1*)

Spinrad, J. (1985). On comparability and permutation graphs, *SIAM J. Comput.*, **14**, pp. 658–670.   (7.1*)

Spinrad, J. P. (1993). Doubly lexical ordering of dense 0-1 matrices, *Inform. Process. Lett.*, **45**, pp. 229–235.   (6.2*)

Spinrad, J., Brandstädt, A. and Stewart, K. (1987). Bipartite permutation graphs, *Discrete Appl. Math.*, **18**, pp. 279–292.   (8.1)

Srinivasa Rao, A. and Pandu Rangan, C. (1989/90) Linear algorithm for domatic number problem on interval graphs, *Inform. Process. Lett.*, **33** pp. 29–33. (5.5*)

Srinivasan, A. and Pandu Rangan, C. (1991). Efficient algorithms for the minimum weighted dominating clique problem on permutation graphs, *Theor. Comput. Sci.*, **91**, pp. 1–21. (8)

Stewart, L. (1985). *Permutation in Graph Structure and Algorithms*, Ph.D. Thesis, Univ. Toronto. (1.1 thesis)

Stone, H. S. and Stone, J. M. (1987). Efficient search techniques–an empirical study of the N-queens problem, *IBM J. Res. Develop.*, **31**, pp. 464–474. (1.1 chess)

Strassen, V. (1969). Gaussian elimination is not optimal, *Numer. Math.*, **13**, pp. 354–356. (7.1*)

Tarjan, R. E. (1976). Maximum cardinality search and chordal graphs, Standford Univ., Unpublished Lecture Notes CS 259. (4.3*)

Telle, J. A. (1994). *Vertex Partition Problems: Characterization, Complexity and Algorithms on Partial k-Trees*, Ph.D. Thesis, Univ. Oregon. (1.1 thesis)

Telle, J. A. (1994). Complexity of domination-type problems in graphs, *Nordic J. Comput.*, **1**, pp. 157–171. (4.4)

Telle, J. A. and Proskurowski, A. (1993). Practical algorithms on partial k-trees with an application to domination-type problems, in Dehne, F., Sack, J. R., Santoro, N. and Whitesides, N., eds., *Proc. Third Workshop on Algorithms and Data Structures (WADS'93), Lecture Notes in Comput. Sci.*, **703**, pp. 610–621, Montréal, Springer-Verlag. (3.7)

Telle, J. A. and Proskurowski, A. (1993). Efficient sets in partial k-trees, *Discrete Appl. Math.*, **44**, pp. 109–117. (3.7)

Topor, R. W. (1982). Fundamental solutions of the eight queens problem, *BIT*, **22**, pp. 42–52. (1.1 chess)

Tripathi, V., Kloks, T., Pandey, A., Paul, K. and Wang, H.-L. (2022). Complexity of paired domination in AT-free and planar graphs, *Theor. Comput. Sci.*, **930**, pp. 53–62. (7.7)

Tripathi, V., Pandey and Maheshwari, A. (2023). A linear-time algorithm for semitotal domination in strongly chordal graphs, *Discrete Appl. Math.*, **338**, pp. 77–88. (6.5)

Tsai, K.-H. and Hsu, W.-L. (1993). Fast algorithms for the dominating set problem on permutation graphs, *Algorithmica*, **9**, pp. 601–614. (8.2*)

Tucker, A. C. (1970). Characterizing circular-arc graphs, *Bull. Amer. Math. Soc.*, **76**, pp. 1257–1260. (5.6*)

Tucker, A. C. (1971). Matrix characterizations of circular-arc graphs, *Pacific J. Math.*, **39**, pp. 535–545. (5.6*)

Tutte, W. T. and Nash-Williams, C. St. J. A. (1984). *Graph Theory*, Addison-Wesley, Menlo Park, CA. (1.2 book)

Ungerer, E. (1996). *Aspects of Signed and Minus Domination in Graphs*, Ph.D. Thesis, Rand Afrikaans Univ. (1.1 thesis)

Valdes, J., Tarjan, R. E. and Lawler, E. L. (1979). The recognition of series parallel digraphs, *STOC'79: Proc. of the Eleventh Annual ACM Symposium on Theory of Computing*, pp. 1–12.   (4.6)

Valdes, J., Tarjan, R. E. and Lawler, E. L. (1982). The recognition of series parallel digraphs, *SIAM J. Comput.*, **11**, pp. 298–313.   (4.6)

van Emde Boas, P. (1977). Preserving order in a forest in less than logarithmic time and linear space, *Inform. Process. Lett.*, **6**, pp. 80–82.   (8.2*)

Voloshin, V. I. (1982). Properties of triangulated graphs (Russian), in *Oper. Research & Progr.* (ed. Shcherbakov, B. A) Shtiintsa, pp. 24–32. (4.1* 4.7*)

Voloshin, V. I. and Gorgos, I. M. (1982). Some properties of 1-simply connected hypergraphs and their applications (Russian), in Graphs, hypergraphs and discrete optimization problems, *Mat. Issled.*, **66**, pp. 30–33.   (4.7*)

Wagner, R. and Geist, R. (1984). Crippled queens placement problem, *Sci. Comput. Programming*, **4**, pp. 221–248.   (1.1 chess)

Walikar, H. B. (1980). *Some Topics in Graph Theory (Contributions to the Theory of Domination in Graphs and its Applications)*, Ph.D. Thesis, Karnatak Univ.   (1.1 thesis)

Wallis, C. K. (1994). *Domination Parameters of Line Graphs of Designs and Variations of Chessboard Graphs*, Ph.D. Thesis, Clemson Univ. (1.1 thesis chess)

Walter, J. R. (1972). *Representations of Rigid Cycle Graphs*, Ph.D. Thesis, Wayne State Univ.   (4.1*)

Walter, J. R. (1978). Representations of chordal graphs as subtrees of a tree, *J. Graph Theory*, **2**, pp. 265–267.   (4.1*)

Wang, C., Chen L. and Lu, C. (2016). k-Power domination in block graphs, *J. Combin. Optim.*, **31**, pp. 865–873.   (3.7)

Weakley, W. D. (1995). Domination in the queen's graph, in Alavi, Y. and Schwenk, A. J. eds., *Graph Theory, Combinatorics, Algorithms and Applications, Proc. Seventh Quad. Internat. Conf. on the Theory and Application of Graphs*, **2**, pp. 1223–1232, Kalamazoo, John Wiley and Sons, Inc.
                                                                            (1.1 chess)

Wegner, G. (1967). *Eigenschaften der Nerven Homologische-einfacher Familien in R^n*, Ph.D. Thesis, Göttingen.   (5.7*)

West, D. B. (2001). *Introduction to Graph Theory*, Second Edition, Prentice Hall, Upper Saddle River, NJ.   (1.2*)

White, K., Farber, M. and Pulleyblank, W. (1985). Steiner trees, connected domination and strongly chordal graphs, *Networks*, **15**, pp. 109–124.
                                                                            (6.3* 6.5*)

Wilf, H. S. (1995). The problem of the kings, *Electron. J. Combin.*, **2**, pp. 1–7.
                                                                            (1.1 chess)

Williams, V. V., Xu, Y., Xu, Z. and Zhou, R. (2023). New bounds for matrix multiplication: from alpha to omega, arXiv: 2307.07970 [cs.DS].   (7.1*)

Wimer, T. V. (1986). Linear algorithms for the dominating cycle problems in series-parallel graphs, partial k-trees and Halin graphs, *Congr. Numer.*, **56**, pp. 289–298.   (3.7)

Wimer, T. V. (1987). *Linear Algorithms for k-Terminal Graphs*, Ph.D. Thesis, Dept. Computer Science, Clemson Univ. (1.1 thesis)

Wimer, T. V., Hedetniemi, S. T. and Laskar, R. (1985). A methodology for constructing linear graph algorithms, *Congr. Numer.*, **50**, pp. 43–60. (3.3)

Yaglom, A. M. and Yaglom, I. M. (1964). *Challenging Mathematical Problems with Elementary Solutions, Volume 1: Combinatorial Analysis and Probability Theory*, Dover Publications, Inc., New York. (1.1 chess)

Xu, G., Kang, L., Shan, E. and Zhao, M. (2006). Power domination in block graphs, *Theor. Comput. Sci.*, **359**, pp. 299–305. (3.7)

Yannakakis, M. and Gavril, F. (1980). Edge dominating sets in graphs, *SIAM J. Appl. Math.*, **38**, pp. 364–372. (3.2, 4.4)

Yeh, H.-G. (1997). *Distance-Hereditary Graphs: Combinatorial Structures and Algorithms*, Ph.D. Thesis, National Chiao-Tung Univ., Hinchu, Taiwan. (1.1 thesis)

Yeh, H.-G. and Chang, G. J. (1997). Algorithmic aspect of majority domination, *Taiwanese J. Math.*, **1**, pp. 343–350. (3.3)

Yeh, H.-G. and Chang, G. J. (1998). Weighted connected domination and Steiner trees in distance-hereditary graphs, *Discrete Appl. Math.*, **87**, pp. 245–253. (9.3*)

Yeh, H.-G. and Chang, G. J. (2001). Weighted k-domination and weighted k-dominating clique in distance-hereditary graphs, *Theor. Comput. Sci.*, **263**, pp. 3–8. (9.3*)

Yen, W. C.-K (2002). Bottleneck domination and bottleneck independent domination on graphs. *J. Inform. Sci. Engin.*, **18**, pp. 311–331. (4.4)

Yen, W. C.-K. (2003). The bottleneck independent domination on the classes of bipartite graphs and block graphs, *Inform. Sci.*, **157**, pp. 199–215. (4.4)

Yen, C.-C. and Lee, R. C. T. (1990). The weighted perfect domination problem, *Inform. Process. Lett.*, **35**, pp. 295–299. (3.2, 4.4)

Yen, C.-C. and Lee, R. C. T. (1996). The weighted perfect domination problem and its variants, *Discrete Appl. Math.*, **66**, pp. 147–160. (3.7, 4.7*)

Yu, M. S., Chen, C. L. and Lee, R. C. T. (1989). Optimal algorithms for the minimum dominating set problem on circular-arc graphs, in *Proc. First Ann. IEEE Symp. on Parallel and Distributed Computing*, pp. 3–10. (5.6*)

Zhao, Y. and Shan, E. (2016). An efficient algorithm for distance total domination in block graphs, *J. Combin. Optim.*, **31**, pp. 372–381. (3.7)

Zhao, Y., Shan, E., Liang, Z. and Gao, R. (2014). A labeling algorithm for distance domination on block graphs, *Bull. Malays. Math. Sci. Soc.*, **37**, pp. 965–970. (3.7)

Zou, Y. H., Liu, J. J., Hsu, C. C. and Wang, Y. L. (2019). A simple algorithm for secure domination in proper interval graphs, *Discrete Appl. Math.*, **260**, pp. 289–293. (5.3)

Zuse, K. (1972). *Der Plankalkül* (in German), Konrad Zuse Internet Archive (http://zuse.zib.de). (3.1*)

Zverovich, I. E. and Zverovich, V. E. (1995). An induced subgraph characterization of domination perfect graphs, *J. Graph Theory*, **20**, pp. 375–395. (4.4)

Index